Enzyme Assays
Essential Developmental Biology
Essential Molecular Biology I and II
Experimental Neuroanatomy
★ Extracellular Matrix
★ Flow Cytometry (2nd Edition)
Gas Chromatography
Gel Electrophoresis of Nucleic Acids (2nd Edition)
Gel Electrophoresis of Proteins (2nd Edition)
★ Gene Probes 1 and 2
Gene Targeting
Gene Transcription
Glycobiology
Growth Factors
Haemopoiesis
Histocompatibility Testing
★ HIV Volumes 1 and 2
Human Cytogenetics I and II (2nd Edition)
Human Genetic Disease Analysis
Immunocytochemistry
In Situ Hybridization
Iodinated Density Gradient Media
★ Ion Channels
Lipid Analysis
Lipid Modification of Proteins
Lipoprotein Analysis
Liposomes
Mammalian Cell Biotechnology
Medical Bacteriology
Medical Mycology
★ Medical Parasitology
★ Medical Virology
Microcomputers in Biology
Molecular Genetic Analysis of Populations
★ Molecular Genetics of Yeast
Molecular Imaging in Neuroscience
Molecular Neurobiology
Molecular Plant Pathology I and II
Molecular Virology
Monitoring Neuronal Activity
Mutagenicity Testing
Neural Transplantation
Neurochemistry
Neuronal Cell Lines
NMR of Biological Macromolecules
★ Non-isotopic Methods in Molecular Biology
Nucleic Acid Hybridization
Nucleic Acid and Protein Sequence Analysis
Oligonucleotides and Analogues
Oligonucleotide Synthesis
PCR 1
★ PCR 2
★ Peptide Antigens
Photosynthesis: Energy Transduction
★ Plant Cell Biology
★ Plant Cell Culture (2nd Edition)
Plant Molecular Biology
★ Plasmids (2nd Edition)

Pollination Ecology
Postimplantation Mammalian Embryos
Preparative Centrifugation
Prostaglandins and Related Substances
★ Protein Blotting
Protein Engineering
Protein Function
Protein Phosphorylation
Protein Purification Applications
Protein Purification Methods
Protein Sequencing
Protein Structure
Protein Targeting
Proteolytic Enzymes
★ Pulsed Field Gel Electrophoresis
Radioisotopes in Biology
Receptor Biochemistry
Receptor–Effector Coupling
Receptor–Ligand Interactions
RNA Processing I and II
Signal Transduction
Solid Phase Peptide Synthesis
Transcription Factors
Transcription and Translation
Tumour Immunobiology
Virology
Yeast

HIV
Volume 2
Biochemistry, Molecular Biology, and Drug Discovery

The Practical Approach Series

SERIES EDITORS

D. RICKWOOD
Department of Biology, University of Essex
Wivenhoe Park, Colchester, Essex CO4 3SQ, UK

B. D. HAMES
Department of Biochemistry and Molecular Biology
University of Leeds, Leeds LS2 9JT, UK

★ indicates new and forthcoming titles

Affinity Chromatography
Anaerobic Microbiology
Animal Cell Culture (2nd Edition)
Animal Virus Pathogenesis
Antibodies I and II
★ Basic Cell Culture
Behavioural Neuroscience
Biochemical Toxicology
★ Bioenergetics
Biological Data Analysis
Biological Membranes
Biomechanics—Materials
Biomechanics—Structures and Systems
Biosensors
★ Carbohydrate Analysis (2nd Edition)
Cell–Cell Interactions
★ The Cell Cycle
★ Cell Growth and Apoptosis
Cellular Calcium
Cellular Interactions in Development
Cellular Neurobiology
Clinical Immunology
Crystallization of Nucleic Acids and Proteins
★ Cytokines (2nd Edition)
The Cytoskeleton
Diagnostic Molecular Pathology I and II
Directed Mutagenesis
★ DNA Cloning 1: Core Techniques (2nd Edition)
★ DNA Cloning 2: Expression Systems (2nd Edition)
★ DNA Cloning 3: Complex Genomes (2nd Edition)
★ DNA Cloning 4: Mammalian Systems (2nd Edition)
Electron Microscopy in Biology
Electron Microscopy in Molecular Biology
Electrophysiology

HIV

A Practical Approach

Volume 2

Biochemistry, Molecular Biology, and Drug Discovery

Edited by

JONATHAN KARN

Medical Research Council
Laboratory of Molecular Biology
Hills Road, Cambridge CB2 2QH, UK

IRL PRESS
at
OXFORD UNIVERSITY PRESS
Oxford New York Tokyo

Oxford University Press, Walton Street, Oxford OX2 6DP
Oxford New York
Athens Auckland Bangkok Bombay
Calcutta Cape Town Dar es Salaam Delhi
Florence Hong Kong Istanbul Karachi
Kuala Lumpur Madras Madrid Melbourne
Mexico City Nairobi Paris Singapore
Taipei Tokyo Toronto
and associated companies in
Berlin Ibadan

Oxford is a trade mark of Oxford University Press

Published in the United States
by Oxford University Press Inc., New York

A catalogue record for this book is available from the British Library

Library of Congress Cataloging-in-Publication Data
HIV : a practical approach / edited by Jonathan Karn. – 1st ed.
(Practical approach series ; v. 156–157)
Includes bibliographical references and index.
Contents: v. 1. Virology and immunology – v.2. Biochemistry, molecular biology, and drug discovery.
1. HIV infections – Research – Methodology. 2. HIV (Viruses) – Research – Methodology. I. Karn, J. (Jonathan) II. Series.
[DNLM: 1. HIV–isolation & purification. 2. HIV Infections – virology. 3. HIV Infections – immunology. 4. HIV Seropositivity. 5. Antiviral Agents. 6. Microbiological Techniques. QW168.5.H6 H6761995]
QR201.A37H55 1995 616.97'9201–dc20 95–15495

ISBN 0 19 963499 8 (v. 2: Hbk)
ISBN 0 19 963498 X (v. 2: Pbk)
ISBN 0 19 963493 9 (v. 1: Hbk)
ISBN 0 19 963492 0 (v. 1: Pbk)
ISBN 0 19 963501 3 (set: Hbk)
ISBN 0 19 963500 5 (set: Pbk)

Typeset by Footnote Graphics, Warminster, Wilts
Printed in Great Britain by Information Press Ltd, Eynsham, Oxon.

Preface

The acquired immunodeficiency syndrome (AIDS) is undeniably one of the most pressing medical emergencies of our era. The World Health Organization's projections indicate that by the year 2000, between 30 and 40 million people world-wide, including five to ten million children will have been infected with the human immunodeficiency virus. It is likely that around half of those infected with HIV will die from AIDS within seven to ten years of being infected. Although the development of improved approaches to clinical care, such as the treatment and prevention of opportunistic infections have improved life expectancy there is clearly an urgent need for antiviral drugs of improved efficacy and prophylactic vaccines.

It is now 12 years after the discovery of HIV, and in spite of an intensive world-wide research effort, we are still a long way from understanding and controlling AIDS. Sadly, we do not even know why HIV infections produce a lengthy asymptomatic phase followed by the catastrophic loss in CD4-positive T cells that leads to the immunodeficiency. Indeed, our present ignorance of the pathogenic mechanisms involved in AIDS has permitted Dr Peter Duesberg and others to argue that HIV is not the causal agent in AIDS. The arguments they have put forward are complex, but can be easily answered. For example, they have argued that there is insufficient virus replication to produce disease. However, the methods for studying HIV infections in patients are largely restricted to analysis of virus growth in the peripheral blood and other relatively accessible tissues, and this can be misleading. It has been shown recently that lymph nodes provide 'hidden' foci for the infection. As a biologist, I find the most compelling argument that HIV infection induces AIDS, stems from the observation that closely related viruses such as the simian immunodeficiency virus (SIV) and the feline immunodeficiency virus (FIV) produce AIDS-like diseases in experimentally infected animals. In these systems it is even possible to correlate loss of pathogenicity with the presence of mutations in the infecting viruses. Because of these, and many other observations, the majority of scientists studying AIDS have agreed that the evidence for a causal link between HIV infections and AIDS is solid and undeniable.

These two volumes are guides to *how* to study HIV in the laboratory. As the reader of these books will quickly see, HIV research is highly multi-disciplinary. Virtually every technique used in modern medical research, ranging from cloning viruses and classical virological studies of their growth properties to the expression of recombinant viral antigens and the analysis of their structure by X-ray crystallography or NMR, to studies of antigens, antibodies, and cellular immune responses could have been legitimately included in these volumes. A complete compendium of all these methods would have been unwieldy and unreadable. Instead, these two volumes focus on the specialized techniques which are unique to studies of HIV.

Volume 1 describes studies of the virus and infected cells. Working with a pathogenic agent such as HIV requires special precautions. Techniques for handling HIV safely and efficiently, as well as methods for propagating laboratory adapted and primary viral isolates in a variety of host cells are described in Part I. The methods used to analyse HIV infection including methods for studying viral entry, viral replication, viral neutralization by antibodies, and viral sequence variation are described in Part II. Part III describes basic immunological techniques used to study HIV infections ranging from methods to detect antibodies to HIV and quantify CD4-positive cells in clinical specimens, to advanced techniques used in cellular immunology to study lymphokine production, helper T cell function, and cytotoxic T cell activity.

Volume 2 explores the biochemistry of HIV using the reductionist approach that is characteristic of modern molecular biology. Methods for cloning and expressing most of the viral proteins are described. Many of the viral proteins, including reverse transcriptase, RNaseH, and integrase, are enzymes, and specific assays have been devised to study their functions *in vitro*. The genetic regulatory proteins, *tat*, *rev*, and *nef* are less well defined and require a combination of cellular and cell-free assays to study their functions. The functions of some of the other proteins, such as the viral envelope, *vpu*, and *vif* can still only be studied in cellular systems.

The detailed studies of HIV molecular biology are providing impressive insights into the molecular machinery used by the virus, but fulfilment of the mandate to discover new antiviral agents necessarily requires studies of the virus itself. Part II adds to the methods described in Volume 1 and presents recent approaches to drug discovery, ranging from methods used for the large scale screening of antiviral compounds to methods used to explore gene therapy approaches to managing HIV infections, including development of ribozymes, antisense oligonucleotides, and inhibitory proteins including dominant negative mutants of the HIV regulatory proteins.

In keeping with the spirit of the Practical Approach series, detailed protocols providing a step by step guide to performing experiments are the central focus of each chapter. The books have been designed so that a novice entering HIV research can find clear instructions about how to perform standard experiments. In addition the books include many novel methods which are described in detail for the first time. I hope that older hands will find these books filled with useful 'tips' which can be incorporated into their own work. We've already modified several of our routines in Cambridge after reading the draft chapters!

Cambridge JONATHAN KARN
July 1995

Contents

Contributors

W. A. BEARD
Sealy Center for Molecular Science, University of Texas Medical Branch, Medical Research Building, J-71 Galveston, Texas 77555–1071, USA.

E. A. BERGER
National Institute of Allery and Infectious Diseases, National Institute of Health, Bldg. 4, Rm 232, 9000 Rockville Pike, Bethesda, Maryland 20892, USA.

E. BERTRAND
Institute Jacques Monod, CNRS, Université Paris 7, Tour 43, 2 Place Jussieu, 75251 Paris Cedex 05, France.

E. BÖHNLEIN
Progenesys, 1501 California Avenue, Palo Alto, CA 94304, USA.

C. C. BRODER
National Institute of Allery and Infectious Diseases, National Institute of Health, Bldg. 4, Rm 232, 9000 Rockville Pike, Bethesda, Maryland 20892, USA.

C. BRUCK
Smith Kline Beecham Biologicals s.a., Rue de l'Institut 89, B-1330 Rixensart, Belgium.

P. J. G. BUTLER
Medical Research Council, Laboratory of Molecular Biology, Hills Road, Cambridge, CB2 2QH, UK.

D. CASTANOTTO
Center for Molecular Biology and Gene Therapy, Mortensen Hall, Loma Linda University, 11058 Campus Drive, Loma Linda, CA 92350, USA.

M. J. CHURCHER
Medical Research Council, Laboratory of Molecular Biology, Hills Road, Cambridge CB2 2QH, UK.

E. A. COHEN
Laboratoire de Rétrovirologie Humaine, Départment de Microbiologie et Immunologie, Faculté de Médicine, Université de Montréal, Cp6 128 Station A, Montréal, Québéc H3C 3J7, Canada.

R. CRAIGIE
Laboratory of Molecular Biology, National Institute for Diabetes and Digestive and Kidney Diseases, National Institutes of Health, Building 5, Room 301, Bethesda, Maryland 20892, USA.

J. CULP
Smith Kline Beecham Laboratories, 709 Swedeland Road, PO Box 1539, King of Prussia, Pennsylvania 19406–0939, USA.

C. DEBOUCK
Smith Kline Beecham Laboratories, 709 Swedeland Road, PO Box 1539, King of Prussia, Pennsylvania 19406–0939, USA.

A. ENGELMAN
Laboratory of Molecular Biology, National Institute for Diabetes and Digestive and Kidney Diseases, National Institutes of Health, Building 5, Room 301, Bethesda, Maryland 20892, USA.

L. FABRY
Smith Kline Beecham Biologicals s.a., Rue de l'Institut 89, B-1330 Rixensart, Belgium.

M. FRANCOTTE
Smith Kline Beecham Biologicals s.a., Rue de l'Institut 89, B-1330 Rixensart, Belgium.

M. J. GAIT
Medical Research Council, Laboratory of Molecular Biology, Hills Road, Cambridge CB2 2QH, UK.

R. GAYNOR
Department of Microbiology, Southwestern Medical School, 5323 Harry Hines Blvd., Dallas, Texas 75235–9048, USA.

H. G. GÖTTLINGER
Dana-Farber Cancer Institute, 44 Binney Street, Boston, Massachusetts 02115, USA.

A. B. HICKMAN
Laboratory of Molecular Biology, National Institute for Diabetes and Digestive and Kidney Diseases, National Institutes of Health, Building 5, Room 301, Bethesda, Maryland 20892, USA.

Z. HOSTOMSKA
Agouron Pharmaceuticals Inc., 3565 General Atomics Court, San Diego, CA 92121–1121, USA.

Z. HOSTOMSKY
Agouron Pharmaceuticals Inc., 3565 General Atomics Court, San Diego, CA 92121–1121, USA.

L. A. IVANOFF
Smith Kline Beecham Laboratories, 709 Swedeland Road, PO Box 1539, King of Prussia, Pennsylvania 19406–0939, USA.

I. JONES
National Environment Research Council, Institute for Virology and Environmental Microbiology, Mansfield Road, Oxford OX1 3SR, UK.

J. KARN
Medical Research Council, Laboratory of Molecular Biology, Hills Road, Cambridge CB2 2QH, UK.

A. KELLEY
Medical Research Council, Laboratory of Molecular Biology, Hills Road, Cambridge CB2 2QH, UK.

M. LASPIA
Department of Microbiology, Dartmouth-Hitchcock Medical Center, Dartmouth Medical School, Borwell Building, Lebanon, New Hampshire 03756, USA.

N. MAHMOOD
Medical Research Council Collaborative Centre, 1–3 Burtonhole Lane, Mill Hill, London NW7 1AD, UK.

M. H. MALIM
Howard Hughes Medical Institute, Department of Microbiology, University of Pennsylvania School of Medicine, Clinical Research Building, 422 Curie Boulevard, Philadelphia, Pennsylvania 19104–6148, USA.

D. A. MANN
Clinical Biochemistry, Faculty of Medicine, University of Southampton, Level D, South Block, Southampton General Hospital, Southampton, SO9 4XY, UK.

W. A. MARASCO
Dana-Farber Cancer Institute, Boston, Massachusetts 02115, USA.

R. MARIANI
James Laboratory, Cold Spring Harbor Laboratory, PO Box 100, Cold Spring Harbor, New York 11724, USA.

O. NUSSBAUM
National Institute of Allery and Infectious Diseases, National Institutes of Health, Bldg. 4, Rm 232, 9000 Rockville Pike, Bethesda, Maryland 20892, USA.

S.-H. I. OU
Department of Microbiology, Southwestern Medical School, 5323 Harry Hines Blvd., Dallas, Texas 75235–9048, USA.

K. RITTNER
Transgene, 11 rue de Molsheim, 67082, Strasbourg, Cedex-France.

J. J. ROSSI
Center for Molecular Biology and Gene Therapy, Mortensen Hall, Loma Linda University, 11058 Campus Drive, Loma Linda, CA 92350, USA.

J. SKOWRONSKI
James Laboratory, Cold Spring Harbor Laboratory, PO Box 100, Cold Spring Harbor, New York 11724, USA.

C. SMITH
Duke University Medical Center, Durham, North Carolina 27710, USA.

C. THIRIART
Smith Kline Beecham Biologicals s. a., Rue de l'Institut 89, B-1330 Rixensart, Belgium.

T. A. TOMASZEK Jr
Smith Kline Beecham Laboratories, 709 Swedeland Road, PO, Box 1539, King of Prussia, Pennsylvania 19406–0939, USA.

DIDIER TRONO
Infectious Disease Laboratory, The Salk Institute for Biological Studies, PO Box 85800, San Diego, California 92186–5800, USA.

O. VAN OPSTAL
Smith Kline Beecham Biologicals s.a., Rue de l'Institut 89, B-1330 Rixensart, Belgium.

S. H. WILSON
Sealy Center for Molecular Science, University of Texas Medical Branch Medical Research Building, J-68 Galveston, Texas 77555–1068, USA.

F. WU
Department of Microbiology, Southwestern Medical School, 5323 Harry Hines Blvd., Dallas, Texas 75235–9048, USA.

Abbreviations

aa	amino acid
ADA	adenosine deaminase
AIDS	acquired immune deficiency syndrome
α-MMP	α-methyl-β-D-mannopyranoside
AZT	3′-azido-3′-deoxythymidine
AZTTP	3′-azido-3′-deoxythymidine 5′-triphosphate
BFA	brefeldin A
BHK	baby hamster kidney
BSA	bovine serum albumin
CAT	chloramphenicol acetyltransferase
CHAPS	3-[(3-cholamidopropyl)-dimethylammonio]-1-propanesulfonate
CHO	chinese hamster ovary
CMC	Critical micellar concentration
CMV	Cytomegalovirus
CPE	cytopathic effect
CPRG	chlorophenol red-β-D-galactopyranoside
CS	calf serum
CTL	cytotoxic T lymphocytes
CV-1	African green monkey kidney cell line
ddC	2′,3′-dideoxycytidine
ddI	2′,3′-dideoxyinosine
DEAE	diethylaminoethyl
DEPC	diethylpyrocarbonate
DHFR	dihydrofolate reductase
DMEM	Dulbecco's modified minimal essential medium
DMSO	dimethyl sulfoxide
DNase	deoxyribonuclease
dNMP	deoxynucleoside monophosphate
dNTP	deoxynucleoside triphosphate
DTE	dithioerythritol
DTT	dithiothreitol
EC_{50}	compound concentration inhibiting virus yield by 50%
EDTA	ethylenediaminetetraacetic acid
EGS	external guide sequence
ELISA	enzyme-linked immunosorbent assay
Env	envelope glycoprotein
env	envelope gene
FACS	flourescence activated cell sorting

FCS (or FBS)	Fetal calf serum (fetal bovine serum)
FDG	fluorescein di-β-D-galactopyranoside
FITC	fluorescein isothiocyanate
FPLC	fast-performance liquid chromatography
G418	Geneticin 418
Gag	group specific antigen
gag	group specific antigen gene
GAL-4	galactose-4 gene (*S. cerevisiae*)
γG-F-CS	gamma-globulin-free calf serum
GNA	*Galanthus nivalis*
gp	glycoprotein
gpt	xanthine-guanine phosphoribosyl transferase gene
GTC	guanidine isothiocyanate
HBS	Hepes-buffered saline
HEPES	N-2-hydroxyethylpiperazine-N′-2-ethanesulfonic acid
HIV-1	human immunodeficiency virus type-1
HIV-2	human immunodeficiency virus type-2
HLB	hydrophilic lipophilic balance
HPLC	high-performance liquid chromatography
HSB	high salt buffer
HSV	herpes simplex virus
HTLV	human T lymphotrophic virus
IE	immediate early
Ig	immunoglobulin
IGS	internal guide sequence
IMDM	Iscove's modified Dulbecco's medium
INS	instability sequence
IPTG	isopropylthio-β-D-galactoside
lacZ	β-galactosidase Z gene (*E. coli*)
LGT	low gelling temperature
LTR	long terminal repeat
mAb	monoclonal antibody
MEM	minimal essential medium
MES	2-(*N*-morpholino)ethanesulfonic acid
mϕ	macrophage
m.o.i.	multiplicity of infection
MOPS	3-(*N*-morpholino)-propanesulfonic acid
MTT	2-(4,5-dimethylthiazol-2-yl)-2,5-di-phenyl-tetrazolium bromide
NCS	newborn calf serum
nef	negative factor gene
neo	neomycin
NF-κB	nuclear factor κB
Ni-NTA	nickel-nitrilo-acetate

NMR	nuclear magnetic resonance
NOE	nuclear Overhauser effect
NP-40	Nonidet P-40
NSI	non-syncitia-inducing strain
nt	nucleotide
OGP	*n*-octyl-β-D-glucopyranoside
ONPG	2-nitrophenyl-β-D-galactopyranoside
OPDA	*ortho*-phenylenediamine
PAGE	polyacrylamide gel electrophoresis
PBL	peripheral blood lymphocytes
PBMC	peripheral blood mononuclear cells
PBS	phosphate-buffered saline
PCR	polymerase chain reaction
PEG	polyethylene glycol
PHA	phytohaemagglutinin
PMA	phorbol-12-myristate-13-acetate
PMSF	phenylmethylsulfonyl fluoride
Pol	polymerase
pol	polymerase (reverse transcriptase) gene
prt	protease gene
PVDF	polyvinyldiformate
QC–PCR	quantitative competitive polymerase chain reaction
R	virus production, infectivity studies
rev	regulator of envelope gene
rIL-2	recombinant interleukin-2
RIPA	radioimmunoprecipitation assay
RNase	ribonuclease
RNase H	ribonuclease H
RNasin	ribonuclease inhibitor
RPMI 1640	Roswell Park Memorial Institute medium 1640
RRE	rev-response element
RSV	rous sarcoma virus
R.T.	retention time
RT	reverse transcriptase
SDS	sodium dodecyl sulfate
SF	syncitia formation
SIV	simian immunodeficiency virus
S-MEM	minimal essential medium for suspension cultures
snRNP	small nuclear ribonuclear proteins
SP-1	simian virus 40 promoter binding protein-1
SSC	sodium chloride, sodium citrate buffer
SV40	simian virus 40
TAR	*trans*-activation-responsive region
Tat	*trans*-activator

tat	*trans*-activator gene
TBE	Tris borate EDTA
TC_{50}	compound concentration inhibiting XTT–Formazan assay by 50%
TCA	trichloroacetic acid
$TCID_{50}$	tissue culture infectious dose (half-maximal)
TE	Tris–EDTA
TFA	triflouroacetic acid
T–P	template–primer pair
Tris	tris(hydroxymethyl)aminomethane
UBP-1	upstream binding protein-1
UV	ultraviolet
vif	viral infectivity factor gene
vpr	viral protein R gene
vpu	viral protein U gene (HIV-1 only)
X-gal	5-bromo-4-chloro-3-indolyl-β-D-galactopyranoside
XTT	2,3-bis[2-methoxy-4-nitro-5-sulfophenyl]-5-[(phenylamino) carbonyl]-2H-tetrazolium hydroxide

Part I

Biochemistry and molecular biology

1

An introduction to the growth cycle of human immunodeficiency virus

J. KARN

1. The growth cycle

HIV and related lentiviruses have growth cycles that are typical of all retroviruses. It is convenient to think of viral growth as comprising four distinct stages:

- infection
- reverse transcription and integration
- viral gene expression
- virus assembly and maturation

The major biochemical steps in virus growth are shown in *Figure 1*. Soon after infection of the cells, the viral RNA genome is reverse transcribed into a DNA copy, transported to the nucleus, and integrated at random sites in the chromosome. Once integrated, the proviral genome is subject to transcriptional regulation by the host cell, as well as its own transcriptional control mechanisms. The later stages of the life cycle involve expression of the viral genes and eventual assembly and release of virus particles.

The following informal summary is intended to provide an introduction to the gene products encoded by HIV and their roles in the viral life cycle.

2. Genome structure

The HIV genome is tightly compressed (*Figure 2*). At least 30 different mRNA transcripts are produced by splicing using the six splice acceptor and two splice donor sequences. The HIV gene products and their functions are listed in *Table 1*.

- HIV encodes a total of nine genes
- processing by cellular and viral enzymes generates 14 protein products

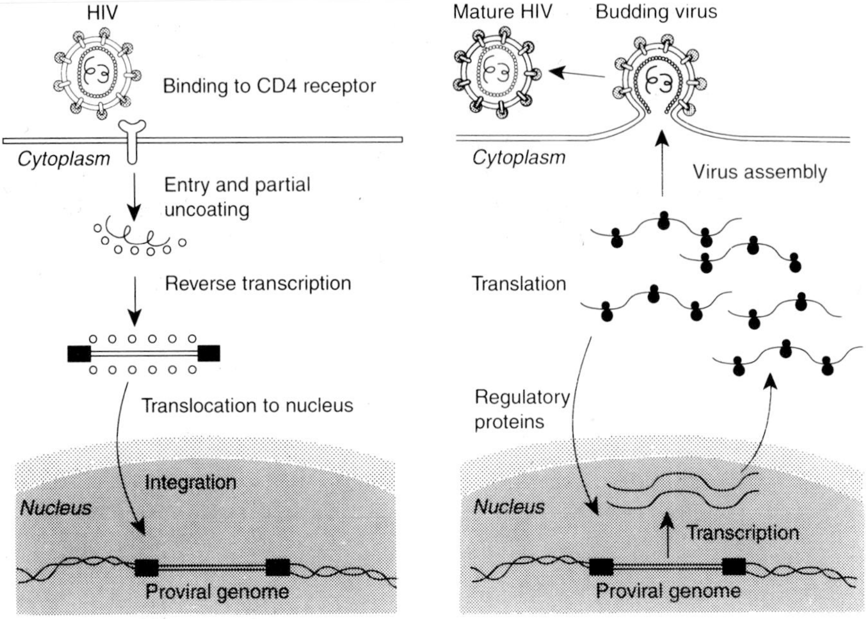

Figure 1. Growth cycle of HIV. *Left*: Infection and pre-integration events. *Right*: Transcription and virus assembly.

2.1 The Gag–Pol processing pathway

The Gag–Pol protein is produced by ribosomal frame shifting. The process is inefficient; only 10% of the translated mRNA gives rise to the Gag–Pol product. The Gag–Pol polyprotein is cleaved by the viral protease (see Chapters 5 and 6) during virus assembly (*Figure 3*):

(a) The protease gene is encoded by the N-terminus of the frame shifted *pol* gene.

(b) Protease is activated in the budding virion particle and leads to cleavage of itself, the Gag proteins, RT, and IN.

2.2 Envelope biosynthesis

The HIV-1 envelope also undergoes a complicated biosynthetic pathway (see Chapters 6 and 7). Initial product is an apoprotein of 88 kDa, which is subsequently glycosylated to form the gp160 precursor molecule.

(a) gp160 is cleaved by cellular proteases to form a complex composed of the amino-terminal gp120 and the carboxy-terminal gp41.

(b) Cleavage of gp160 into gp120 and gp41 is essential for virus infectivity, but not for virus assembly, since mutation of the tryptic-like cleavage site produces uninfectious particles.

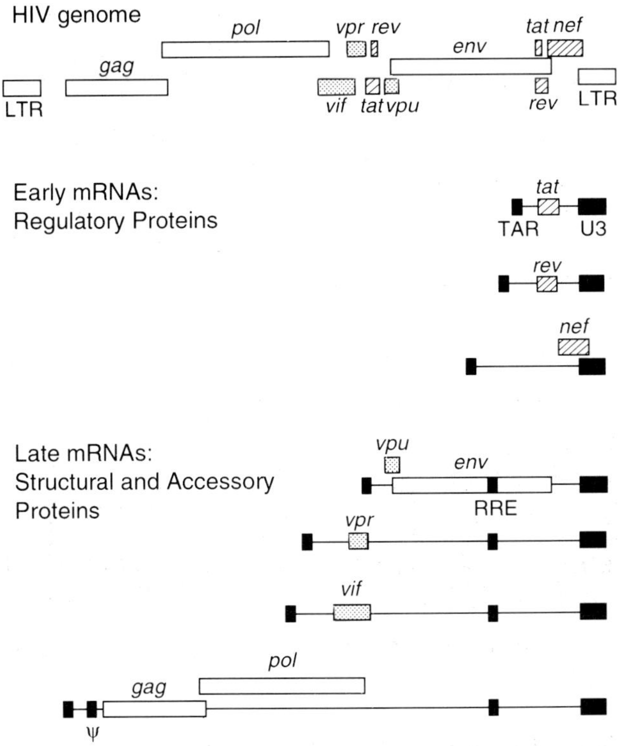

Figure 2. HIV-1 genome structure and mRNA expression. *Top*: Genome map. *Middle*: Early gene expression is mainly composed of 1.8 kb transcripts containing doubly spliced mRNAs encoding the regulatory proteins *tat, rev,* and *nef.* Each mRNA begins with the TAR sequence at the 5′ end and the U3 region of the viral LTR at its 3′ end. *Bottom*: Late gene expression includes the 4.3 kb mRNA encoding *vpu* and *env.* Messengers for *vpr, vif,* and the first exon of *tat* are also approximately 4.3 kb (not shown). Each of the late mRNAs carries the RRE sequence located within the coding region of the *env* gene. The 9.2 kb virion RNA also acts as the mRNA for *gag–pol.* This RNA carries the retroviral packaging signal, Ψ, located upstream of the start of the *gag* gene. Both *env* and *gag–pol* undergo proteolytic processing to generate the mature viral structural proteins.

3. Infection of cells

3.1 Envelope–CD4 interactions

One of the first clinical symptoms exhibited by patients suffering from HIV disease is a selective depletion of CD4-bearing (helper/inducer) T lymphocytes, suggesting that HIV might show selective growth in this population of cells.

(a) HIV grows preferentially in T cells that express CD4 (see Volume 1).

(b) Infection and syncytia formation can be blocked by monoclonal antibodies directed against either the CD4 molecule itself, or the viral envelope glycoprotein gp120 (see Volume 1).

Table 1. The proteins encoded by HIV-1 and HIV-2

Gene	Protein	Function
Virion proteins		
gag	MA (p17)	Matrix protein; membrane binding; virus assembly
	CA (p24)	Capsid protein; virus assembly
	NC (p9)	Nucleocapsid protein; RNA binding; virus assembly
pol	PR (p11)	Protease; virus maturation
	RT (p66/p51)	Reverse transcriptase; virus replication
	IN (p32)	Integrase; virus replication
env	SU (gp120)	Viral envelope; CD4 receptor binding; virus infectivity
	TM (gp41)	Transmembrane envelope glycoprotein; virus infectivity
Accessory proteins		
vpr	Vpr	Viral protein R; translocation of preintegration complex to nucleus
vif	Vif	Viral infectivity factor; preintegration events
vpx	Vpx	Viral protein X; viral structural protein found in HIV-2 only
vpu	Vpu	Viral protein U; envelope maturation and virus release; found in HIV-1 only
Regulatory proteins		
tat	Tat	*Trans*-activator protein; stimulates transcriptional elongation
rev	Rev	Regulator of late gene expression; stimulates appearance of unspliced mRNAs in cytoplasm
nef	Nef	Down-regulates CD4; regulates T cell signalling pathways

The gp120 molecule binds to the CD4 receptor with an affinity constant on the order of 1 nM. Mutational studies have defined the critical residues in both proteins which are required for this interaction. Recent research suggests that binding to the viral envelope induces a conformational change in gp120 which in turn unmasks a specific fusion domain located in gp41. This process is conveniently studied by monitoring syncytium formation between cell lines that express the viral envelope and CD4 molecules (see Chapters 6–8).

3.2 Virus uptake

The CD4 molecule was the first retroviral receptor to be identified. However, simple binding between gp120 and CD4 is not sufficient to allow infection or cell fusion. For example, expression of human CD4 on murine cells does not allow infection, suggesting that additional components are present in human

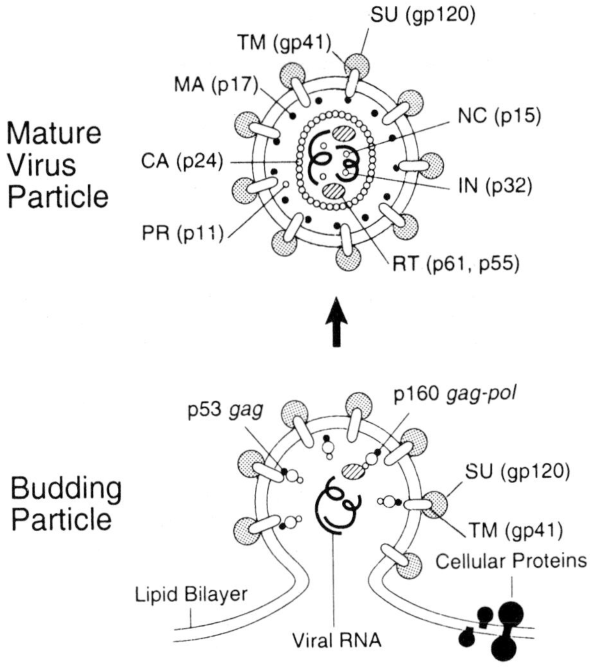

Figure 3. Assembly and maturation of viral particles. *Top*: Structure of the HIV virion. *Bottom*: HIV structural proteins accumulate on the innner surface of the cytoplasmic membrane and begin to bud as immature particles. The budding particles incorporate dimerized genomic RNA molecules and their associated tRNA primers, the myristoylated *gag* and *gag–pol* precursor molecules, and the envelope glycoproteins. Dimerization of the protease activates the enzyme and allows cleavage of the *gag* and *gag–pol* precursors. This is accompanied by structural rearrangements leading to the assembly of a mature core structure.

cells. The search for these accessory proteins is a major focus of current research.

The mechanism of HIV entry into cells is only poorly understood. HIV enters cells in a pH-independent manner, but the process does require endocytosis since mutant CD4 receptors with deletions in their cytoplasmic domains function as receptors even though they are not normally internalized. The viral accessory protein Vif also appears to be required for efficient infection of cells by HIV (see Chapter 15).

4. Reverse transcription and integration

4.1 Reverse transcription

Immediately after entry of HIV into the cytoplasm of a susceptible cell, the viral RNA is transcribed into a double-stranded proviral copy. The reaction is catalysed the viral reverse transcriptase (see Chapter 2). The reverse transcriptase has been recently crystallized and shown to be a heterodimer containing the p55 and p61 sub-units. The two subunits not only differ in their configurations, but also differ because p61 carries the catalytic domain for RNase H. This enzyme is required to remove the RNA template strand during reverse transcription, and has been extensively studied on its own (see Chapter 3).

4.2 Integration

Reverse transcription takes place in the cytoplasm. After synthesis of the proviral DNA, a nucleoprotein complex containing the MA protein, integrase, and possibly other viral proteins, is translocated to the nucleus. The transport signals for this appear to be encoded by the p17 MA protein, but it has also recently been suggested that the Vpr protein might also assist in this process.

The proviral genome is inserted at random into regions of actively transcribed chromatin. The reaction is catalysed by the viral integrase protein which directs both cleavage events and strand exchange (see Chapter 4).

5. Transcriptional control

Once the HIV provirus is integrated into the host chromatin, it is subject to transcriptional control by cellular DNA binding proteins (see Chapter 12). The integrated provirus carries duplicated structures, called the long terminal repeats (LTR) at each end (*Figure 4*):

- the 5′ LTR functions as a promoter element
- the 3′ LTR supplied a polyadenylation signal

Transcription of the HIV genome is regulated by *cis*-acting viral regulatory proteins (*Figure 4*). These two regulatory proteins play complementary roles in the HIV life cycle:

(a) Tat stimulates transcription from the viral long terminal repeat (LTR).
(b) Rev is required for the efficient cytoplasmic expression of the mRNAs encoding the structural proteins of the virus (see Chapters 10 and 11).

Remarkably, both Tat and Rev are RNA binding proteins that exert their effects through specific *cis*-acting viral RNA regulatory sequences (see Chapter 9). Previously studied retroviral regulatory proteins all acted through DNA elements.

5.1 Control of transcriptional elongation by Tat

Tat interacts with a regulatory element located downstream of the initiation site for transcription between residues +1 and +79—the *trans*-activation-responsive region (TAR).

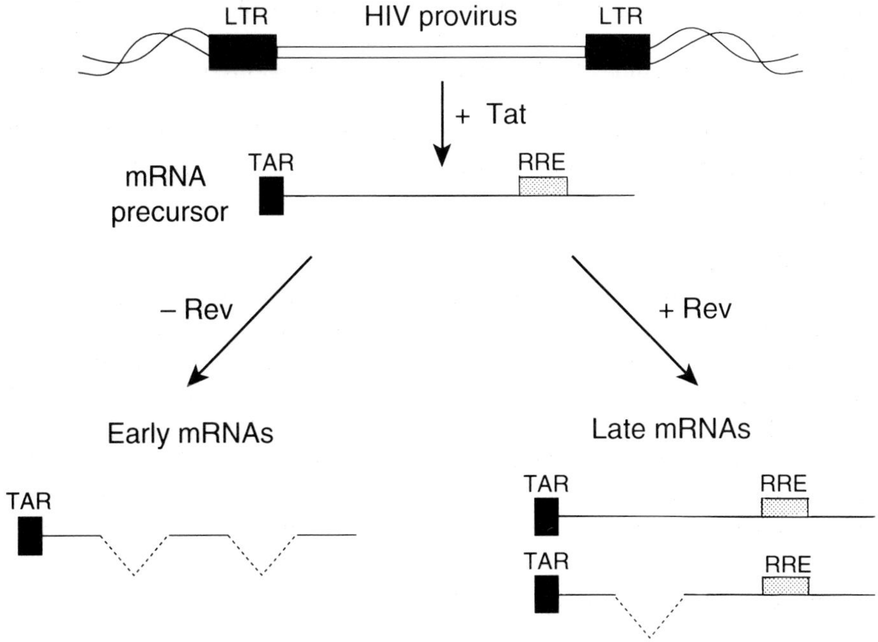

Figure 4. Regulatory circuits in HIV gene expression. Early gene expression produces the regulatory proteins Tat and Rev from doubly spliced mRNAs. Initially, Tat stimulates transcription from the viral long terminal repeat (LTR) by interacting with the *trans*-activation response region (TAR). Once Tat and Rev levels are sufficiently high, Rev then binds to the Rev-response element (RRE) and stabilizes the mRNAs for the virion proteins.

(a) TAR is only functional when it is placed 3′ to the HIV promoter and in the correct orientation. These observations suggested that TAR does not act as an ordinary DNA element, but instead encodes a functionally important RNA sequence.

(b) Tat binds with high affinity to a site located near the apical tip of the TAR RNA sequence (see Chapter 9).

How does Tat stimulate transcription? The HIV LTR characteristically gives rise to two populations of transcripts: short non-polyadenylated RNAs terminating at, or near, the end of the TAR stem–loop, and longer polyadenylated mRNAs. Nuclear run-on experiments have shown that these short transcripts arise because of a strong transcriptional polarity. In the absence of Tat, the majority of the transcriptional complexes formed at the HIV promoter stall or disengage near TAR, but Tat is able to increase the density of RNA polymerases found downstream of the promoter.

The observations described above have led to the working hypothesis that Tat acts as a generalized elongation factor. According to this proposal, soon after transcription through TAR, RNA polymerase either pauses or falls off the template. If Tat is present in the cell, it can associate with TAR RNA, cellular cofactors, and the transcribing RNA polymerase. This produces a modified transcription complex which is then able to transcribe the remainder of the HIV genome efficiently. This type of anti-termination mechanism appears closely analogous to the mechanism used by the bacteriophage λN protein. The recent development of cell-free transcription systems that respond to Tat is now allowing direct tests of the *trans*-activation model (see Chapter 10).

5.2 Rev control of late mRNAs

Transcription of the HIV genome is also characterized by a progressive shift in mRNA production:

(a) Multiply spliced mRNAs encoding the viral regulatory proteins (Tat, Rev, and Nef) are the first mRNAs produced.

(b) Singly spliced mRNAs (Env, Vpu, Vif, and Vpr) are produced at intermediate times.

(c) mRNA production switches to full-length, unspliced transcripts which act both as the virion RNA and the mRNA for the Gag–Pol polyprotein late in infection.

This process is controlled by the Rev protein. Only the short, fully spliced viral mRNAs appear in the cytoplasm of Rev-minus cells (see Chapter 11).

Rev activity requires both positive and negative regulatory sequences. These are located in the regions of the *gag* and the mRNAs which are removed by splicing:

(a) The Rev-responsive element (RRE) acts as a positive control element that is absolutely required for Rev activity. The RRE is located within the *env* reading frame and acts as a binding site for the Rev protein (see Chapter 9).

(b) The Rev-dependent mRNAs also carry negative regulatory elements, now called instability (INS) sequences. In the absence of either Rev, or a functional RRE sequence, Rev-dependent mRNAs are highly unstable in the cytoplasm. Their cytoplasmic levels rise dramatically if Rev is added in *trans* or when the INS sequences are inactivated by a series of mutations.

The mechanism of action of Rev is still unknown. It was originally believed that Rev activity was coupled to splicing. However, Rev can also function in the absence of splicing, due to its ability to overcome the effects of the INS sequences. There is still a debate over whether Rev acts to stimulate the export of partially spliced viral mRNAs from the nucleus or simply to stabilize these mRNAs in the cytoplasm. Although it is probable that Rev initially binds RRE-containing mRNAs in the nucleus, recent immunoprecipitation experiments have shown that Rev is associated with mRNAs in the cytoplasm of infected cells. Rev itself is able to shuttle between the nucleus and cytoplasm.

5.3 Nef regulation

The Nef protein is the third viral protein synthesized early during infection from spliced mRNAs. It was named Nef, or negative factor, because of early reports that deletion of Nef from recombinant viruses stimulated replication three- to tenfold. However, these observations were probably due to a number of experimental artefacts:

(a) The *nef* gene overlaps the viral LTR sequence. In some cases mutations in *nef* gene produced polar effects on promoter function.
(b) Recombinant viruses used to assay Nef function in these early experiments are now known to have carried mutations in the *vpr* and *vpu* genes.

Although Nef does not appear to directly regulate transcription from the viral LTR, it undoubtedly plays a significant role in the viral growth cycle. In recent experiments studying the pathogenesis of SIV*mac*239, have provided an important demonstration that Nef is essential for viral growth *in vivo*. SIV viruses rescued from clones carrying premature stop codons in the *nef* gene are able to grow normally *in vitro*. However, when these same *nef*-minus viruses were used to infect monkeys, they quickly reverted to a *nef*-positive genotype.

One biochemical function of Nef is to down-regulate the cell surface expression of the CD4 receptor (see Chapter 13):

(a) Normal CD4 mRNA levels are maintained in presence of Nef.
(b) The interaction between CD4 and Nef appears to be direct. A complex between the two proteins can be immunoprecipitated from bacculovirus-infected cells which express high levels of CD4 and Nef.

However, the activity of Nef is unlikely to be limited to the regulation of CD4 levels. There is increasing evidence that Nef may interfere with normal T cell signalling pathways, perhaps by blocking the interactions between CD4 and cellular protein kinases.

6. Assembly and release of virus particles

The latter stages of the HIV life cycle involve the assembly of viral components into particles and release from the cell via budding. Although the details of assembly are not well understood for any retrovirus, it does seem clear that budding is a self-assembly process (*Figure 3*).

(a) Gag precursor molecules accumulate on the inner surface of the cellular membrane by virtue of N-terminal myristoylation and protein–protein interactions.

(b) The Gag proteins in turn bind retroviral RNA. This interaction appears to involve the binding of RNA packaging signal sequences by the zinc-figure nucleic acid binding domain of the NC (nucleocapsid) protein.

(c) Prior to budding, envelope glycoproteins must also accumulate at the site of virus assembly. This process has never been properly studied but it presumably requires an interaction between the envelope glycoproteins and MA (matrix) proteins.

6.1 Control of assembly by protease

The viral protease is required to cleave the Gag precursors after the virus has budded (see Chapter 5). Mutants of HIV which carry inactive protease show normal assembly and budding but are unable to form mature infectious virions. Similarly, treatment of HIV-infected cells with inhibitors of protease leads to the accumulation of immature and non-infectious virus particles.

6.2 Control of assembly by Vpu

In HIV-1, envelope biosynthesis and particle release appears to be regulated by a specialized maturation protein, Vpu (see Chapter 14). The HIV envelope glycoprotein is able to bind to the CD4 receptor in cells and this traps the protein and reduces its levels. Vpu reduces the formation of envelope–CD4 complexes by inducing the rapid degradation of the CD4 protein. HIV-1 viruses carrying mutations in the *vpu* gene also show an accumulation of characteristic virion particles with an 'immature' morphology and an unusual pattern of budding. This suggests that Vpu also plays a role in virion release.

7. Drug discovery

The AIDS crisis has stimulated an unprecedented joint effort between research groups in the academic world and the pharmaceutical industry aimed at drug discovery. The detailed understanding of the mechanisms of virus growth outlined above have provided a wealth of new targets. Indeed virtually every step in the virus life cycle can potentially be converted into a drug target. In addition rapid and reliable screening methods have been developed (see Chapter 16).

7.1 Drug resistance

Unfortunately, the problem of drug discovery in the HIV field is even more complicated than the already formidable task of finding small molecules that specifically and effectively inhibit virus replication. Because HIV mounts a systemic infection and mutates rapidly, drug resistant mutants rapidly appear in patients treated with anti-reverse transcriptase inhibitors, such as the nucleosides AZT, ddI, and ddC, or against peptidic anti-protease inhibitors.

Clearly new and innovative approaches will be needed to circumvent the problem of resistance.

Within the small molecule field several approaches can be taken to counter drug resistance using 'cocktails of multiple drugs':

(a) Treatment with multiple drugs against different targets, such as reverse transcriptase and protease might offer some clinical advantages in the short-term. However, recombination between resistant mutants will probably limit the value of this approach in the longer-term.
(b) Treatment with multiple drugs directed against different sites on the same protein can potentially limit resistance. However, this approach will not work if one site is notably more susceptible to resistance.
(c) Drugs specifically directed to resistant mutants appears to offer the best hope to limit resistance. Unfortunately this approach requires lengthy research into the nature of resistance, as well as new drug discovery programmes.

A related approach that deserves more attention is to direct drugs against non-viral targets, such as human cofactors. For example, it might be possible to design CD4-directed drugs that inhibit gp120 binding, but not the normal function of CD4. The regulatory proteins, Tat and Rev, also appear to interact with cellular cofactors. It might be possible to specifically inhibit these interactions once the cellular factors have been identified.

7.2 Drug design

The three-dimensional structures of an impressive number of molecules involved in HIV replication are now known. These include, gp120 binding domain of CD4, reverse transcriptase, RNase H, protease, and the MA (p17) protein. High-resolution structural information will soon be available about the RNA binding sites for Tat and Rev. The structure of the DNA-binding domain integrase is also known. Hopefully, this wealth of structural information will make a substantial contribution to drug design programmes. It is probably premature to use structural information for *de novo* drug design, although recent progress in the field has been impressive. In spite of this, knowledge of the structures and the structures of lead compounds complexed to the target molecules finds a useful place in modern drug discovery programmes. This information provides useful insights into new leads, and perhaps most importantly, quickly rules out avenues which will not work because of steric clashes.

7.3 Ribozymes

Ribozymes and antisense constructs (see Chapter 17) are nucleic acid moieties that can specifically inhibit HIV. The great advantage of this approach is that specificity can be easily achieved through hybridization between the target molecule and appropriate flanking sequences in the ribozyme. The major disadvantage of the approach is that cellular and intracellular targeting of the ribozymes is difficult to achieve. Resistance mutations are also possible. None the less, selective inhibition of HIV *in vitro* is now routinely achieved using ribozymes.

7.4 Gene therapy

Gene therapy approaches can also employ delivery of an inhibitory protein (see Chapter 18). Numerous strategies are possible, including the delivery of antibodies and mutant viral proteins which interfere with replication. Once again, progress has been impressive, and the first gene therapy trials using dominant mutants of the regulatory protein Rev will begin shortly. The major disadvantage of this approach is that the gene therapy protocols are all extremely complicated and expensive. Unusually, repeated bone marrow transplantation is required, since the expression of heterologous genes in this tissue is short-lived.

8. Future prospects

Since the discovery of HIV in 1983 there has been an explosion of research into the virus life cycle. More is now known about HIV than any other retrovirus, and if the present pace of research is maintained, we shall soon know as much about HIV as any other virus.

The information generated by the basic research effort is enhancing drug discovery efforts, but sadly the goal of finding an effective anti-viral therapy remains elusive. The design of novel protease inhibitors has shown that highly selective compounds can be rapidly developed by exploiting structural information. Unfortunately these compounds have only achieved limited clinical success because of problems of resistance. None the less, there is cause for optimism. As long as the HIV research effort continues at its present pace, it can not be too long before a new generation of effective anti-viral compounds is discovered.

Acknowledgements

I thank my colleagues at LMB and in the MRC AIDS Directed Programme for their advice and support.

2

Reverse transcriptase

W. A. BEARD and S. H. WILSON

1. Introduction

The reverse transcriptase (RT) of human immunodeficiency virus type-1 (HIV-1) converts the single-stranded (+) viral RNA genome into double-stranded proviral DNA prior to its integration into the host genomic DNA. The HIV-1 RT has therefore been the target for anti-viral drug design (1). The cloning of the RT in biologically active form has made large quantities of enzyme available for biophysical and kinetic studies. *In vivo*, the heterodimeric form (p66/p51) of RT is believed to be the result of carboxy-terminal proteolytic cleavage of one subunit of the homodimer (p66/p66) (2). Crystal structures of RT complexed with DNA (3) and a non-nucleoside inhibitor (4) have recently been published. In addition, pre-steady state kinetic studies are now available (5–7). This detailed structural and kinetic information is providing the prerequisite information for rational drug design of specific and potent inhibitors.

HIV-1 RT is a multifunctional enzyme with three recognized enzymatic activities:

- RNA-dependent polymerase
- DNA-dependent polymerase
- ribonuclease H

The RNase H is responsible for hydrolysis of the RNA template after first strand DNA synthesis. Both polymerase activities catalyse template-directed phosphodiester bond formation in the 5′→3′ direction. The DNA polymerase reaction occurs by means of a nucleophilic displacement of pyrophosphate at the α-phosphorous of an incoming deoxynucleoside triphosphate (dNTP) by the 3′-OH terminus of the growing primer strand. Steady state kinetic studies indicate that DNA synthesis proceeds by an ordered mechanism in which template–primer (T–P) binds to free heterodimeric RT to form the first complex in the reaction pathway (8).

The measurement of deoxynucleoside monophosphate (dNMP) incorporation into DNA by DNA polymerases is straightforward, but the conditions of the assay will influence the detailed interpretation of the results significantly.

In this chapter, after outlining the standard protocols for following RNA- and DNA-dependent polymerase activity, there is a detailed discussion of interpretation of typical results. Since steady state kinetic parameters (K_M, K_i, k_{cat}) are not always easily interpreted, we outline methods for determining individual rate constants which may be important in the catalytic cycling of the reverse transcriptase. Once all of the rate constants describing substrate (product) binding, chemistry, and enzyme•substrate (product) complex isomerizations have been accurately measured, a kinetic model which is able to predict the steady state behaviour of the reverse transcriptase under a variety of conditions can be derived.

2. Steady state characterization

Much of the enzymological data on reverse transcriptase is derived from initial velocity measurements using low concentrations of enzyme. The steady state kinetic parameters that are measured under these conditions provide useful parameters in gauging the effects of altered substrates, inhibitors, or enzymes.

2.1 Reaction conditions

The choice of reaction conditions should be considered in some detail. Generally, it is appropriate to choose reaction conditions that approximate physiological conditions:

- pH $\approx$ 7.4
- temperature = 37 °C
- ionic strength $\approx$ 150 mM

We routinely employ a minimal reaction mix consisting of: 50 mM Tris–HCl pH 7.4, 100 mM KCl, and 5 mM $MgCl_2$. Variation of these conditions can illuminate mechanistic detail. For example, the identification of ionizable groups participating in ligand binding, or catalysis, can be deduced from a systematic pH study of the enzymatic reaction. Additionally, it may be advantageous to work at lower temperatures if the reaction is too fast to measure reliably, or if the enzyme or substrates are not stable at an elevated temperature.

In some instances, 5 mM DTT and 50–100 μg/ml BSA are included in the reaction mixture to stabilize the enzyme. DTT is useful for protecting protein sulfhydryl groups from oxidation, whereas BSA serves to stabilize dilute solutions of reverse transcriptase.

(a) Include DTT when making enzyme dilutions.

(b) DTT can be omitted from the final reaction mixture without a deleterious effect on activity.

(c) Omit DTT from inhibition studies employing 3′-azido-3′-deoxythymidine 5′-triphosphate (AZTTP) since DTT is known to reduce AZTTP to 3′-amino-3′-deoxythymidine 5′-triphosphate (9).

(d) The pH of Tris buffers should be measured at the appropriate temperature using a 400 μl 'mock-up' reaction mixture, since the pK_a of the Tris buffer is very sensitive to temperature.

(e) Use sterile buffers, reagents, and glassware.

2.2 Substrates and enzyme

2.2.1 Template–primer pairs

Homopolymeric template–primer pairs have been commonly used in the past to assay RT. The use of these template–primer pairs in assays to measure specific steps in the reaction pathway has several advantages:

- widely available and inexpensive
- insensitive to the effects of competing deoxynucleotides since only a single dNTP is required in the polymerase reaction
- permits comparison of results from numerous laboratories employing a variety of enzymes and techniques
- eliminates the effect that a heterogeneous sequence might have on kinetic constants

The disadvantage of these template–primer pairs are:

- short primers can stack on the template if care is not taken to anneal them properly, therefore precluding accurate calculation of the concentration of 3′ primer termini (10)
- the physiological relevance of these template-primers can be questioned

Heteropolymeric template–primers can also be used in RT assays. However, it should be recognized that a heteropolymeric nucleotide sequence can significantly influence the enzymatic reaction, as shown by the characteristic pattern of termination sites observed with RT using a M13 DNA template (11–13) or a heteropolymeric RNA sequence (12). To minimize this problem, single nucleotide incorporation can be assayed. In this case, only the correct deoxynucleoside triphosphate, corresponding to the complementary base on the template, need be included in the reaction mix.

The quantification of the concentration of 3′ primer termini in the reaction is a critical parameter in defining binding constants as well as in the interpretation of the results. It is therefore meaningless to express homopolymeric T–P concentrations as 260 nm absorbance units/ml or μg/ml. The concentration of commercially available homopolymeric oligonucleotides can be determined from the extinction coefficients supplied by the manufacturer. Since homopolymeric templates are often heterogeneous in size, we typically express the

Table 1. Extinction coefficients at 260 nm (ε, 1/mM/cm)[a] for DNA and RNA oligonucleotides (14)

Nucleotide		**Deoxynucleotide**	
A	15.4	dA	15.4
C	7.2	dC	7.4
G	11.5	dG	11.5
U	9.9	dT	8.7

First deoxynucleotide	**Second deoxynucleotide**			
	A	**C**	**G**	**U**
A	13.7	10.5	12.5	12.0
C	10.5	7.1	8.9	8.1
G	12.6	8.7	10.8	10.6
U	12.3	8.6	10.0	9.8
	dA	**dC**	**dG**	**dT**
dA	13.7	10.6	12.5	11.4
dC	10.6	7.3	9.0	7.6
dG	12.6	8.8	10.8	10.0
dT	11.7	8.1	9.5	8.4

[a]25 °C and neutral pH.

concentration of template as the concentration of template nucleotide relative to primer. A template to primer nucleotide ratio of ten would correspond to 200 template nucleotides per 20-mer primer (i.e. ≈ 180 nucleotides of single-stranded nucleic acid per primer). The concentration of primer (3′ termini) would also need to be specified as determined spectrophotometrically.

Heteropolymeric RNA and DNA oligonucleotides are readily available commercially, as well as through DNA synthesis facilities in private laboratories. The extinction coefficients can be calculated from the values compiled in *Table 1* (14). For example, the extinction coefficient for the hypothetical 26-mer oligonucleotide ABCD ... YZ can be calculated from the formula:

$$\epsilon_{ABCD \ldots YZ} = [2(\epsilon_{AB} + \epsilon_{BC} + \epsilon_{CD} + \ldots + \epsilon_{YZ}) - \epsilon_B - \epsilon_C - \epsilon_D - \ldots - \epsilon_Y)].$$

Oligonucleotides can be stored lyophilized at −20 °C for long periods of time. They are stable when reconstituted in TE buffer (10 mM Tris–HCl pH 7.4, 1 mM EDTA) and stored at −20 °C for several months.

The procedure for annealing heteropolymeric template–primers is outlined in *Protocol 1*. The optimal temperature for annealing short oligonucleotides (< 100 nucleotides) is generally between 50–70 °C. The precise optimum can be determined by doing several hybridizations at 5 °C intervals in this temperature range. If annealing is poor, the duplex can be purified from the single-stranded oligonucleotides by non-denaturing 20% polyacrylamide gel electrophoresis (15, 16). Efficiency of hybridization can be determined by measuring the level of incorporation of a single [^{32}P]dNTP complementary to the first single-stranded template base. Extended duplexes can then be collected by filter binding (see *Protocol 2*).

Protocol 1. Heteropolymeric template–primer annealing

Equipment and reagents

- 650 μl Eppendorf tube
- Oligonucleotides in TE buffer (10 mM Tris–HCl pH 7.4, 1 mM EDTA)
- Variable temperature water-bath or heating block
- TE buffer

Method

1. Mix equal concentrations (1–10 μM) of oligonucleotides in TE buffer.
2. Incubate the tube for 3 min in a beaker with ≈ 750 ml water heated to 90 °C.
3. Place the tube in a water-bath set at 50–70 °C for an additional 30 min.
4. Cool the hybridization mixture to room temperature slowly, 1–2 h.
5. Store annealed template–primers at −20 °C.

Annealing of homopolymeric template–primers is similar to that outlined for heteropolymeric oligonucleotides. The homopolymeric primer (1–20 μM) is mixed with template (template to primer nucleotide ratio of at least ten) and heated to 90 °C. In this case, however, the reaction mixture is allowed to cool to room temperature over a 3–4 h period after removing the heat source to the beaker of water.

2.2.2 Deoxynucleoside triphosphate

The highest quality dNTP reagents available should be used. These can be purchased lyophilized or as a lightly buffered 100 mM dNTP solution (Pharmacia). A working solution of 1–3 mM is made in 10 mM Tris–HCl pH 7.4. If the product DNA will be radioactively labelled by the incorporation of ^{3}H or ^{32}P-labelled deoxynucleotides, then radioactively labelled deoxynucleoside triphosphate is added to this working solution so that the specific activity is ≈ 5000 d.p.m./pmol dNTP.

2.2.3 Active site titration

A critical parameter in defining the activity of an enzyme preparation is the concentration of active enzyme. The active form of RT is a dimer with polypeptides of molecular mass of 66 kDa and 51 kDa (i.e. heterodimer) or two 66 kDa polypeptides (i.e. homodimer). The homodimer forms less readily than the heterodimer (17, 18). However, it has recently been demonstrated that the homo- and heterodimeric forms of RT derived from HXB2R proviral DNA are kinetically equivalent and that nucleic acids stabilize the dimeric form of the enzyme (19). Homodimeric enzyme is available commercially from Worthington Biochemical Corporation. Bacterial overexpression and purification of RT has also been described (20).

Enzyme concentration can be estimated from a standard dye-binding protein assay (21) with a purified protein preparation. However, unless the dye-binding assay has been calibrated with a preparation of gravimetrically determined purified enzyme, the estimate would only serve as a reference concentration with similar preparations. Likewise, UV absorbance may not reflect the active concentration of enzyme (5).

Enzyme dilutions are made in a cold (4°C) freezing buffer: 50 mM Tris–HCl pH 7.4 (adjusted at 25°C), 500 mM KCl, 50% glycerol, and 5 mM DTT. The dimeric form of RT is stabilized by Mg^{2+} (22), so that 5 mM $MgCl_2$ may also be included. The DTT is added immediately before use from a 500 mM stock stored at −20°C. Concentrated enzyme (>1 μM) is stable for several months when stored at −70°C. Dilute enzyme stocks (nM) are stable for several hours when stored on ice.

If product release is the rate-limiting step (i.e. slowest step) during catalytic cycling, then a burst of product formation is observed upon initiation of the reaction (*Figure 1a* and *b*). The amount of product in the burst is equivalent to the amount of active enzyme. When nucleotide incorporation is limited to a single dNMP, then a burst of product formation is observed indicating that incorporation is rapid and product release is slow (6, 23). Incorporation can be limited to a single deoxynucleotide by using a heteropolymeric T–P and including only the complementary template deoxynucleoside triphosphate to be incorporated at the 3′ primer terminus. If a homopolymeric T–P is used, such as poly(rA)–p(dT)$_{20}$, then incorporation can be limited to a single incorporation with dideoxythymidine triphosphate, ddTTP, or AZTTP. Active site titrations can be performed by following the time course of radioactive nucleotide incorporation (*Protocol 2*) or by following the extension of a radioactively labelled primer (n) to form extended primer (n + 1) (*Protocol 4*). Increasing the enzyme concentration should give a proportional increase in the burst amplitude (ordinate-intercept; *Figure 1b* and *c*). With some DNA templates or enzyme preparations, it has been found that $k_3 \approx k_2$ (*Figure 1a*) so that the burst amplitude does not reflect the active site concentration (7).

2.3 Product analysis

DNA synthesis is followed primarily by the incorporation of [α^{32}P] or [^{3}H]dNTPs into DNA. The amount of synthesis is directly proportional to the amount of radioactivity in the product DNA. Alternatively, the primer can be 5′ ^{32}P-labelled with polynucleotide kinase and [γ^{32}P]ATP. In this case, the extended labelled primers are separated by gel electrophoresis according

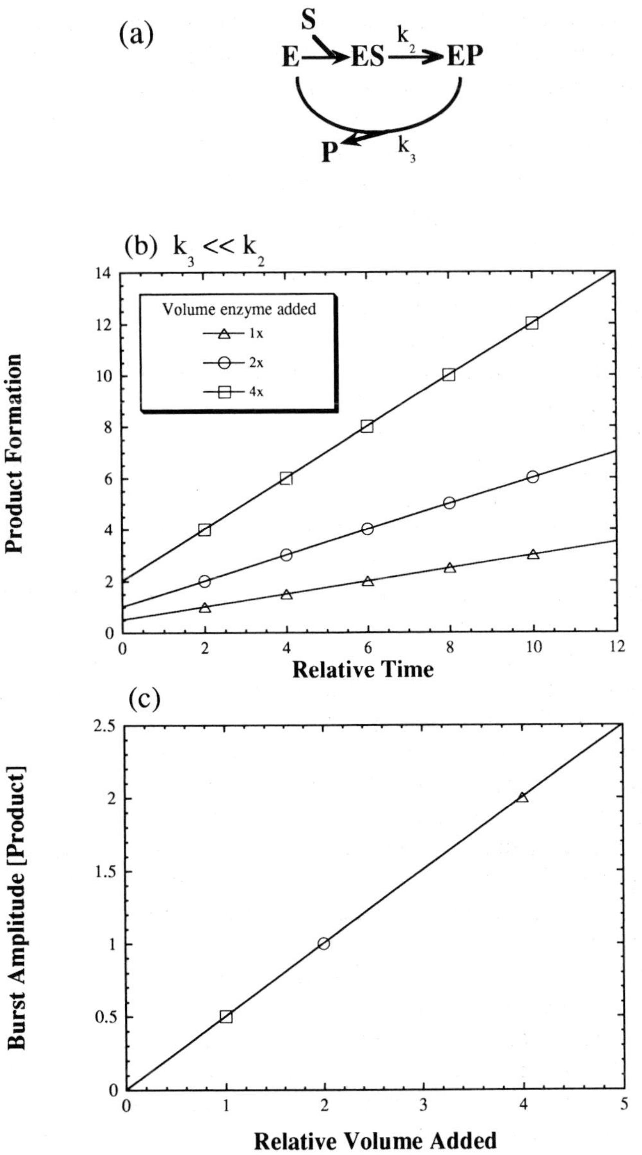

Figure 1. Reverse transcriptase active site titration. (a) Kinetic model illustrating substrate (S) binding, catalysis (k_2), and product (P) dissociation (k_3). (b) When product release is the slowest step ($k_3 \ll k_2$), the time course of product formation is biphasic. A rapid phase, which is too rapid to measure by manual mixing and sampling techniques, and a linear steady state phase. The amount of product formed in the 'burst' phase can be determined by extrapolating the linear phase to the ordinate-intercept and is equivalent to the product formed in the first turnover of active enzyme. The burst amplitude, and the slope of the linear phase, are proportional to the concentration of active enzyme. The dependence of the ordinate-intercept on active enzyme from (b) is re-plotted in (c).

to size. The product is visualized by autoradiography, excised from the gel, and the amount of radioactivity determined by scintillation counting (24).

2.3.1 Filter binding

Protocol 2 describes a filter binding assay for RT. DNA binds avidly to positively charged DEAE filters. Radioactively labelled DNA can be collected on these filters and the unincorporated radioactively labelled dNTPs can be removed by washing.

Protocol 2. Reverse transcriptase polymerization assay

Equipment and reagents

- 650 μl Eppendorf tubes
- 5 ×–10 × RT buffer components: Tris–HCl pH 7.4, KCl, $MgCl_2$
- Radioactively labelled dNTP
- T–P pair
- RT
- Water-bath
- Heat lamp
- 25 mm DE-81 filters
- 0.5 M EDTA pH 8
- Small glass dish or beaker
- Shaker
- Filter washing solutions: 300 mM ammonium formate pH 8 and 95% ethanol
- Glass scintillation vials and scintillation cocktail

Method

1. Assemble in a 650 μl Eppendorf tube held in a water-bath at the appropriate temperature (e.g. 37 °C), a 120 μl reaction mixture containing:
   - RT buffer: 50 mM Tris–HCl pH 7.4, 100 mM KCl, 5 mM $MgCl_2$
   - 100 nM or more T–P (expressed as 3′ primer termini)
   - 25 μM of the appropriate radioactively labelled dNTP

   This reaction mixture should be made from concentrated stocks of the components and brought up to 120 μl with water.
2. Allow the temperature of the reaction mixture to equilibrate.
3. Initiate reaction by addition of a small volume (e.g. 5 μl) of diluted enzyme. If any of the components of the freezing buffer need to be omitted from the reaction, then intermediate enzyme stocks can be made up in 10 mM Tris–HCl pH 7.4 and used immediately.
4. Remove 20 μl aliquots from the reaction mixture at time intervals and add to Eppendorf tubes containing 10 μl of 0.5 M EDTA pH 8.
5. Number Whatman DE-81 filters (25 mm diameter) with a pencil.
6. Spot quenched reaction aliquots on separate filters.
7. Spot 5 μl of the reaction mixture on one filter if ^{32}P-labelled dNTP incorporation is followed. It is put in a scintillation vial with 5 ml of

scintillation fluid. This sample will be used to determine the specific activity (c.p.m./pmol dNTP) of the reaction mixture. Note: Do not wash this filter.

8. Dry the filters in air or with a heat lamp.
9. Wash filters to remove unincorporated radioactive deoxynucleoside triphosphates in a beaker or glass dish.
   (a) Wash filters four times with gentle agitation for 5 min/wash. Include approx. 5 ml of 0.3 M ammonium formate pH 8 per filter. The pH of the ammonium formate is adjusted with ammonium hydroxide. Alternatively, the filters can be washed five times with 125 mM Na_2HPO_4 pH 7.
   (b) Wash filters with 70–95% ethanol. This step can be omitted if the filters are initially washed with Na_2HPO_4.
10. Dry filters and place in glass scintillation vials with 5 ml scintillation fluid for scintillation counting.

Follow time courses of product formation at different enzyme concentrations (e.g. 1–20 nM). This will demonstrate that the velocity is proportional to enzyme concentration and determine the time interval over which the reaction velocity is linear. The pmol of [^{32}P]dNMP incorporated can be calculated from the c.p.m. of a sample and the specific activity (c.p.m./pmol) determined from an unwashed filter. [^{3}H]dNMP is counted with a higher efficiency when incorporated into DNA than as a free deoxynucleoside triphosphate on DE-81 filters (25). Incorporated [^{3}H]dNMP bound to DE-81 filters can be used as a measure of specific activity. In order to incorporate all of the ^{3}H-labelled deoxynucleoside triphosphate, reactions should be carried out with excess RT and T–P and limiting dNTP for a prolonged period to ensure complete incorporation of the labelled dNTP. No difference in c.p.m. between washed and unwashed filters is taken as an index of total incorporation of the deoxynucleoside triphosphate.

2.3.2 DNA precipitation

An alternative method for separating unincorporated radioactively labelled dNTPs from the radioactive DNA product is precipitation of the DNA with trichloroacetic acid (TCA) followed by collection on nitrocellulose or glass fibre filters (*Protocol 3*). ^{3}H-labelled DNA is counted with a very low counting efficiency. The efficiency can be improved by solubilizing the nitrocellulose filters or the precipitated DNA with the glass fibre filters (26). For this reason, we recommend using [^{32}P]dNTPs which can be counted on dry filters with a 25% counting efficiency (Cerenkov radiation determined in the ^{3}H channel of a scintillation counter) or near 100% with a toluene-based scintillation cocktail.

Protocol 3. TCA precipitation of DNA

Equipment and reagents

- Whatman GF/A glass fibre or Millipore nitrocellulose filters (25 mm)
- Vacuum manifold (Millipore)
- 10% trichloroacetic acid and 20 mM pyrophosphate

Method

1. Initiate reaction as in *Protocol 2*.
2. Stop reactions by adding aliquots to 1 ml of cold 10% TCA and 20 mM pyrophosphate.
3. Collect precipitated DNA by vacuum filtering the solution through glass fibre or nitrocellulose filters. Wash four times with 5 ml 10% TCA, then twice with 5 ml 95% ethanol.
4. Place dried filters in scintillation vials with scintillation fluid for counting.
5. Spot a separate filter with 5 μl of reaction mixture and count this unwashed filter to determine the specific activity of the radioactive product.

2.3.3 Gel electrophoresis

A more sensitive and informative measure of primer extension is to 5′ ^{32}P-label the primer, and after extension of the primer with non-labelled dNTPs (*Protocol 4*), separate the product DNA by gel electrophoresis corresponding to its size. The ability to visualize the product distribution (variety of extension products) offers an advantage over collection of the total products by filter binding. In addition, very low levels of incorporation can be measured requiring very little reagent. The primer is labelled by T4 polynucleotide kinase and [γ-^{32}P]ATP as suggested by the manufacturer or as outlined by Sambrook *et al.* (16). The unincorporated [^{32}P]ATP is separated from the 5′-labelled DNA by gel filtration with centrifugation (i.e. spin column; e.g. Bio-Spin-30 column from Bio-Rad). Detailed protocols describing gel electrophoresis of nucleic acids and their detection has recently appeared in this series (15).

Protocol 4. Gel electrophoresis

Equipment and reagents

- Gel apparatus
- Polyacrylamide slab gel
- Electrophoresis buffer (10 × TBE): 0.89 M Tris base, 0.89 M borate, 25 mM EDTA
- Gel loading buffer: 80% formamide, 50 mM Tris–borate, 1 mM EDTA, 0.1% bromophenol blue, 0.1% xylene cyanol
- Kodak XAR-2 X-ray film
- 5′ ^{32}P-labelled primer

Method

1. Assemble gel apparatus with polyacrylamide gel.
2. Add a small aliquot of labelled primer to the hybridization reaction, ≈ 3000 c.p.m./pmol (see *Protocol 1*).
3. Initiate reaction as in *Protocol 2.*
4. Stop reactions by withdrawing aliquots (5 μl) and adding to an equal volume of 2 × gel loading buffer.
5. Heat samples to 90°C for 3 min and then quickly put on ice.
6. Apply samples, 4–5 μl, to 15–20% denaturing (7 M urea) polyacrylamide gels.
7. Electrophorese.
8. Expose the gel to Kodak XAR-2 film at −70°C.
9. Quantify the autoradiograph by densitometry or use it as a template to cut out the product bands prior to scintillation counting.

2.4 Kinetic analysis

In the past, kinetic equations have been linearized and initial velocity data routinely analysed by graphical methods. However, the popularity of laboratory computers and the availability of published computer programs (27, 28) and commercially available software packages (Jandel), has made analyses more rapid and less tedious. On the other hand, it has also separated the scientist from the data so they are less aware of what the analyses represent.

2.4.1 Michaelis constants

Initial velocity data can be collected as described in *Protocol 2* or determined by an end-point assay (single time point) provided product formation is linear over this time interval. The data can then be fitted to the Michaelis–Menton equation, where S is the varied substrate concentration, k_{obs} and k_{cat} are the initial and maximum velocities normalized for enzyme concentration (i.e. k = velocity/[E]), and K_{M} is the Michaelis constant:

$$k_{\text{obs}} = \frac{S \times k_{\text{cat}}}{S + K_{\text{M}}} \qquad (1)$$

The precise value for these kinetic constants will strongly depend on reaction conditions, choice of T–P, as well as the strain of virus from which the reverse transcriptase is derived (19). At 37°C with poly(rA)–oligo(dT), a commonly used T–P, the kinetic constants are generally reported to be: k_{cat}, 1–5/sec; $K_{\text{M,dTTP}}$ = 1–10 μM; $K_{\text{M,T-P}} < 50$ nM. If the K_{M} for dNTP is to be measured by following the incorporation of radioactively labelled dNMP, then time courses determined at different dNTP concentrations offers the advantage that a correction for the background radioactivity binding to the filters is not required, as for an end-point assay.

2.4.2 Processive versus distributive dNTP incorporation

The reverse transcriptase from HIV is a processive polymerase in that it can incorporate many deoxynucleoside triphosphates with a single T–P encounter. A simple model describing single nucleotide incorporation is illustrated below where E represents reverse transcriptase, T–P is the template–primer, N is the deoxy- or dideoxynucleoside triphosphate, and $T–P_{+1}$ is the template–primer after a single nucleotide incorporation.

$$E + T\text{–}P \underset{k_{-1}}{\overset{k_1}{\leftrightarrow}} E_{T\text{–}P} + N \underset{k_{-2}}{\overset{k_2}{\leftrightarrow}} E^{N}_{T\text{–}P} \overset{k_3}{\rightarrow} E_{T\text{–}P+1} \overset{k_4}{\rightarrow} E + T\text{–}P_{+1} \qquad \text{(Scheme I)}$$

If incorporation is limited to a single deoxynucleoside monophosphate (i.e. distributive synthesis), then the steady state kinetic constants reflect this change in catalytic cycling. *Table 2* summarizes the kinetic constants for dNTP under distributive and processive polymerizing conditions. Note that the K_M for dNTP, as well as k_{cat}, is dependent on the sequence of steps in

Table 2. Summary of steady state kinetic constants as defined by individual rate constants for the mechanism illustrated in Scheme I[a]

		Incorporation[b]			
		Single nucleotide		Processive	
Approach	Assumptions	K_M	k_{cat}	K_M	k_{cat}
Steady state[c]					
	$k_3 \ll k_4$	$(k_{-2} + k_3)/k_2$	k_3	Not applicable	
	$k_3 \gg k_4$	$k_4(k_{-2} + k_3)/k_2k_3$	k_4	$(k_{-2} + k_3)/k_2$	k_3
Rapid equilibrium ($k_{-2} \gg k_3$)					
	$k_3 \ll k_4$	K_d	k_3	Not appicable	
	$k_3 \gg k_4$	K_dk_4/k_3	k_4	K_d	k_3

[a] Template–primer saturating.
[b] Single nucleotide incorporation refers to a situation where only a single nucleotide is incorporated per polymerase encounter with template–primer such as when the template–primer is a heteropolymer and only the correct incoming deoxynucleoside triphosphate is present or when a homopolymeric T–P is used and a dideoxynucleoside triphosphate is incorporated. Processive incorporation refers to the situation when $E_{T\text{–}P+1}$ would not accumulate (i.e. $E_{T\text{–}P+1} = E_{T\text{–}P}$).
[c] The full expressions for k_{cat} and K_M, when incorporation is limited to a single nucleoside triphosphate, are $k_3k_4/(k_3 + k_4)$ and $[k_4/(k_3 + k_4)][(k_{-2} + k_3)/k_2]$, respectively. The expressions for k_{cat} and K_M when incorporation is processive are as given.

the catalytic cycle. The rate-determining steps for a processive polymerase is different for single nucleotide and processive incorporation. For single nucleotide incorporation, the rate-determining step is the release of the extended primer, T–P_{+1}, from E (i.e. k_4), whereas during processive incorporation, it is k_3 and the release of the extended primer is not on the catalytic pathway.

2.4.3 Inhibition

The reverse transcriptase is a popular therapeutic target. However, there is a need for a wide variety of new and potent inhibitors due to the problem of side effects and the development of viral resistance with existing inhibitors. The most recognized inhibitor of reverse transcriptase is AZTTP which is the phosphorylated form of AZT. It belongs to a general group of nucleoside inhibitors which are 2′,3′-dideoxynucleosides. Among these are 2′,3′-dideoxycytidine (ddC) and 2′,3′-dideoxyinosine (ddI). Their mechanism of inhibition is discussed in detail below.

A number of structurally divergent compounds have recently been described (29–31) and are generally termed non-nucleoside inhibitors. They appear to bind to a distinct site on the RT as they can compete with one another for this site, but dNTPs or T–P afford no protection to binding (29, 32, 33). Inhibition by one of these non-nucleoside inhibitors, nevirapine (BI-RG-587), has been shown to be non-competitive with respect to T–P and dNTPs (30).

Inhibitors are typically classed according to the effect they have on steady state kinetic parameters, k_{cat} and K_M. Simple competitive inhibitors increase K_M without affecting k_{cat}, whereas non-competitive inhibitors decrease k_{cat} without influencing K_M. If both parameters are effected to the same extent, then the inhibition is uncompetitive, and is termed mixed if they are affected differentially. Steady state analysis to determine the type of inhibition, as well as, an apparent inhibition constant, K_i, are outlined in most basic enzymology texts.

The nucleoside inhibitors discussed above are substrate analogues for reverse transcriptase. In this situation, and assuming rapid equilibrium binding of the analogue, K_i should be equivalent to K_M. AZTTP has been found to be as efficiently incorporated as the natural dNTP, deoxythymidine triphosphate, when the template base is riboadenosine (6, 23). Inhibition of dTTP incorporation by AZTTP is competitive since these substrates are competing for the same form of the enzyme (i.e. $E_{T\text{–}P}$). However, the low K_i determined for AZTTP, typically in the low nanomolar range with a RNA template, is a consequence of the catalytic pathway. Incorporation of AZTMP results in chain termination because of the lack of a 3′-OH. Since this limits the reaction to a single nucleotide addition, and the reverse transcriptase is processive on this template, K_M, and therefore K_i, is determined by the T–P dissociation rate constant (k_4 in Scheme I) which is now the rate-determining step ($k_{cat} = k_4$). As pointed out in *Table 2*, K_M underestimates the dissociation constant by a factor equivalent to the processivity of the polymerase.

Therefore, the K_i for AZTTP is observed to increase on a substrate where the dissociation rate constant for $T\text{–}P_{+1}$ does not limit catalytic cycling as observed when poly(dA)–$(dT)_{20}$ is the T–P (23). As expected, the K_i is approximately equivalent to its K_M (low micromolar range) which is similar to the equilibrium dissociation constants measured for dNTPs on a DNA substrate (5, 6).

3. Microscopic kinetic constants for substrate and inhibitor binding

From the discussion above, it should be evident that steady state kinetic parameters are not always easily interpreted. This is because they are often a composite of several rate constants. Depending on the situation, a particular rate constant may dominate the kinetic expression. In addition, if this rate constant changes because of a site-directed alteration in the enzyme, it is difficult to predict the magnitude of the effect on an observed steady state kinetic parameter. There are several methods which have been used to measure substrate binding parameters, equilibrium and rate constants.

3.1 Processivity

The processivity of a polymerase refers to the average number of deoxynucleotides incorporated per T–P encounter (association and dissociation). The processivity of RT can be examined with a challenge assay that has been used previously with *E. coli* Pol I (34) and RT (12, 19, 35) with homopolymeric template–primers (*Protocol 5*). This simple assay takes advantage of the processive nature of these polymerases to amplify the signal for a polymerase/T–P interaction. For every T–P binding event there are numerous incorporations of radioactive dNMP that can be easily measured. RT is first pre-incubated with T–P to allow $RT_{T\text{–}P}$ complex formation. A challenger, such as heparin, acts as a trapping agent that competes with T–P for free RT. When heparin and Mg^{2+}/[α-^{32}P]dNTP are added to a mixture of RT and T–P, the $RT_{T\text{–}P}$ complex binds dTTP and undergoes processive dTMP incorporation, while the heparin binds free RT as well as enzyme that has dissociated from the T–P. As a result, dNMP incorporation is limited to a single processive cycle. Under these conditions, deoxynucleotide incorporation is initially rapid, but becomes progressively slower at longer time intervals (*Figure 2a*). The effectiveness of the challenge is demonstrated by including the trapping agent in the pre-incubation phase of the reaction and repeating the time course. Under this condition, no incorporation of dTMP should be observed. The time course for processive polymerization can be described by a model where deoxynucleotide incorporation results from a competition between incorporation and dissociation of the extended primer from the ternary complex with bound dNTP and T–P complex:

$$E + T\text{–}P_{+1} \xleftarrow{k_{off}^{pol}} E^{N}_{T\text{–}P} \xrightarrow{k_{cat}} E^{N}_{T\text{–}P+1} \qquad \text{(Scheme II)}$$

This model predicts that the time course should follow the relationship:

$$\frac{N}{E} = \left(\frac{k_{cat}}{k_{off}^{pol}}\right)\left(1 - \exp(k_{off}^{pol}t)\right) \qquad (2)$$

where N/E is the number of dTMP nucleotides incorporated per dimer of RT, k_{cat} is the steady state rate constant observed in the absence of trapping agent, and k_{off}^{pol} is the dissociation rate constant for T–P from the ternary substrate complex during processive polymerization. After a sufficiently long time period (i.e. $t \gg 1/k_{off}^{pol}$), N/E represents the average number of nucleotides incorporated during a single processive cycle and is equivalent to the ratio of the competing pathways (i.e. incorporation versus dissociation; k_{cat}/k_{off}^{pol}).

Protocol 5. Heparin challenged time course to determine processivity

Equipment and reagents

- See *Protocol 2*
- Heparin

Method

1. Assemble reaction mixtures in 650 μl Eppendorf tubes. Final concentrations would be 15 nM RT, 75 nM poly(rA)–$(dT)_{20}$, 30 μM dTTP, 50 mM Tris–HCl pH 7.4, 10 mM $MgCl_2$, and 1 mg/ml heparin.
2. Add 110 μl $MgCl_2$, heparin, and [α-^{32}P]dTTP to 110 μl of RT which has been pre-incubated with a saturating concentration of T–P for 10 min.
3. Remove 20 μl aliquots at various times from this mixture.
4. Quench aliquots with 10 μl of 0.5 M EDTA pH 8.
5. Determine incorporation of radioactive dTMP by DE-81 filter binding (see *Protocol 2*).

3.2 Template–primer binding affinity

3.2.1 Equilibrium binding

In general, RNA–DNA template–primers bind with a higher affinity than their DNA–DNA homologues (7). The nucleotide sequence and the length of the primer can also influence affinity. Short primers annealed to poly(rA), such as $p(dT)_{10}$ bind to RT much weaker than primers greater than 14 nucleotides (19, 35).

Incubation of RT with increasing concentrations of poly(rA)–$(dT)_{16}$

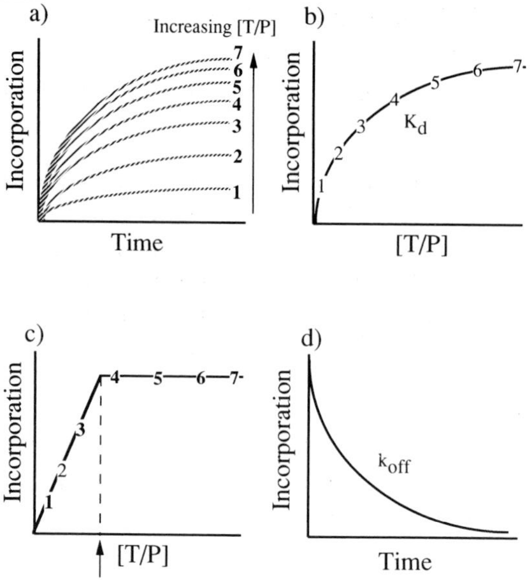

Figure 2. Template–primer binding to reverse transcriptase. (a) Challenged time courses of nucleotide incorporation. Enzyme is pre-incubated with increasing concentrations of T–P (1–7) to form $E_{T\text{–}P}$ complex and assayed by adding a challenger, such as heparin, to bind free reverse transcriptase and enzyme after dissociation of T–P (see *Protocol 5*). Incorporation is therefore limited to a single processive cycle, and the magnitude of the incorporation, after the activity has decayed, is proportional to $RT_{T\text{–}P}$. (b) If binding is weak, the equilibrium dissociation constant can be determined by fitting the T–P concentration dependence of this maximum incorporation to Eqn 3. (c) If T–P binding is tight, then the inflection point in the binding isotherm will represent the binding site concentration (*arrow*). (d) The breakdown of the $RT_{T\text{–}P}$ complex can be followed by challenging the complex with heparin, to trap RT dissociating from T–P, and assaying for the concentration of complex remaining after increasing periods of challenge. If binding of RT to T–P is a simple equilibrium, a monophasic exponential decrease in incorporation is expected representing the dissociation rate constant, k_{off} (see *Protocol 7*).

(*Protocol 6*) results in an apparent increase in the total incorporation of dTMP during a processive cycle of incorporation (*Figure 2a*). Since total incorporation is proportional to the concentration of $RT_{T\text{–}P}$ complex formed in the pre-incubation phase of the reaction, an equilibrium dissociation constant can be determined from a plot of total incorporation versus T–P concentration (*Figure 2b*). The data can be fitted to the equation:

$$I = \frac{I_{max} \times S}{K_d + S} \qquad [3]$$

where I is the amount of deoxynucleotide incorporation, I_{max} is the incorporation at infinite S, and S is the T–P concentration. If binding is tight, so that

K_d is similar to the RT concentration, then the quadratic form of the binding equation should be used to determine K_d,

$$[E_{T-P}] = 0.5(K_d + E_t + T\text{-}P_t) - \sqrt{0.25\,(K_d + E_t + T\text{-}P_t)^2 - (E_t \times T\text{-}P_t)} \quad [4]$$

where E_t and $T\text{-}P_t$ refer to the total concentration of ligand. If ligand binding is very tight so that $K_d \ll$ [RT], then the binding isotherm can be used to titrate the number of T–P binding sites. With this situation, as the T–P concentration exceeds the RT dimer concentration, the incorporation measured after a time interval in the presence of heparin, does not increase further (*Figure 2c*). The inflection point in this isotherm represents the number of binding sites (*arrow* in *Figure 2c*).

Protocol 6. Challenge assay to measure equilibrium binding of template–primer

Equipment and reagents

- See *Protocol 5*

Method

1. Assemble reactions in 650 µl Eppendorf tubes:
   - 40 µl solution of 10–20 nM enzyme (expressed as concentration of dimer)
   - 50 mM Tris–HCl pH 7.4
   - 0.2 K_d to 5 K_d nM poly(rA)–$(dT)_x$
2. Incubate for 10 min.
3. Add 10 µl of 5 mg/ml heparin, 150 µM [α-^{32}P]dTTP, and 10 mM $MgCl_2$.
4. Incubate 10 min.
5. Stop reaction by adding 20 µl of 0.5 M EDTA pH 8.
6. Determine total incorporation by DE-81 filter binding (see *Protocol 2*).

As discussed above, when the dissociation of extended T–P from RT is the rate-limiting step, as is the case for single nucleotide incorporation on a RNA template, then there is a burst of product formation. When the RT is saturated with T–P, then the amount of product formed in the burst is equivalent to the active enzyme concentration (*Figure 1*). Since the breakdown of the RT_{T-P} complex is much slower than incorporation when dNTP is saturating, the burst amplitude reflects the concentration of this complex. The concentration of RT_{T-P} complex can be determined by extrapolation of the steady state rate of product formation to the ordinate (*Figure 1b*) when

RT is pre-incubated with varying concentrations of T–P. The data can then be analysed using Eqn 3 or 4 as outlined above.

3.2.2 Dissociation rate constant

Since association rate constants are in many instances diffusion controlled, the dissociation constant (i.e. K_d) is often dependent on the rate constant for the breakdown of the complex (i.e. k_{off}). Examination of the dissociation rate constant for the RT_{T-P} complex is therefore a sensitive measure of the nature of the interactions occurring between RT and T–P. The breakdown of the RT_{T-P} complex can be followed by challenging the complex with heparin, to trap RT dissociating from T–P, and assaying for the concentration of complex remaining after increasing periods of challenge. If binding of RT to T–P is a simple equilibrium ($RT + T\text{–}P \longleftrightarrow RT_{T-P}$, a monophasic exponential decrease in incorporation is expected representing the dissociation rate constant, k_{off} (*Figure 2d*). Data can be fitted to a single exponential model:

$$I_t = Ae^{-kt} \quad (5)$$

where I_t = pmol of dNMP incorporated at time t, A = amplitude, and $k = k_{off}$. In some instances, the decay of incorporation has multiple phases (12, 19, 35). This suggests that there may be multiple RT_{T-P} forms or multiple kinetic intermediates in T–P binding. A reaction should be run in parallel where heparin is pre-incubated with enzyme, before addition of T–P, to demonstrate the effectiveness of the challenge. If the primer is ^{32}P-labelled, then the complex can be challenged with unlabelled T–P. In this case, radioactive dNTP would be omitted. *Protocol 7* uses radioactive dNMP incorporation to follow RT_{T-P} complex decay.

Protocol 7. Determination of the T–P dissociation rate constant (k_{off})

Equipment and reagents

- See *Protocol 5*

Method

1. Assemble reaction components in a single 650 μl Eppendorf tube.
2. Pre-incubate enzyme with T–P for 10 min before challenging.
3. Add heparin. After heparin addition, the reagent concentrations are 100 nM RT, 80–150 nM T–P (expressed as primer 3′ termini), 50 mM Tris–HCl pH 7.4, and 2 mg/ml heparin.
4. Remove 10 μl aliquots at time intervals after adding challenge.
5. Mix with 10 μl of 20 mM $MgCl_2$, 60 μM [^{32}P]dTTP, and 50 mM Tris–HCl pH 7.4 in another Eppendorf tube.

6. Incubate the reaction mixture. Allow the reaction to proceed through a processive cycle of incorporation (≈ 10 min).
7. Stop reaction with the addition of 10 μl of 0.5 M EDTA pH 8.
8. Determine incorporation by DE-81 filter binding (see *Protocol 2*).

3.3 Deoxynucleoside triphosphate binding

Binding of dNTPs to DNA polymerases is assumed to be in rapid equilibrium. The observed rate of the incorporation step (k_3 in Scheme I) is probably much slower than the dissociation of dNTP from the ternary complex with T–P and RT (k_{-2} in Scheme I). This dissociation rate constant is too fast to measure by rapid-quench-flow techniques.

3.3.1 Steady state dNTP incorporation

As summarized in *Table 2*, the dNTP dissociation constant can be determined by a steady state approach under certain situations. When the T–P dissociation rate constant is rapid, relative to the rate of nucleotide incorporation (i.e. $k_3 \ll k_4$), the polymerase is effectively distributive and, $K_M = K_d$. Since RT is less processive on DNA than on a hybrid RNA–DNA T–P, the K_M determined for AZTTP incorporation on a DNA homopolymeric T–P or a dNTP on a DNA heteropolymeric T–P is a more accurate reflection of the dissociation constant than the K_M determined on the corresponding RNA template. The K_M determined under highly processive conditions should also be equivalent to K_d. This approach assumes that the simple reaction mechanism illustrated in Scheme I does not include potential rate-determining isomerizations which have been postulated to occur with other DNA polymerases (5, 36, 37). Additionally, steady state measurements will be influenced by competing pathways. Although it appears straightforward to determine K_M and k_{cat} on a homopolymeric T–P, a rapidly polymerizing processive polymerase may quickly terminate at a downstream primer so that it will need to dissociate before continuing incorporation. Since it is processive, k_4 is slow and a kinetically significant fraction of enzyme will be in a non-productive complex during steady state. This will result in an underestimate of k_3 and K_d.

3.3.2 Rate of burst

Another approach to determine the dissociation constant for dNTP is by means of rapid-quench-flow techniques. Manual mixing and sampling techniques are not fast enough to measure the rate of the first turnover (i.e. burst; see *Figure 1*). There are a number of commercially available rapid-quench-flow instruments (KinTek Instruments, Molecular Kinetics, Hi-Tech Scientific) which can initiate and stop a reaction after only 1 msec and can therefore be used to measure the early part of a reaction time course. The rate of the first turnover is dependent on the concentration of ternary complex with dNTP and T–P bound to RT and follows the relationship:

$$k_{\text{obs}} = \frac{[\text{dNTP}]k_3}{K_\text{d} + [\text{dNTP}]}. \qquad (6)$$

An exponential increase in product formation during the first turnover can be measured with different concentrations of dNTP. The increase in rate constant describing the burst (k_{obs}) should follow a hyperbolic relationship and can be used not only to determine the dissociation constant for dNTP, but also the limiting rate of incorporation when $E_{T\text{–}P}$ is saturated with dNTP, k_3. It is not clear if this limiting rate of nucleotide incorporation represents rate-limiting bond formation or a conformational change followed by very rapid chemistry (5, 7).

4. Kinetic properties of site-directed mutants

The reverse transcriptase from HIV has been the focus of an extensive amount of cellular, clinical, molecular, biophysical, and kinetic study. With constantly improving structural information, rational site-directed mutagenesis is now possible. Modelling of a RNA–DNA T–P into its putative binding site has suggested that enzyme residues which confer AZT resistance may interact with the template strand (4). The observation that mutagenesis of these residues translate into *in vitro* resistance in some instances (38) and not in others (39) emphasizes our lack of understanding of the reactions catalysed by RT.

Mutagenesis of p66 results in a homodimer with two mutations (one per polypeptide). A heterodimer may have one or two mutations depending if one of the altered side chain is processed out during maturation of the p51 from p66. The structure determined from crystallography has shown that the RT is an asymmetric heterodimer (4). Although the catalytic residues appear to be associated with p66, p51 does contribute to T–P binding. Mutagenesis studies can distinguish the function of subunit-specific residues by taking advantage of the greater stability of the heterodimer than homodimer (17, 18). If a mutation in p66 does not influence dimerization, mutant p66 polypeptides mixed with wild-type p51 should form heterodimers with a single alteration. This asymmetry in amino acid function in the heterodimeric RT has been noted for D110 (40), and D185–D186 (41) which are critical for polymerase activity in p66, but not in p51, and L289 which is important for dimerization in p66, but not p51 (42). The identity of other protein side chain and substrate and inhibitor interactions remain to be determined. Kinetic characterization of HIV RTs from other sources, as well as RTs altered by site-directed mutagenesis, will lead to a better understanding of the low fidelity of this polymerase (11), inhibition, and enzymology of reverse transcription.

References

1. Mitsuya, H., Yarchoan, R., and Broder, S. (1990). *Science*, **249**, 1533.
2. Chandra, A., Gerber, T., Kaul, S., Wolf, C., Demirhan, I., and Chandra, P. (1986). *FEBS Lett.*, **200**, 327.
3. Jacobo-Molina, A., Ding, J., Nanni, R. G., Clark, A. D. Jr., Lu, X., Tantillo, C., *et al.* (1993). *Proc. Natl Acad. Sci. USA*, **90**, 6320.
4. Kohlstaedt, L. A., Wang, J., Friedman, J. M., Rice, P. A., and Steitz, T. A. (1992). *Science*, **256**, 1783.
5. Kati, W. M., Johnson, K. A., Jerva, L. F., and Anderson, K. S. (1992). *J. Biol. Chem.*, **267**, 25988.
6. Reardon, J. E. (1992). *Biochemistry*, **31**, 4473.
7. Reardon, J. E. (1993). *J. Biol. Chem.*, **268**, 8743.
8. Majumdar, C., Abbotts, J., Broder, S., and Wilson, S. H. (1988). *J. Biol. Chem.*, **263**, 15657.
9. Kedar, P. S., Abbotts, J., Kovacs, T., Lesiak, K., Torrence, P., and Wilson, S. H. (1990). *Biochemistry*, **29**, 3603.
10. Mesner, L. D. and Hockensmith, J. W. (1992). *Proc. Natl Acad. Sci. USA*, **89**, 2521.
11. Bebenek, K., Abbotts, J., Roberts, J. D., Wilson, S. H., and Kunkel, T. A. (1989). *J. Biol. Chem.*, **264**, 16948.
12. Huber, H. E., McCoy, J. M., Seehra, J. S., and Richardson, C. C. (1989). *J. Biol. Chem.*, **264**, 4669.
13. Abbotts, J., Bebenek, K., Kunkel, T. A., and Wilson, S. H. (1993). *J. Biol. Chem.*, **268**, 10312.
14. Fasman, G. D. (1975). In *Handbook of biochemistry and molecular biology*, Vol. 1—Nucleic Acids, CRC Press, Cleveland, OH.
15. Rickwood, D. and Hames, B. D. (ed.) (1990). *Gel electrophoresis of nucleic acids: a practical approach*. Oxford University Press, Oxford.
16. Sambrook, J., Fritsch, E. F., and Maniatis, T. (ed.) (1989). *Molecular cloning, a laboratory manual*. Cold Spring Harbor Laboratory Press, NY.
17. Restle, T., Müller, B., and Goody, R. S. (1990). *J. Biol. Chem.*, **265**, 8986.
18. Becerra, S. P., Kumar, A., Lewis, M. S., Widen, S. G., Abbotts, J., Karawya, E. M., *et al.* (1991). *Biochemistry*, **30**, 11708.
19. Beard, W. A. and Wilson, S. H. (1993). *Biochemistry*, **32**, 9745.
20. Becerra, S. P., Kumar, A., and Wilson, S. H. (1993). *Protein Expression and Purification*, **4**, 187.
21. Bradford, M. M. (1976). *Anal. Biochem.*, **72**, 248.
22. Divita, G., Restle, T., and Goody, R. S. (1993). *FEBS Lett.*, **324**, 153.
23. Reardon, J. E. and Miller, W. H. (1990). *J. Biol. Chem.*, **265**, 20302.
24. Detera, S. D. and Wilson, S. H. (1982). *J. Biol. Chem.*, **257**, 9770.
25. Altman, S. and Lerman, L. S. (1970). *J. Mol. Biol.*, **50**, 235.
26. Schrier, B. K. and Wilson, S. H. (1975). *Methods Cell Biol.*, **13**, 105.
27. Barshop, B. A., Wrenn, R. F., and Frieden, C. (1983). *Anal. Biochem.*, **130**, 134.
28. Cleland, W. W. (1979). In *Methods in enzymology*, (ed. D. L. Purich), Vol. 63, pp. 103–38. Academic Press, San Diego.
29. Goldman, M. E., Nunberg, J. H., O'Brien, J. A., Quintero, J. C., Schleif, W. A., Freund, K. F., *et al.* (1991). *Proc. Natl Acad. Sci. USA*, **88**, 6863.

30. Merluzzi, V. J., Hargrave, K. D., Labadia, M., Grozinger, K., Skoog, M., Wu, J. C., *et al.* (1990). *Science*, **250**, 1411.
31. Pauwels, R., Andries, K., Desmyter, J., Schols, D., Kukla, M. J., Breslin, H. J., *et al.* (1990). *Nature*, **343**, 470.
32. Dueweke, T. J., Kezdy, F. J., Waszak, G. A., Deibel, M. R. Jr., and Tarpley, W. G. (1992). *J. Biol. Chem.*, **267**, 27.
33. Wu, J. C., Warren, T. C., Adams, J., Proudfoot, J., Skiles, J., Raghavan, P., *et al.* (1991). *Biochemistry*, **30**, 2022.
34. Bryant, F. R., Johnson, K. A., and Benkovic, S. J. (1983). *Biochemistry*, **22**, 3537.
35. Reardon, J. E., Furfine, E. S., and Cheng, N. (1991). *J. Biol. Chem.*, **266**, 14128.
36. Dahlberg, M. E. and Benkovic, S. J. (1991). *Biochemistry*, **30**, 4835.
37. Patel, S. S., Wong, I., and Johnson, K. A. (1991). *Biochemistry*, **30**, 511.
38. Martin, J. L., Wilson, J. E., Haynes, R. L., and Furman, P. A. (1993). *Proc. Natl Acad. Sci. USA*, **90**, 6135.
39. Lacey, S. F., Reardon, J. E., Furfine, E. S., Kunkel, T. A., Bebenek, K., Eckert, K. A., *et al.* (1992). *J. Biol. Chem.*, **267**, 15789.
40. Hostomsky, Z., Hostomska, Z., Fu, T.-B., and Taylor, J. (1992). *J. Virol.*, **66**, 3179.
41. Le Grice, S. F. J., Naas, T., Wohlgensinger, B., and Schatz, O. (1991). *EMBO J.*, **10**, 3905.
42. Goel, R., Beard, W. A., Kumar, A., Casas-Finet, J. R., Strub, M. P., Stahl, S. J., *et al.* (1993). *Biochemistry*, **32**, 13012.

3

Ribonuclease H

Z. HOSTOMSKA and Z. HOSTOMSKY

1. Introduction

The ribonuclease H (RNase H) is an essential activity associated with HIV reverse transcriptase (RT). RNase H cleaves the RNA in hybrid RNA–DNA intermediates during minus-strand DNA synthesis initiated from the $tRNA^{Lys,3}$ primer. In the course of this ordered degradation of the genomic RNA template, the enzyme also generates a specific RNA fragment in the polypurine tract (ppt) region, which is then used as a primer for initiation of the plus-strand DNA synthesis. RNase H is also involved in the specific removal of both tRNA and ppt primers so that synthesis of the double-stranded DNA of the provirus can be completed. These well-defined functions of RNase H during reverse transcription can not be substituted by cellular enzymes. Since inhibition of any of these functions would block virus replication, RT-associated RNase H represents a valid target for anti-retroviral therapy (1).

The RNase H domain is located in the C-terminal portion of the p66 subunit in the heterodimer of HIV-1 RT (*Figure 1*). Although this domain

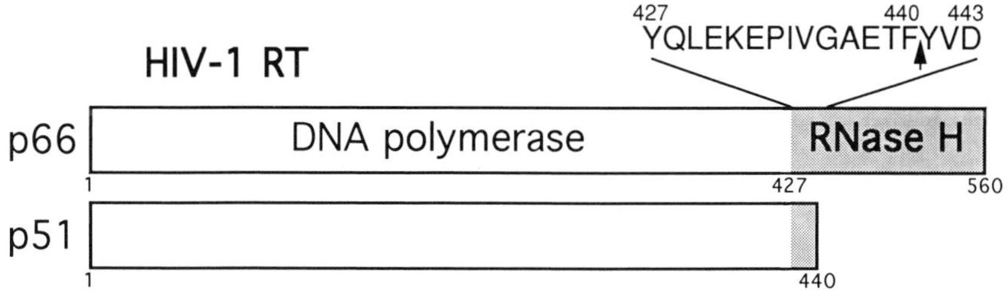

Figure 1. Subunit composition of HIV-1 RT. The RNase H domain is shaded. Note that the cleavage by HIV-1 protease (*arrow*) to generate the C-terminus of p51 at Phe440 incapacitates the second RNase H domain, as it occurs within the RNase H region, only two residues from the catalytically important Asp443. To obtain a stable properly folded protein, the N-terminus of the recombinant RNase H domain should extend at least to Tyr427.

can exist separately from RT as a folded protein with an intact catalytic site, its enzymatic activity requires specific interactions with the polymerase portions of the heterodimer which mediate substrate binding (2–4). By contrast the RNase H domain of the Moloney murine leukaemia virus RT is active when separated from the rest of polymerase.

In this chapter we describe the expression, purification, and crystallization of the RNase H domain of HIV-1 RT. The section on enzymatic activity, however, will concentrate on the RNase H activity of the heterodimeric RT as the only form of RNase H relevant for the process of reverse transcription during HIV replication.

2. Expression of the RNase H domain

Although there is no evidence for a role in the HIV life cycle of an RNase H domain non-associated with RT, study of the HIV RNase H domain as an isolated protein has been of considerable importance since it enables more detailed characterization of the RNase H catalytic site—a potential anti-retroviral drug target.

The mature RT heterodimer (p66/p51) arises from the proteolysis of the p66/p66 homodimer precursor by HIV-1 protease, which removes a polypeptide of 120 amino acids from one of the subunits. This cleaved polypeptide corresponds to a substantial portion of the RNase H domain and it was therefore proposed that the site of protease cleavage defines the N-terminus of RNase H (*Figure 1*). However, initial attempts to express the HIV-1 RNase H starting with residue Tyr441 failed to produce a stable protein in *E. coli*. The explanation for this instability has been provided by the three-dimensional structure of the RNase H domain which revealed that the Phe440–Tyr441 cleavage site is located in the middle strand of a five-stranded central beta sheet and would be accessible to HIV-1 protease only if the domain is unfolded. Removal of three residues from the N-terminal beta strand is incompatible with a native conformation and results in further degradation of the improperly folded polypeptide (*Figure 2A*). Stable RNase H domains require a C-terminal region of the p66 subunit starting at Tyr427.

The purification of the RNase H domain as an isolated form, rather than as a part of HIV RT, provided protein for high-resolution structural studies. The suitability of protein prepared for this purpose can vary, depending on expression and purification protocols.

The following strategies have been applied successfully for expression of a stable recombinant RNase H domain in *E. coli*:

(a) The RNase H domain is covalently attached to a stable *E. coli* protein, such as dihydrofolate reductase (DHFR).
(b) The RNase H domain is expressed with varying N-terminal extensions, mostly derived from the polymerase domain of HIV-1 RT.

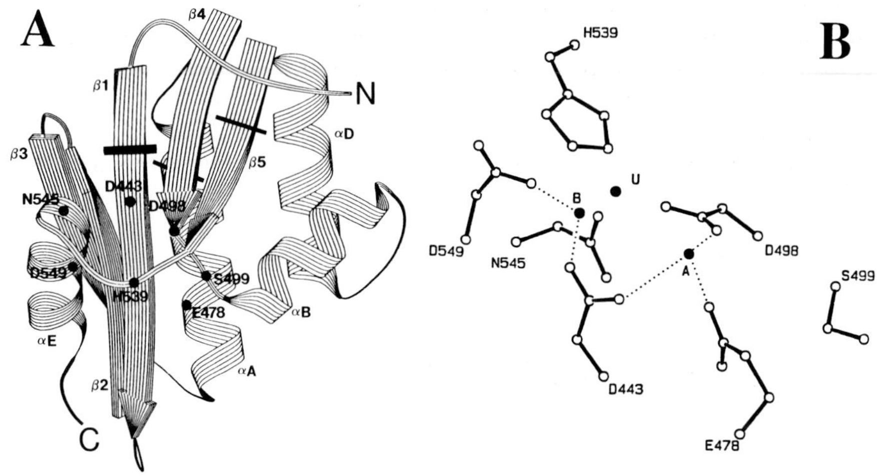

Figure 2. A. Overall folding of the RNase H domain of HIV-1 RT. Positions of the seven residues conserved in the sequences of bacterial and retroviral RNases H are marked as black dots in the ribbon representation of the structure. These residues (Asp443, Glu478, Asp498, Ser499, His539, Asn545, and Asp549) constitute the RNase H catalytic site. The perpendicular bars indicate sites of cleavage by HIV-1 protease occurring during maturation of HIV-1 RT (18). The thicker of these bars indicates cleavage site between Phe440–Tyr441 which generates the carboxy-terminus of the p51 subunit. None of these sites is accessible to HIV-1 protease when the RNase H domain is properly folded. Alpha helices and beta strands are labelled, as are the carboxy-(C) and amino-(N) termini. B. Detailed view of the catalytic site of HIV-1 RNase H, highlighting the side chains of the seven invariant residues. Coordination of the two metal ions in positions A and B by four carboxylates (Asp443, Glu478, Asp498, and Asp549) is indicated by dotted lines. Binding of uranyl pentafluoride (U) occurs in the presumed phosphate binding site. To avoid overlap of the residues in the two-dimensional representation, the catalytic site is viewed from a slightly different angle than that in *Figure 2A*.

(c) The RNase H domain is expressed with an oligo-histidine tag for affinity purification.

The RNase H domain expressed in *E. coli* under the control of a T7 promoter as a C-terminal portion of a fusion protein with bacterial dihydrofolate reductase can be released by various proteolytic enzymes, including HIV-1 protease (5). The released domain is resistant to further proteolysis and produced well-ordered crystals diffracting X-rays to high resolution. By contrast, when HIV-1 RNase H was expressed in *E. coli* as a separate domain under the control of the P_L promoter (6), unstable crystals were obtained which diffracted X-rays only to moderate resolution (about 3 Å). Interestingly, when the same protein preparation was used in NMR studies, NOE connectivities from the C-terminal portion of the RNase H domain were missing, indicating that the region corresponding to αE (*Figure 2A*) was disordered in this protein (7).

Forms of the RNase H domain purified via an oligo-histidine tag have also been crystallized, but these crystals are smaller, less well-ordered, and in general less suitable for high-resolution structural studies compared to those described above.

The possibility of disorder in certain regions of a heterologously expressed protein means there may be significant benefit in using proteolytic enzymes, especially proteinases with broad sequence specificities, to release domains from fusion proteins. Controlled or limited proteolysis performs a dual function in the preparation of protein for crystallization:

- release a stable domain from a larger multidomain protein
- eliminate misfolded polypeptides, generating a homogeneous population of the properly folded protein molecules

Appropriate proteinases will preferentially hydrolyse the disordered and solvent accessible regions found in the conformationally heterogeneous protein population leaving only the tightly folded subpopulations resistant to further proteolysis. The RNase H domain is better ordered when expressed as a C-terminal fusion protein than as a single, isolated domain, perhaps because the RNase H domain is located at the C-terminus of the RT, which is itself synthesized as part of a larger polyprotein precursor.

2.1 Purification of the RNase H domain for structural studies

A procedure for providing uniformly folded protein suitable for crystallization studies includes purification of the DHFR–RNase H fusion protein (*Protocol 1*) followed by limited proteolysis of the fusion protein (*Protocol 2*). HIV-1 protease or α-chymotrypsin can be used for the cleavage. Both proteases cleave the fusion protein in an accessible interdomain linker (5). Although the DHFR–RNase H fusion protein could, in principle, be purified by affinity chromatography on methotrexate–Sepharose, the procedure using ion-exchange chromatography described in *Protocol 1* gives the best results.

Protocol 1. Purification of the DHFR–RNase H fusion protein[a]

Equipment and reagents

- Sonicator or microfluidizer
- Buffer A: 25 mM potassium phosphate pH 6.5, 50 mM NaCl, 1 mM dithiothreitol, 1 mM phenylmethylsulfonyl fluoride (PMSF)
- DEAE–Sephacel or DEAE–Sepharose
- Sephacryl S-100
- FPLC Mono Q
- 16% SDS–PAGE

Method

1. Take 10 g (wet weight) of induced *E. coli* cells and resuspend them in 30 ml of buffer A.

2. Sonicate cells or break by passing through a microfluidizer (Microfluidics Corp.).
3. Centrifuge for 30 min, at 4°C, 15000 *g*.
4. Discard pellet and load supernatant on a DEAE–Sephacel column equilibrated in buffer A.
5. Elute the fusion protein with a linear gradient of 0.05–1 M NaCl in buffer A.
6. Pool fractions containing the fusion protein (0.4–0.5 M NaCl).
7. Load on a Sephacel S-100 equilibrated in buffer A.
8. Pool fractions with the fusion protein.
9. Load on FPLC Mono Q column and elute with gradient of 0.05–0.8 M NaCl.

[a] This method can be scaled-up proportionally.

Protocol 2. Cleavage of fusion protein with HIV-1 protease and purification of RNase H domain[a]

Equipment and reagents

- DHFR–RNase H fusion protein (1 mg/ml; see *Protocol 1*)
- Cleavage buffer: 50 mM sodium acetate pH 5, 1 mM dithiothreitol, 100 mM NaCl
- HIV-1 protease (0.1 mg/ml; see Chapter 5)
- Buffer B: 25 mM potassium phosphate pH 7.0
- FPLC Mono Q in buffer B
- FPLC Mono S in buffer B
- Centricon 3, ultrafiltration device (Amicon)

Method

1. Prepare 20 ml of the DHFR–RNase H fusion protein at 1 mg/ml in the cleavage buffer.
2. Add HIV-1 protease to a final concentration of 0.01 mg/ml to the fusion protein.
3. Adjust the final pH to 5.
4. Mix gently and incubate on ice for 3–16 h. The time depends on the extent of the digestion.
5. Apply the reaction mixture to FPLC Mono Q and FPLC Mono S columns linked in tandem.
6. Collect RNase H domain in the flow-through fractions. The uncleaved fusion protein is retained on Mono Q, HIV-1 protease is retained on Mono S column.

Protocol 2. *Continued*

7. Concentrate the protein using the Centricon 3 to 8–10 mg/ml for crystallization trials.

[a] This protocol describes the use of HIV-1 protease as a general agent for cleaving interdomain linker regions. It can be applied for cleavage of many other multidomain proteins.

2.2 Purification of a separately expressed RNase H domain

Unfused RNase H domains expressed in *E. coli* can be purified as described in *Protocol 3*. The procedure includes one step on DEAE–Sephacel to remove nucleic acids in *E. coli* extracts, followed by a gel filtration step on Sephacryl S-100. RNase H runs as a monomer of about 18 kDa on S-100. To remove contaminating activities it is advisable to run the sample through the tandem of Mono Q and Mono S columns. Mono S efficiently removes *E. coli* RNase H.

Protocol 3. Purification of a separately expressed RNase H domain

Equipment and reagents

- Buffer C: 50 mM Tris–HCl pH 8.0, 100 mM NaCl, 1 mM PMSF
- DEAE–Sephacel
- Sephacryl S-100
- Tandem columns Mono Q and Mono S in buffer B (see *Protocol 2*)

Method

1. Lyse 10 g (wet weight) of induced *E. coli* cells in 30 ml of buffer C.
2. Load soluble proteins on a DEAE–Sephacel column equilibrated in buffer C.
3. Recover RNase H in the flow-through and load the sample on S-100 in buffer C.
4. Evaluate purity of RNase H-containing fractions on 16% SDS–PAGE.
5. Dialyse RNase H into buffer B (*Protocol 2*).
6. Run tandem columns. RNase H domain is in the flow-through fractions.

2.3 Purification of RNase H with an oligo-histidine tag

Addition of an oligo-histidine affinity label at the amino- or C-termini allows simple, usually one-step purification by metal chelating chromatography (8). The oligo-histidine tag facilitates the purification protocol but has the disadvantage that the tag may modulate enzymatic properties of the protein (cf. section 3.4). It is important to include a protease-sensitive site into the construct to be able to release an authentic RNase H domain. Proteins with the histidine tag are bound to nickel-nitrilo-acetate resin (Ni-NTA agarose), and eluted specifically with imidazole-containing buffers or by a change of pH to 4.5. Further details on the expression systems and purification procedures are described in the 'QIAexpressionist' manual distributed by QIAGEN Corp.

Some forms of the HIV RNase H domain are occasionally expressed in *E. coli* as insoluble proteins. Although these proteins can be solubilized in 8 M urea and purified by metal chelating chromatography, it is our experience that refolded forms of RNase H are unsuitable for structural studies.

2.4 Crystallization of the RNase H domain

The best diffracting crystals (2.2 Å resolution) were obtained from protein expressed as a DHFR–RNase H fusion which was subsequently cleaved with HIV-1 protease. Conformationally homogeneous material is obtained by limited proteolysis. Pure isolated RNase H in 50 mM potassium phosphate buffer pH 7.0 is concentrated to 8–10 mg/ml. Crystals are grown by the hanging drop vapour diffusion method using 24-well Linbro plastic tissue culture plates (ICN Biomedicals). 5 μl of the protein is mixed with 5 μl of the reservoir containing 0.15 M sodium/potassium tartarate, 20% polyethylene glycol 8000, 0.1 M sodium citrate pH 5.2. Drops are spotted on coverslips and suspended over wells containing 750 μl of the reservoir solution. Crystals appear in several days and grow over the course of two weeks to their maximum dimensions (0.8 × 0.4 × 0.3 mm).

Structure of the isolated RNase H domain at 2.4 Å resolution containing an intact catalytic site is available (2). This high-resolution structure represents a target for structure-based drug design, second in detailed molecular characterization only to HIV-1 protease. Although RNase H is part of the crystal structure of HIV-1 RT solved at 3.0 Å resolution, crystals of the small isolated RNase H domain could be more practical for use in repeated cycles of co-crystallization with inhibitors binding to the catalytic site compared to the large multidomain HIV RT.

3. Detection of RNase H activity

Ribonuclease H is defined as an enzyme that specifically cleaves RNA only when it is in duplex with a complementary DNA sequence. In studying this activity, careful assays are necessary to control for RNA hydrolysis by non-enzymatic mechanisms or by contaminating nucleases. This is especially important in attempts to distinguish low, but specific RNase H activity, over a variable background of RNA degradation.

It is advisable to study RNase H activity using several independent techniques. In addition to following degradation of a generic RNA–DNA substrate, characterization of RNase H activity should ideally include several of the following:

- activity studies with an RNA–DNA substrate of defined length and sequence
- analysis of degradation products, e.g. by gel electrophoresis
- determination of cleavage site(s) using RNA sequencing
- correlation of activity with the size of the protein using an *in situ* activity gel assay
- use of catalytically inactive forms of an enzyme (knock-out mutants) as negative controls

3.1 Preparation of substrates

Most types of substrates for RNase H assays are not commercially available and they have to be prepared in individual laboratories.

Protocol 4. Preparation of RNase H substrate

Equipment and reagents

- *In vitro* transcription kit (e.g. Ambion, Stratagene, etc.)
- Denaturing polyacrylamide gel electrophoresis system
- Diethylpyrocarbonate (DEPC)-treated water for all reagents
- RNase H assay buffer: 50 mM Tris–HCl pH 8.0, 50 mM KCl, 7 mM $MgCl_2$, 2 mM dithiothreitol

Method

1. Synthesize a radioactively labelled RNA as a run-off transcript from a linearized plasmid with sequence of interest cloned downstream from the T7 promoter. The *in vitro* transcription system typically contains 20 μM [α-^{32}P]UTP with 500 μM each ATP, CTP, and UTP.
2. Add 5 μl of 95% formamide dye, boil for 3 min, and load on a 10% polyacrylamide sequencing gel containing 7 M urea.
3. Purify the electrophoretically separated RNA: autoradiograph wet gel; excise a gel slice containing the identified band; place band into a plastic tube (Falcon 2059) with 600 μl 0.5 M ammonium acetate; freeze–thaw; incubate at 37°C with gentle shaking for 2 h; precipitate the eluted RNA with 3 vol. of ethanol; wash with 80% ethanol; air dry.
4. Dissolve RNA in 100 μl RNase H assay buffer.
5. Add a > tenfold molar excess of a complementary DNA (synthetic oligonucleotide or single-stranded DNA from a M13 phage or phagemid subclone).

6. Heat to 65°C in a water-bath.
7. Cool slowly (several hours) to room temperature.
8. Store at −20°C.

Protocol 4 describes preparation of a uniformly labelled RNA transcript of defined length and sequence. The following variations are also possible:

(a) To obtain an end-labelled substrate, the 5′ end of an unlabelled RNA transcript is dephosphorylated by calf intestine alkaline phosphatase, and phosphorylated by T4 polynucleotide kinase and [γ-^{32}P]ATP. Alternatively, 5′-end-labelled substrate can be obtained by using 20 μM [γ-^{32}P]GTP with 500 μM each ATP, CTP, and UTP in the initial transcription mixture. The yield of the end-labelled RNA in the latter case is relatively low, however, as high levels of GTP are required for efficient initiation of transcription by T7 RNA polymerase.

(b) Synthetic RNA of at least 25 nt can be used instead of RNA transcript and labelled directly by polynucleotide kinase and [γ-^{32}P]ATP.

(c) Instead of gel purification (*Protocol 4*, steps 2–4), the reaction mixture after labelling can be phenol extracted and passed through a Sephadex G-50 spin column, or a NucTrap column (Stratagene) to remove unincorporated radiolabelled NTPs.

3.2 RNase H activity assays

RNase H activity is determined traditionally using high molecular weight hybrid substrates in which RNA strands are radioactively labelled. The extent of RNA hydrolysis is derived from measurements of radioactivity remaining in the substrate, which is separated from the lower molecular weight products by precipitation with trichloroacetic acid, perchloric acid, or ethanol. The acid or alcohol soluble radioactivity of the supernatant can also be used as a measure of the reaction products. Below are listed some alternative techniques which are being used to assay RNase H activity:

- gel electrophoresis assay (*Protocol 5*)
- *in situ* gel assay (*Protocol 6*)
- filter binding assay (*Protocol 7*)
- scintillation proximity assay

3.2.1 Gel electrophoretic assay

Using a 5′ end-labelled substrate of the defined sequence enables precise determination of the RNase H cleavage sites (9).

Protocol 5. Gel electrophoresis assay for RNase H cleavage

Equipment and reagents

- Denaturing polyacrylamide sequencing gel containing 7 M urea
- Formamide dye
- RNase H assay buffer (*Protocol 4*)
- Labelled RNA–DNA substrate (*Protocol 4*)

Method

1. Incubate 10 μl reaction mixtures in RNase H assay buffer containing 5–50 nM labelled RNA–DNA substrate and RT at 37 °C for 5–20 min.
2. Stop reactions with 2 μl of formamide dye.
3. Boil samples for 2 min.
4. Load on a denaturing polyacrylamide sequencing gel.
5. Autoradiograph the gel.

3.2.2 *In situ* gel assay

The *in situ* gel assay (sometimes called activated gel assay *Protocol 6*) allows measurement of RNase H activity directly in the polyacrylamide gel containing radioactively labelled RNA–DNA substrate. After electrophoretic separation, proteins are renatured in the gel by incubating in the assay buffer. Refolded proteins digest the substrate co-polymerized in the gel and low molecular weight digestion products are released into the washing buffer. The digested area is free of radioactivity. The autoradiography of the dried gel reveals clear bands on a dark background. After activity development separated proteins can be stained with Coomassie blue and size of the protein can be correlated with the activity band. This approach helps in detecting nuclease activities present in the sample which are unrelated to the protein of interest. Purified samples as well as cell lysates can be analysed by *in situ* gel assay. To minimize the size of the gel and thus reduce the amounts of the radioactive substrate needed, a small electrophoretic chamber (e.g. from Hoeffer) can be used.

As incubation in these assays takes several days at room temperature, a very weak activity can be amplified as a detectable cleared zone. This assay is semi-quantitative—size of the cleared zone in a certain range correlates with the hydrolytic activity of the protein. In studying various types of mutants using the *in situ* gel assay—especially mutations caused by deletions or insertions—it is important to realize that activity in this assay is dependent on successful refolding of a protein species. Negative results may thus reflect, besides impairment of catalytically important residues, an inability of certain types of mutated proteins to refold properly.

Protocol 6. *In situ* gel assay for RNase H activity

Equipment and reagents

- 16% separating gel contains 375 mM Tris–HCl pH 8.8, 2 mM EDTA, 0.02% SDS, 1.10^6 c.p.m. of the uniformly ^{32}P-labelled hybrid (see *Protocol 4*)—best results are obtained with short (about 100 nt) transcripts hybridized to a M13-derived ssDNA
- 4% stacking gel contains 66.7 mM Tris–HCl pH 6.8, 2 mM EDTA, 0.02% SDS
- Sample buffer: 65 mM Tris–HCl, pH 7.0, 2 mM EDTA, 4% SDS, 20% glycerol, 0.6 mM β-mercaptoethanol
- Electrode buffer: 3.03 g Tris–HCl, 14.4 g glycine, 0.2 g SDS in 1 litre
- RNase H assay buffer: 50 mM Tris–HCl pH 8.0, 50 mM KCl, 2 mM dithiothreitol, 7 mM $MgCl_2$ (or $MnCl_2$—see section 3.4)

Method

1. Prepare 16% SDS–polyacrylamide gel containing RNA–DNA substrate (with uniformly labelled RNA strand) polymerized into the gel.
2. Take 10 μl of the protein sample, add 5 μl of sample buffer, and boil for 1 min at 100°C.
3. Run the gel for approx. 45 min at 4°C.
4. Shake gently the gel in the RNase H assay buffer at room temperature with several changes of buffer for 48–66 h.
5. Stain and destain the gel with Coomassie blue.
6. Dry the gel and expose to the X-ray film.

3.2.3 Filter binding assay

Protocol 7. Filter binding of RNA–DNA hybrids

Equipment and reagents

- 24 mm DEAE membranes NA45 (Schleicher & Schuell)
- 0.5 M Na_2HPO_4
- 0.3 M ammonium formate pH 8
- Liquid scintillation counter

Method

1. Apply aliquots of reaction mixture in RNase H buffer (as in *Protocol 5*) on 24 mm DEAE membrane disks NA45.
2. Wash disks twice with 0.5 M Na_2HPO_4; once with 0.3 M ammonium formate pH 8; once with ethanol; once with ether.
3. Air dry disks.
4. Measure radioactivity either by liquid scintillation counting when the RNA is ^3H-labelled, or by Cerenkov radiation when the RNA is ^{32}P-labelled.

3.2.4 Scintillation proximity assay

The scintillation proximity assay developed by Amersham Corporation may soon become more widely used for detection of RNase H activity. The assay contains fluoromicrospheres with bound radioactive RNA–DNA substrate. Since excitation of the fluor by the radioactive energy of the released reaction products is unlikely, no separation of the product from the remaining substrate is necessary. Although it uses a generic RNA–DNA substrate, this commercially available kit could be helpful as a reference for standardizing levels of RNase H activity detected by different laboratories.

3.3 Contaminating activities

Since it is almost exclusively recombinant forms of HIV enzymes that are used for *in vitro* activity studies, the possibility of contamination by activities originating in the *E. coli* host should be borne in mind. This issue has become of importance especially in connection with attempts to demonstrate catalytic activity of the isolated RNase H domain of HIV, when highly concentrated preparations of recombinant protein are incubated with a labelled RNA–DNA substrate. Besides highly efficient *E. coli* RNase H I, whose specific activity is two orders of magnitude higher than the RNase H activity associated with HIV-1 RT, at least 20 other distinct ribonucleases have been identified in *E. coli*. For example, a contamination by *E. coli* RNase III in preparations of recombinant RT was erroneously interpreted as a new activity, double-stranded RNA-specific ribonuclease ('RNase D'), associated with HIV-1 RT.

3.4 Altered substrate specificities of RNase H

It has been observed that besides hydrolysing RNA in an RNA–DNA duplex, the HIV-1 RT can also cleave double-stranded RNA when Mg^{2+} is replaced by Mn^{2+} in an *in situ* activity gel assay (*Protocol 6*). This inherent Mn^{2+}-dependent activity seems to reflect a relaxation of substrate specificity of RT-associated RNase H. It has been proposed to name this activity RNase H*, by analogy with a relaxed 'star' activity of certain restriction endonucleases, which can be detected under non-standard assay conditions (1, 19).

Interestingly, various recombinant forms of the HIV-1 RNase H domain were found to be active on an RNA–DNA substrate only in the presence of Mn^{2+}, but not Mg^{2+}, in an *in situ* gel assay (10). The activity of these forms of the RNase H domain, however, was dependent on an oligo-histidine tag (cf. section 2.3), which presumably facilitates substrate binding in the absence of the polymerase domain.

3.5 Reconstitution of RNase H activity

The isolated RNase H domain (p15) displays no detectable activity under standard assay conditions (see section 3.2.1). However, addition of the purified native N-terminal polymerase domain p51 of HIV-1 RT leads to the reconstitution of RNase H activity (11). Although the reconstituted activity produces RNase H cleavage patterns identical to that of wild-type RT, about ten times more molar equivalents of both p51 and p15 are needed than the wild-type p66/p51 RT to see the same extent of cleavage.

Based on a detailed comparison of the three-dimensional structures of *E. coli* RNase H I with the RNase H domain of HIV-1 RT, it is now clear that the HIV-1 domain lacks the basic protrusion ('handle region'), implicated in substrate binding (2). The crystal structure of mature heterodimeric HIV-1 RT reveals that the polymerase domains, in addition to providing substrate binding determinants, are also important for directly stabilizing portions of the RNase H domain (3, 4). Comparative structural analysis thus offers a plausible explanation for the loss of activity of this domain upon separation from the heterodimer as well as for the reconstitution of an active weakly associated complex.

3.6 Coupled assay of DNA synthesis/RNA degradation by HIV RT

When a short DNA or RNA primer hybridized to a labelled RNA is incubated with RT in the presence of dNTPs, DNA synthesis can be followed simultaneously with the RNase H catalysed degradation of the RNA template. A challenged reaction, in which RT is effectively sequestered after one round of processive DNA synthesis in the presence of excess unlabelled RNA (12) or heparin (13), allows measurement of the number of nucleolytic cleavages by RNase H activity of RT during a single encounter with the primer–template.

The products of these reactions can be distinguished visually using different labelling of RNA template and nascent DNA. For example, RNA can be labelled at the 5′ end with ^{32}P, while the nascent DNA can be labelled by incorporation of ^{35}S-containing nucleotides. The assignment of bands to RNA or DNA is based on differential exposure of the gel. Alternatively, a short DNA primer can be ^{32}P-labelled at its 5′ end so that primer extension can be followed in the presence of unlabelled dNTPs. The number of nucleotides between the 3′ end of the primer and the cleavage site in the RNA template reflects the distance between the polymerase and RNase H catalytic sites in HIV RT (14).

4. Mutagenesis of the catalytically important residues in HIV RNase H

Computer analysis of amino acid sequences of retroviral and bacterial RNases H revealed seven highly conserved amino acid residues. These residues are Asp443, Glu478, Asp498, Ser499, His539, Asn545, and Asp549 (*Figure 2A*).

When residues Asp443, Glu478, and Asn498 were changed by site-directed mutagenesis independently to Asn443, Gln478, and Asn498, respectively, they were shown to be essential for the catalytic activity (15, 16). Crystallographic analysis of RNase H showed all seven invariant residues clustered near the catalytic site. Further, analysis of crystals of RNase H with divalent metals, which are required for the activity, located two metals approximately 3.9Å apart in close proximity to four acidic amino acid residues Asp443, Glu478, Asp498, and Asp549 (*Figure 2B*). These four residues are important in forming the two metal binding sites. It has been proposed that hydrolysis of RNA–DNA hybrids occur by a two metal ion mechanism, based on the striking similarity in the geometrical arrangement of the corresponding metal–carboxylate clusters in RNase H and the 3′–5′ exonuclease domain of *E. coli* DNA polymerase I (2). The uranyl pentafluoride anion (*Figure 2B*), which was used for isomorphous replacement, binds in the catalytic site of RNase H at the position analogous to the phosphate binding site in the 3′–5′ exonuclease domain. The role of conserved His539 is less clear, as mutations of this residue lead to modulation of RNase H activity (17).

Acknowledgements

The authors wish to thank Dr David A. Matthews for valuable discussions and critical reading of the manuscript. This work was supported in part by NIH grant 5 UO1 AI33380.

References

1. Hostomsky, Z., Hostomska, Z., and Matthews, D. A. (1993). In *Nucleases* 2nd edn. (ed. R. J. Roberts, S. M. Linn, and R. S. Lloyd), pp. 341–76. Cold Spring Harbor Laboratory Press, Cold Spring Harbor, NY.
2. Davies, J. F., Hostomska, Z., Hostomsky, Z., Jordan, S., and Mathews, D. A. (1991). *Science*, **252**, 88.
3. Kohlstaedt, L. A., Wang, J., Freidman, J. M., Rice, P. A., and Steitz, T. A. (1992). *Science*, **256**, 1783.
4. Jacobo-Molina, A., Ding, J., Nanni, R. G., Clark, A. D. Jr., Lu, X., Tantillo, C., *et al.* (1993). *Proc. Natl Acad. Sci. USA*, **90**, 6320.
5. Hostomska, Z., Mathews, D. A., Davies, J. F., Nodes, B. R., and Hostomsky, Z. (1991). *J. Biol. Chem.*, **266**, 14697.
6. Becerra, S. P, Clore, G. M., Gronenborn, A. M., Karlstrom, A. R., Stahl, S. J., Wilson, S. H., *et al.* (1990). *FEBS Lett.*, **270**, 76.
7. Powers, R., Clore, G. M., Bax, A., Garrett, G. S., Stahl, S. J., Wingfield, P. T., *et al.* (1991). *J. Mol. Biol.*, **221**, 1081.
8. Arnold, F. H. (ed.) (1992). *Methods. A companion to methods in enzymology*, Vol. 4, No. 1. Academic Press, San Diego, Calif.
9. Mizrahi, V. (1989). *Biochemistry*, **28**, 9088.
10. Smith, J. S. and Roth, M. J. (1993). *J. Virol.*, **67**, 4037.

11. Hostomsky, Z., Hostomska, Z., Hudson, G. O., Moomaw, E. W., and Nodes, B. R. (1991). *Proc. Natl Acad. Sci. USA*, **88**, 1148.
12. DeStefano, J. J., Buiser, R. G., Mallaber, L. M., Myers, T. W., Bambara, R. A., and Fay, P. J. (1991). *J. Biol. Chem.*, **266**, 7423.
13. Gopalakrishnan, V., Peliska, J. A., and Benkovic, S. J. (1992). *Proc. Natl Acad. Sci. USA*, **89**, 10763.
14. Furfine, E. S. and Reardon, J. E. (1991). *J. Biol. Chem.*, **266**, 406.
15. Mizrahi, V., Usdin, M. T., Harington, A., and Dudding, L. R. (1990). *Nucleic Acids Res.*, **18**, 5359.
16. Schatz, O., Cromme, F., Naas, T., Lindemann, D., Mous, J., and Le Grice, S. F. J. (1989). *FEBS Lett.*, **257**, 311.
17. Wöhrl, B. M., Volkmann, S., and Moelling, K. (1991). *J. Mol. Biol.*, **220**, 801.
18. Chattopadhyay, D., Evans, D. B., Deibel, M. R. Jr., Vosters, A. F., Eckenrode, F. M., Einspahr, H. M., *et al.* (1992). *J. Biol. Chem.*, **267**, 14227.
19. Hostomsky, Z., Hughes, S. H., Goff, S. P., and Le Grice, S. F. J. (1994). *J. Virol.*, **68**, 1970.

4

Integrase

R. CRAIGIE, A. B. HICKMAN, and A. ENGELMAN

1. Introduction

The linear DNA copy of the HIV genome made by reverse transcription must be integrated into a chromosome of the infected cell in order to be stably maintained and efficiently transcribed. DNA integration is therefore a necessary step in the viral replication cycle. HIV DNA is inserted into the host genome by a specialized DNA recombination reaction in which the viral integrase protein is the key player. The mechanism of this reaction is similar to that used by many transposable elements to move from one location to another in DNA, but is dissimilar to any reaction known to be involved in the normal functioning of the host cell. The specialized nature of the HIV DNA integration reaction suggests that it may be a good target in the search for specific inhibitors that block viral replication without disrupting normal cellular processes.

Integrase has been shown to carry out the central steps of the integration reaction with synthetic DNA substrates that model the ends of HIV DNA. These *in vitro* assays have provided a powerful tool to probe the molecular mechanism of integration. In addition, the availability of *in vitro* assays for integrase activity may facilitate the identification of compounds that block HIV replication at the step of DNA integration.

In this chapter we describe methodologies for the expression and purification of HIV integrase protein, and assays for its biochemical activities.

2. Expression of HIV integrase

2.1 Selection of expression system

HIV integrase protein has been expressed in heterologous systems, including *Escherichia coli*, insect cells (Sf9) infected with recombinant baculovirus, and yeast. No differences have been reported in the enzymatic or physical properties of integrase purified from these sources. We routinely choose *E. coli* as the expression host for reasons of simplicity, economy, and ease of scale-up.

2.2 Plasmids for expression of HIV integrase in *E. coli*

Vectors based on the T7 system originally developed by Studier and colleagues (1, 2) are suitable for high level expression of integrase and are commercially available from Novagen. These vectors, which confer ampicillin resistance, place a promoter for T7 RNA polymerase 5′ of the cloned gene. Expression is induced by IPTG, which induces expression of a chromosomal T7 RNA polymerase gene under the control of the *lac*UV5 promoter. Since HIV integrase is normally made as part of a polyprotein precursor that is cleaved during assembly of the virus, a translation initiation codon must be added to the 5′ end of the coding sequence; the cloned protein therefore has a methionine at the amino-terminus. It can be convenient to add additional amino acids to the amino- or carboxyl-terminus, such as a series of His residues, to enable rapid purification by affinity chromatography. The His-tag can be removed after purification if a protease cleavage site is engineered between the tag and the integrase polypeptide. The activities of full-length HIV-1 integrase with and without an amino-terminal His-tag are indistinguishable in the assays described here; however, the presence of the His-tag does quantitatively affect the activities of some deletion derivatives of integrase. Fusion of integrase to other polypeptides has also been successfully used as an aid in purification.

We use the plasmid pINSD (3) for expression of HIV-1 integrase with a methionine at the amino-terminus and pINSD.His for expression of integrase with a His-tag preceded by a methionine at the amino-terminus. pINSD contains the coding sequence of HIV-1 integrase inserted between the *Nde*I and *Bam*HI sites of pET-2c (2). In this construct an adventitious ribosome binding site spanning nucleotides 443 to 447 of the coding sequence has been mutated to suppress a shorter expression product that results from internal translation initiation; the amino acid sequence of the expressed protein is not changed by this mutation. pINSD.His contains the same *Nde*I to *Bam*HI fragment inserted into pET-15b (Novagen), which results in expression of a protein with a His-tag at the amino-terminus and a thrombin cleavage site located between the His-tag and the integrase sequence. We note that integrase proteins of several other retroviruses and many mutant derivatives of HIV integrase have been successfully expressed using similar plasmid constructions.

2.3 Induction of integrase expression

HIV integrase is expressed by transformation (4) of *E. coli* strain BL21(DE3) with the integrase expression plasmid and subsequent induction of expression with IPTG. The cells should be from a fresh transformation, or from a glycerol stock prepared as described in *Protocol 1*. It is also important to maintain antibiotic selection during growth of the transformed BL21(DE3).

Protocol 1. Preparation of glycerol stock

Equipment and reagents

- Super Broth: tryptone 12 g/litre, yeast extract 24 g/litre, glycerol 5 g/litre, KH_2PO_4 3.8 g/litre, K_2HPO_4 12.5 g/litre
- LB agar plates containing 100 μg/ml ampicillin
- 25 mg/ml ampicillin in H_2O
- 80% glycerol in H_2O
- Integrase expression plasmid (e.g. pINSD) and *E. coli* strain BL21 (DE3)[a]
- 37 °C incubator and air shaker
- Spectrophotometer
- −70 °C freezer

Method

1. Transform BL21(DE3) with an integrase expression plasmid.
2. Plate on to LB agar plates containing 100 μg/ml ampicillin.
3. Incubate the plates overnight at 37 °C.
4. Inoculate a single colony into 125 ml Super Broth containing 100 μg/ml ampicillin and incubate at 37 °C in a rotary shaker at 250 r.p.m. until an $A_{600\ nm}$ of 0.8 is reached. This will require approximately 5 h.
5. Add glycerol to a final concentration of 15% (w/v) and freeze portions at −70 °C.

[a] pINSD is available from the NIAID AIDS Research and Reference Reagent Program. Cat. No. 2820.

Protocol 2. Induction of integrase expression

Equipment and reagents

- Glycerol stock (see *Protocol 1*)
- Super Broth (see *Protocol 1*)
- 100 mM IPTG in H_2O
- Air shaker
- Centrifuge
- Equipment for SDS–PAGE
- Liquid nitrogen and a −70 °C freezer

Method

1. Thaw the glycerol stock and dilute 1:100 (v/v) with Super Broth containing 100 μg/ml ampicillin.
2. Incubate in a rotary shaker at 37 °C until an $A_{600\ nm}$ of 0.8 is reached. This will take approximately 3 h.
3. Remove a 0.5 ml sample of the culture for SDS–PAGE. Centrifuge for 1 min in a microcentrifuge. Discard the supernatant and resuspend the cell pellet in 75 μl of SDS–PAGE sample buffer. Store at −20 °C.
4. Add IPTG to a final concentration of 0.4 mM and incubate for a further 3 h. Before harvesting the cells, save a sample for SDS–PAGE as described in step 3.

Protocol 2. *Continued*

5. Harvest the cells. Centrifuge at 6000 *g* for 10 min. Discard the supernatant. Resuspend the cell pellet in ice-cold 25 mM Hepes pH 7.5, 1 mM EDTA (1 : 1 w/v).[a] Freeze the cell suspension in liquid nitrogen. Store at −70 °C.
6. Analyse 10 μl of the samples saved in steps 3 and 4 by SDS–PAGE and staining with Coomassie blue to check the level of integrase expression. Integrase, which has a molecular weight of approximately 32 kDa, should be a prominent protein in the crude lysate of induced cells.

[a] Omit EDTA if the expressed protein contains a His-tag and is to be purified by Ni^{2+} affinity chromatography.

2.4 Large scale expression of integrase

Expression of integrase can be easily scaled-up if a fermentor is available to monitor and regulate dissolved oxygen and pH. The following medium works well:

- 15 g/litre tryptone
- 20 g/litre yeast extract
- 10 g/litre glycerol
- 1% (w/v) dextrose
- 0.1 M potassium phosphate pH 7.2
- 10 μM each $(NH_4)_6Mo_7O_{24}\cdot 4H_2O$, $CoCl_2\cdot 6H_2O$, $CuSO_4\cdot 5H_2O$, H_3BO_3, $MnCl_2\cdot 4H_2O$, and $ZnCl_2$
- 0.1 mM $FeCl_3\cdot 6H_2O$, 0.5 mM $CaCl_2\cdot 2H_2O$, and 1 mM $MgSO_4\cdot 7H_2O$
- 100 μg/ml ampicillin
- antifoam agent, such as Ucaferm Adjuvant 27 at 0.75 ml/litre (Union Carbide Chemicals and Plastics Co. Inc.)

A 'starter' culture should be grown in LB broth containing 100 μg/ml ampicillin. It is important that this culture should not be allowed to reach stationary phase. Pellet the cells and resuspend them in fresh LB broth before inoculating the fermentor; this step removes most of the β-lactamase from the inoculum and helps prevent rapid depletion of the ampicillin. Maintain dissolved oxygen at 25%, and pH at 7.2 by addition of 5 M ammonium hydroxide. Induce by addition of IPTG when $A_{600\ nm} \approx 10$. Harvest at 3 h after induction.

3. Purification of HIV-1 integrase

3.1 Selection of purification procedure

Purification of HIV-1 integrase is relatively simple and should not present undue difficulties to those familiar with basic techniques in protein purification. The major problem is the limited solubility of integrase, especially at low ionic strength. Integrase can be purified either under native conditions, or under denaturing conditions, and subsequently refolded in the presence of detergent. No differences have been detected in the activities of the proteins purified by the alternative approaches of *Protocols 3, 4,* and *5.* Purification under native conditions is generally more convenient and results in a purer protein preparation. However, we have found purification under denaturing conditions to be especially useful for deletion derivatives of integrase that are not efficiently solubilized by extraction with 1 M NaCl.

Protocol 3. Purification of HIV-1 integrase under native conditions[a]

Equipment and reagents

- A column chromatography system capable of running buffer gradients—an FPLC-type system is desirable, but not essential
- Sonicator
- Centrifuge: Beckman J2-21M and Sorvall RC-5B are suitable examples (for large scale preparations a batch rotor such as a Beckman Ti15 (requires a Beckman ultracentrifuge) is very convenient)
- Homogenizer: for small scale preparations a hand-held pestle homogenizer will suffice; large scale preparations are facilitated by use of a motor-driven device; a Tekmar SDT-1810 Tissumizer works well for this purpose
- Butyl–Sepharose 4B (Pharmacia 17–0960–01)
- Heparin–Sepharose CL-6B (Pharmacia 17–0467–01)
- Lysozyme
- Ammonium sulfate
- HED buffer: 20 mM Hepes pH 7.5, 1 mM EDTA, 1 mM DTT
- 50 ml of resuspended cells (25 g of wet cell pellet) from *Protocol 2*—volumes of buffers and column beds should be adjusted in proportion to the mass of cells used

Method

1. Lyse the cell suspension from *Protocol 2.* Thaw on ice. Add an equal volume of HED buffer containing 0.2 M NaCl and 0.4 mg/ml lysozyme. Incubate on ice with occasional stirring until the suspension becomes viscous (about 30 min).
2. Sonicate in short bursts until the lysate is no longer viscous. It is important to sonicate on ice and to carefully monitor the temperature of the lysate between bursts of sonication. Do not let it rise above 8°C.
3. Extract and solubilize the integrase. Centrifuge the sonicated lysate at 40 000 *g* for 45 min. Discard the supernatant and homogenize the pellet in 50 ml HED buffer containing 100 mM NaCl. Stir for 1 h. Centrifuge at 40 000 *g* for 45 min. Discard the supernatant and homogenize the pellet in 100 ml HED buffer containing 1 M NaCl. Stir for 1 h. Centrifuge at 40 000 *g* for 45 min. Retain the supernatant which contains the solubilized integrase. The yield of integrase may be improved by repeating the 1 M NaCl extraction and combining the supernatants.

Protocol 3. *Continued*

4. Further purify the solubilized integrase by ammonium sulfate precipitation. Add solid ammonium sulfate to 2 M (31.4 g/100 ml) with continuous stirring. Stir for 1 h—stirring may be continued overnight if this is convenient. Centrifuge at 10 000 *g* for 30 min. Discard the supernatant and homogenize the pellet in 100 ml HED buffer containing 1 M NaCl and 1 M ammonium sulfate. Stir for 30 min. Centrifuge at 10 000 *g* for 30 min and discard the supernatant. Homogenize the pellet in 100 ml HED buffer containing 1 M NaCl to extract the integrase. Stir for 1 h. Centrifuge at 10 000 *g* for 30 min. Retain the supernatant which contains the solubilized integrase.
5. Add 0.24 ml of saturated ammonium sulfate per millilitre of supernatant, with stirring (0.8 M final concentration). This is necessary to raise the ionic strength prior to loading the butyl–Sepharose column, but may result in some precipitation of integrase. This precipitate can be removed by a brief centrifugation before loading the column.
6. Butyl–Sepharose chromatography. Load on to a 6 ml butyl–Sepharose column equilibrated with HED buffer containing 200 mM NaCl and 0.8 M ammonium sulfate. Wash with about five to ten column volumes of equilibration buffer. Elute with a gradient of the same buffer to HED buffer containing 80 mM NaCl, 0.3 M ammonium sulfate, and 10% (w/v) glycerol. Integrase elutes as a broad peak.
7. Prepare the eluted integrase for heparin–Sepharose chromatography. Identify the fractions that contain integrase by SDS–PAGE. Pool the peak fractions. Add two volume equivalents of HED buffer containing 10% glycerol with stirring. This step is necessary to lower the ionic strength prior to loading the protein on to the heparin–Sepharose column, but may result in some precipitation of integrase. Remove any precipitate by a brief centrifugation before loading the heparin–Sepharose column.
8. Heparin–Sepharose chromatography. Load on to a 6 ml heparin–Sepharose column equilibrated with HED buffer containing 0.2 M NaCl and 10% glycerol. Wash with several column volumes of the same buffer. Elute with a gradient of 0.2 M NaCl to 1.0 M NaCl in HED buffer containing 10% glycerol. Pool the peak fractions. Dialyse against HED buffer containing 1 M NaCl and 20% glycerol. About 4 mg of purified integrase should be obtained from 25 g of wet cells using this procedure.
9. Aliquot the protein, freeze in liquid nitrogen, and store at −70°C. Integrase is stable to multiple cycles of freezing and thawing.

[a] This procedure is a modification of that described in ref. 5. All steps should be carried out at 4°C or on ice.

Some integrase protein is likely to be lost due to precipitation after adding ammonium sulfate to the protein prior to loading on to butyl–Sepharose (*Protocol 3*, step 5). Some precipitate can be tolerated provided it does not block the flow of buffer. Precipitation can be reduced by increasing the volume of the applied sample by further dilution with a buffer that maintains the correct ionic strength for adsorption to the column. However, the best solution is to adjust the ionic strength immediately before applying the sample to the column using a pair of pumps. For example, omit step 5 and load the solubilized integrase from step 4 on to the butyl–Sepharose through a mixer. Simultaneously pump HED buffer containing 200 mM NaCl and 2.4 M ammonium sulfate to the mixer at one-third the flow rate of the integrase sample. With this modification, the ionic strength is adjusted immediately before the sample is applied to the column and integrase is adsorbed to the column before significant precipitation occurs. The same methodology can be used to lower the ionic strength of the sample for heparin–Sepharose chromatography (*Protocol 3*, step 7).

Protocol 4. Purification of His-tagged HIV-1 integrase under native conditions[a]

Equipment and reagents

- Equipment is the same as that required for *Protocol 3*
- Chelating Sepharose FF (Pharmacia 17-0575-01)
- Lysis buffer: 20 mM Tris pH 8, 0.1 mM EDTA, 2 mM 2-mercaptoethanol, 0.5 M NaCl, 5 mM imidazole, 0.2 mg/ml lysozyme
- TNM buffer: 20 mM Tris pH 8, 2 M NaCl, 2 mM 2-mercaptoethanol
- 50 mM $NiSO_4$

Method

1. Prepare the integrase extract as follows: Centrifuge the cell suspension from *Protocol 2* at 6000 *g* for 10 min. Resuspend the cells in 40 ml lysis buffer. Lyse the cells as described in *Protocol 3*, steps 2 and 3. Centrifuge at 40 000 *g* for 45 min. Discard the supernatant and homogenize the pellet in 80 ml TNM buffer containing 5 mM imidazole. Stir for 30 min. Centrifuge at 40 000 *g* for 45 min. Save the supernatant.
2. Prepare a 2 ml column of chelating Sepharose FF by washing the column sequentially with 6 ml H_2O, 6 ml of 50 mM $NiSO_4$, 10 ml H_2O, 10 ml TNM buffer containing 5 mM imidazole.[b]
3. Filter the supernatant from step 1 through a 0.45 μm filter (optional) and load it on to the chelating Sepharose column.
4. Wash and elute the column as follows. > 20 ml of TNM buffer containing 5 mM imidazole. > 20 ml of the same buffer containing 60 mM imidazole. Elute with a gradient of 60 mM to 800 mM imidazole in TNM buffer. Pool the fractions that contain integrase and add EDTA to a final

Protocol 4. *Continued*

concentration of 5 mM. Dialyse against 20 mM Hepes pH 7.5, 1 mM EDTA, 2 mM 2-mercaptoethanol, 1 M NaCl, and 20% (w/v) glycerol. Dialyse against the same buffer containing 1 mM DTT in place of 2-mercaptoethanol.[c]

5. Aliquot the protein, freeze in liquid nitrogen, and store at −70°C.

[a] All steps should be carried out at 4°C or on ice. This protocol is for 50 ml of resuspended cells (25 g of wet cell pellet) from *Protocol 2*. Volumes of buffers and column beds should be adjusted in proportion to the mass of cells used.
[b] The column bed may transiently develop a slight red colour upon equilibration with TNM buffer, presumably due to complexing of free Ni^{2+} with 2-mercaptoethanol.
[c] Mixing DTT with Ni^{2+} results in the formation of a red precipitate. Such a precipitate sometimes forms when the eluted protein is directly dialysed against buffer containing DTT; presumably some Ni^{2+} leaches from the column during elution of the protein. Addition of EDTA prior to dialysis and omission of DTT from the first dialysis buffer avoids this problem.

Protocol 5. Purification of His-tagged HIV-1 integrase under denaturing conditions

Equipment and reagents

- Equipment is the same as that required for *Protocol 3*
- Chelating Sepharose FF (Pharmacia 17–0575–01)
- TEM buffer: 20 mM Tris pH 8, 0.1 mM EDTA, 2 mM 2-mercaptoethanol
- TG buffer: 20 mM Tris pH 8, 6 M guanidine–HCl
- 50 mM $NiSO_4$

Method

1. Prepare the integrase extract as follows (see footnote[a] of *Protocol 4*). Centrifuge the cell suspension from *Protocol 2* at 6000 *g* for 10 min. Resuspend the cells in 40 ml TEM buffer containing 0.5 M NaCl, 5 mM imidazole, and 0.2 mg/ml lysozyme. Lyse the cells as described in *Protocol 3*, steps 1 and 2. Centrifuge at 40 000 *g* for 45 min. Discard the supernatant and homogenize the pellet in 80 ml TEM buffer containing 0.5 M NaCl and 5 mM imidazole. Stir for 30 min. Centrifuge at 40 000 *g* for 30 min. Discard the supernatant and homogenize the pellet in 40 ml TG buffer containing 0.5 M NaCl and 5 mM imidazole. Stir for 30 min. Centrifuge at 40 000 *g* for 30 min. Retain the supernatant.
2. Prepare a 4 ml column of chelating Sepharose FF by washing the column sequentially with 12 ml H_2O, 12 ml of 50 mM $NiSO_4$, 20 ml H_2O, 20 ml of TG buffer containing 0.5 M NaCl and 5 mM imidazole.
3. Filter the supernatant from step 1 through a 0.45 μm filter (optional) and load it on to the chelating Sepharose column.

4. Wash and elute the column as follows: > 40 ml of TG buffer containing 0.5 M NaCl and 5 mM imidazole. > 25 ml TG buffer containing 0.5 M NaCl and 20 mM imidazole. Elute with a gradient of 20 mM to 600 mM imidazole in TG buffer containing 0.5 M NaCl. Pool the fractions that contain integrase.[a] Add EDTA to a final concentration of 5 mM. Dialyse the pooled fractions against 6 M guanidine–HCl, 20 mM Hepes pH 7.5, 2 mM EDTA, 2 mM 2-mercaptoethanol, followed by dialysis against the same buffer containing 10 mM DTT instead of 2-mercaptoethanol. Aliquot the purified denatured integrase and store at −70°C.

[a] The integrase may be further purified by a second column of chelating Sepharose FF (2 ml) after dialysis against TG buffer containing 0.5 M NaCl and 5 mM imidazole.

3.2 Refolding of integrase purified under denaturing conditions

Integrase purified by *Protocol 5* must be refolded to restore activity to the protein (*Protocol 6*). This can be accomplished by sequential dialysis against buffers that are progressively less denaturing.

Protocol 6. Refolding denatured integrase

Equipment and reagents

- Buffer 1: 1 M NaCl, 20 mM Hepes pH 7.5, 2 mM EDTA, 10 mM DTT
- Buffer 2: 2 M urea, 0.5 M NaCl, 20 mM Hepes pH 7.5, 0.1 mM EDTA, 1 mM DTT, 15 mM CHAPS
- Buffer 3: 1 M NaCl, 20 mM Hepes pH 7.5, 0.1 mM EDTA, 1 mM DTT, 15 mM CHAPS, 10% glycerol

Method

1. Adjust the concentration of the denatured integrase from *Protocol 5*, step 4 to less than 1 mg/ml ($A_{280\ nm}$ of a 1 mg/ml integrase solution is 1.6) by dilution with 6 M guanidine–HCl, 20 mM Hepes pH 7.5, 2 mM EDTA, and 10 mM DTT. Add an equal volume of buffer 1 with stirring.
2. Dialyse for at least 12 h against buffer 2.
3. Dialyse for at least 12 h against buffer 3.[a]
4. Centrifuge at 12 000 *g* for 10 min.
5. Freeze aliquots of the supernatant in liquid nitrogen and store at −70°C. The protein can be concentrated by ultrafiltration, but CHAPS is also concentrated by this procedure; the CHAPS concentration can be decreased by extensive dialysis.

[a] The presence of CHAPS facilitates a good yield of the refolded protein. Presumably integrase that has been refolded in the presence of CHAPS can be exchanged into the storage buffer of *Protocol 3*, but this has not been tested.

3.3 Removing the His-tag

The His-tag can be removed by cutting with thrombin at the cleavage site engineered between the His-tag and the integrase polypeptide. Note that four amino acids (GSHM) remain linked to the amino-terminus of integrase after cleavage.

Protocol 7. Removing the His-tag using thrombin

Equipment and reagents

- Thrombin (Sigma T-6884)
- Benzamidine–Sepharose 6B (Pharmacia 17–0568–01)
- Chelating Sepharose FF (Pharmacia 17–0575–01)

Method

1. Add thrombin to the integrase protein from *Protocol 4,* step 5 or *Protocol 6,* step 4 to a final concentration of 40 NIH U/mg of integrase.[a]
2. Incubate at 25°C for 1 h.
3. Remove thrombin from the preparation by adsorption to a column of benzamidine–Sepharose 6B. The sodium chloride concentration should be adjusted to 0.5 M prior to this step. 10 mM CHAPS can be included in the buffer if necessary to keep the protein soluble.
4. Remove the cleaved His-tag from the integrase preparation by adsorption to a column of chelating Sepharose FF (see *Protocol 4*) or by gel filtration through a Superdex 75 column (Pharmacia).

[a] In our experience the activity of thrombin is not greatly influenced by NaCl concentration in the range of 0.2 M to 1 M. Thrombin also tolerates CHAPS at 15 mM, but it may be less active in its presence. The extent of cleavage should be checked by SDS–PAGE and the thrombin concentration and/or incubation time should be adjusted if necessary.

4. Integrase assays

4.1 *In vitro* activities of integrase

Integration of HIV DNA into host DNA occurs by a defined set of DNA cutting and joining reactions (*Figure 1*).

(a) Prior to integration, two nucleotides are removed from the 3′ ends of the flush-ended viral DNA made by reverse transcription.

(b) The 3′ processing reaction exposes the 3′ ends of the viral DNA that are covalently linked to host DNA in a subsequent DNA strand transfer reaction.

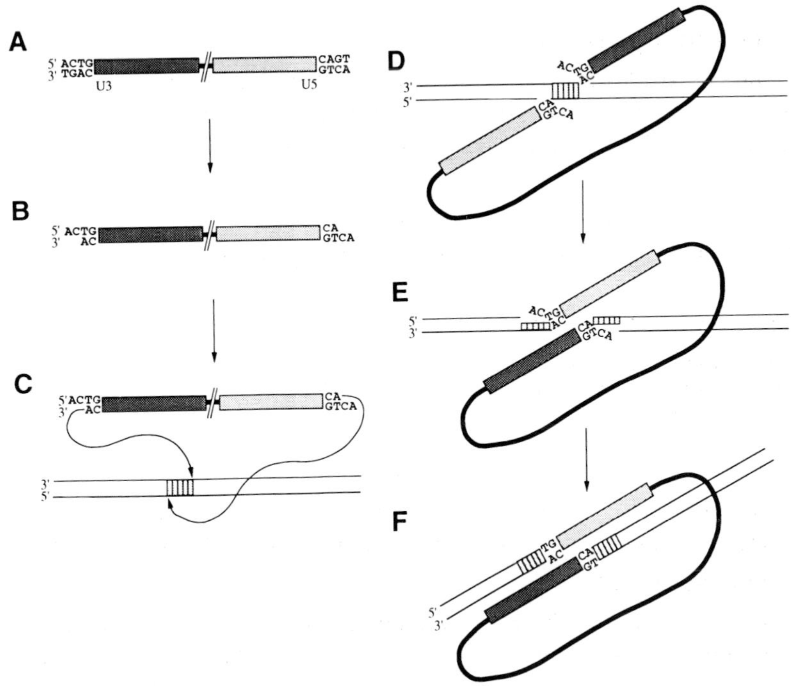

Figure 1. Mechanism of HIV DNA integration. A. The product of reverse transcription is a linear double-stranded DNA molecule with flush ends. The U3 and U5 ends of the viral DNA are distinguished by different shades of stippling. B. Two nucleotides are removed from the 3′ termini at both the U3 and the U5 ends of the viral DNA. This 3′ end processing reaction exposes the CA–$_{OH}$-3′ termini that are joined to target DNA in the next step of the integration process. C. A pair of polynucleotidyl transfer reactions (DNA strand transfer) inserts the pair of viral DNA ends into the target DNA (thin lines). The sites of insertion on the two target DNA strands are staggered by five nucleotides, with a 5′ overhang. D. In the DNA strand transfer product, the 3′ ends of the viral DNA are covalently joined to the 5′ ends of the target DNA at the site of insertion. The 3′ ends of the target DNA that are generated by the pair of DNA strand transfers remain unjoined. E. The five nucleotides between the sites of insertion on the two target DNA strands melt. F. Integration is completed by removal of the two unpaired nucleotides at the 5′ ends of the viral DNA and repair of the single-strand gaps between viral and target DNA. Cellular enzymes are likely to carry out these steps. The overall process results in the loss of two nucleotides from each end of the viral DNA and duplication of five base pairs of host DNA at the site of integration.

(c) The DNA strand transfer step comprises a pair of transesterification reactions that cleave each strand of the target DNA at the site of insertion and join the 3′ ends of the viral DNA to the 5′ ends of the target DNA at this site.

(d) The pair of insertions are staggered by five nucleotides on the two target DNA strands.

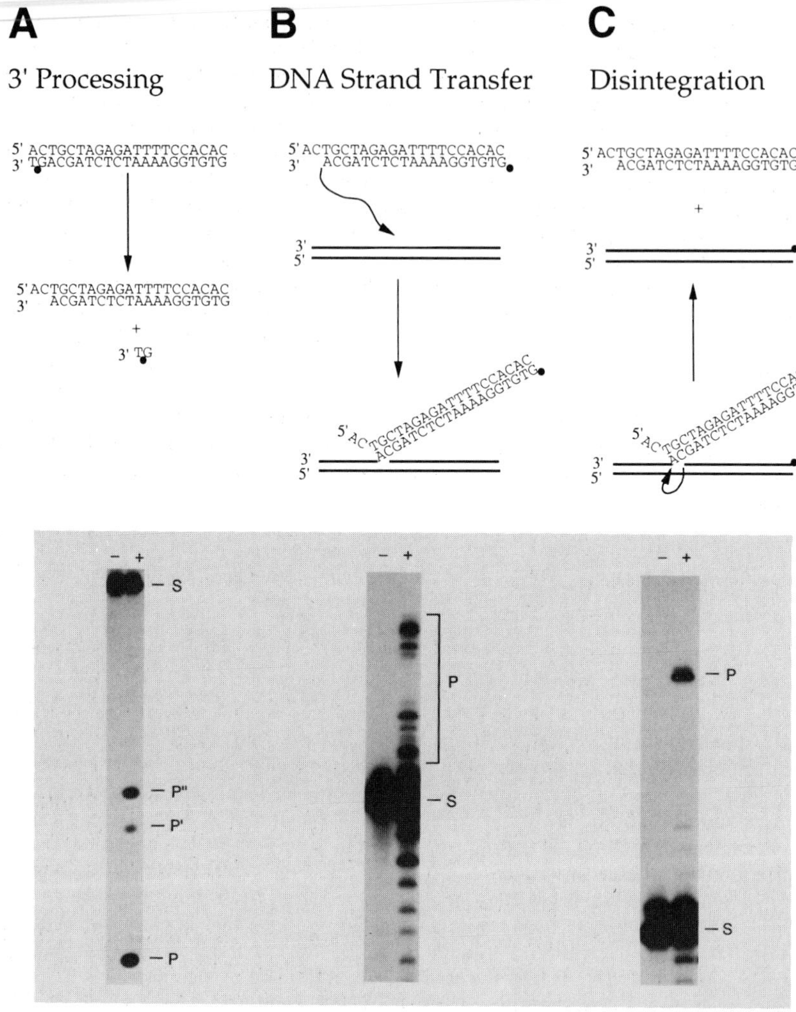

(e) The final step of the integration process, cleavage of the unpaired nucleotides at each 5′ end of the viral DNA and repair of the single-strand gaps between the viral and target DNA, is thought to be completed by cellular enzymes.

Although integration *in vivo* is mediated by a nucleoprotein complex that is likely to include other proteins in addition to integrase, purified integrase possesses the enzymatic activities of 3′ processing and DNA strand transfer.

The *in vitro* activities of integrase are depicted in *Figure 2*.

(a) 3′ processing: Integrase removes two nucleotides from the 3′ end of a DNA molecule that mimics either the U3 or U5 end of HIV DNA. This

Figure 2. Activities of HIV-1 integrase *in vitro*. A. 3′ processing. The duplex oligonucleotide substrate matches the U5 end of HIV DNA. ^{32}P-label (filled circle) is located at the phosphate bridging the two nucleotides that are cleaved off by integrase. The lower panel is an autoradiogram showing the dinucleotide product (P), separated from unreacted substrate (S) by gel electrophoresis; a cyclic (P′) and a glycerol adduct (P″) form of the dinucleotide product are also generated (11, 12). Integrase was included (+) or omitted (−) from reaction mixtures as indicated above each lane. B. DNA strand transfer. The duplex oligonucleotide substrate matches the U5 end of HIV-1 DNA after removal of the terminal dinucleotide. The DNA strand that is joined to target DNA (thick lines) is labelled with ^{32}P (filled circle) at its 5′ end. The lower panel is an autoradiogram showing the labelled strand of the DNA strand transfer products (P) separated from unreacted substrate (S) by gel electrophoresis. The strand transfer products are heterogeneous in length because insertion can occur at any location on either strand of the target DNA; however, insertion at some sites is preferred, as indicated by the uneven intensity of the product bands. C. Disintegration. The substrate matches the DNA strand transfer product shown in *B*, except that the ^{32}P-label is located at the 5′ end of the target DNA strand into which the viral DNA end is joined. The lower panel is an autoradiogram showing the labelled strand of the disintegration product (P) separated from unreacted substrate (S) by gel electrophoresis.

---

reaction exposes the CA-$_{OH}$-3′ end that is to be inserted into another DNA molecule (target DNA).

(b) DNA strand transfer: The DNA strand transfer reaction inserts a processed viral DNA end into the target DNA; the 3′ end of the viral DNA substrate is joined to the 5′ end of the target DNA at the site of insertion. Since there is very little target site specificity, essentially any DNA can serve as a target, including the viral end DNA substrate itself. Most of the DNA strand transfer products in reactions with purified integrase correspond to insertion of a single viral DNA end into a single-strand of the target DNA; the factor(s) required for efficient insertion of pairs of ends have not yet been identified.

(c) Disintegration: Integrase can resolve a substrate that mimics the product of the DNA strand transfer reaction. This apparent reversal of the DNA strand transfer reaction liberates the viral DNA end part of the substrate and seals the target DNA. This activity of integrase may not be biologically important.

Suitable DNA substrates for each of these reactions can be made by annealing short oligonucleotides, one strand of which is radiolabelled; prior to labelling and annealing, the oligonucleotides should be purified by electrophoresis in a denaturing polyacrylamide gel to remove contaminating shorter strands. Reaction products are resolved from unreacted substrate by gel electrophoresis and detected by autoradiography.

4.2 Assay for 3′ processing

The DNA substrate is a duplex oligonucleotide corresponding to the U5 (or U3) end of HIV DNA, labelled with ^{32}P within the dinucleotide that is cleaved off by integrase. [α-^{32}P]TTP is incorporated by a DNA polymerase using, as a template, a duplex oligonucleotide that lacks the terminal T nucleotide at the 3′ end of the DNA strand that is cut by integrase. The substrate is prepared as described in *Protocol 8*, for use in the reaction described in *Protocol 9*.

Protocol 8. Preparation of the substrate for the 3′ processing reaction

Equipment and reagents

- Oligo A (U5 plus-strand) 5′-GTGTGGAAAATCTCTAGCAG
- Oligo B (U5 minus-strand) 5′-ACTGCTAGAGATTTTCCACAC
- 5 × Sequenase reaction buffer (United States Biochemical 70702)
- [α-^{32}P]TTP, at a specific activity of 3000 Ci/mmol (New England Nuclear NEG-015H)
- 1 mM TTP
- Sequenase version 2.0 (United States Biochemical 70775)
- Enzyme dilution buffer: 10 mM Tris–HCl pH 7.5, 5 mM DTT, 0.5 mg/ml BSA
- 0.1 M DTT
- 0.25 M EDTA pH 8.0
- G-25 spin column (Boehringer Mannheim 100402)

Method

1. Mix equal quantities of oligonucleotide A and oligonucleotide B together in a microcentrifuge tube to yield a final DNA concentration of 0.1 mg/ml in 0.1 M NaCl.
2. Heat the DNA to 85°C in a water-bath.
3. Slowly cool the DNA to room temperature. Store at −20°C.
4. Mix the following in a microcentrifuge tube:

• H_2O	2.0 μl
• 5 × reaction buffer	4.0 μl
• 0.1 M DTT	2.0 μl
• DNA mixture	2.0 μl
• [α-^{32}P]TTP	20.0 μl
• 1:8 (v/v) dilution of Sequenase in enzyme dilution buffer	4.0 μl

5. Incubate the mixture for 10 min at 37°C.
6. Add 2.0 μl of 1 mM TTP. Incubate for 5 min at 37°C.
7. Stop the reaction by adding 2.0 μl of 0.25 M EDTA pH 8.0.
8. Adjust the volume to 50 μl with H_2O and separate the labelled substrate from the unincorporated [α-^{32}P]TTP by passing the mixture through the G-25 spin column as recommended by the manufacturer. Store the labelled substrate at −20°C.

Protocol 9. Assay for 3′ processing

Equipment and reagents

- 4 μM (128 μg/ml) integrase (see *Protocol 3, 4, or 5*)
- ^{32}P-labelled DNA substrate (see *Protocol 8*)
- 0.5 M MOPS–NaOH pH 7.2
- 1 mg/ml BSA
- 0.1 M 2-mercaptoethanol
- 150 mM $MnCl_2$
- 80% (w/v) glycerol
- Sequencing stop buffer: 95% formamide, 10 mM EDTA, 0.003% xylene cyanol, 0.003% bromophenol blue
- Sequencing gel electrophoresis equipment

Method

1. Mix the following in a microcentrifuge tube:

		[Final]
• H_2O	9.0 μl	
• 0.5 M MOPS pH 7.2	1.0 μl	25 mM
• 1 mg/ml BSA	2.0 μl	100 μg/ml
• 0.1 M 2-mercaptoethanol	2.0 μl	10 mM
• 80% glycerol	2.25 μl	10% [a]
• 150 mM $MnCl_2$	1.0 μl	7.5 mM
• DNA substrate	1.75 μl	25 nM (7 ng)
• integrase	1.0 μl	200 nM (128 ng)
•		50 mM NaCl (from the integrase storage buffer) [b]

2. Incubate for 1 h at 37 °C.
3. Add 20 μl sequencing stop buffer.
4. Heat to 100 °C for 2 min.
5. Electrophorese 1–2 μl in a 15% or 20% denaturing polyacrylamide gel.
6. Visualize the reaction products by autoradiography.

[a] Includes glycerol contributed by the integrase storage buffer.
[b] The reaction is most efficient at low ionic strength and is strongly inhibited by NaCl concentrations greater than approximately 100 mM.

4.3 Assay for DNA strand transfer

In the reactions described here, a single duplex oligonucleotide species serves as both the viral end DNA substrate and the target DNA substrate. The reaction conditions for DNA strand transfer are identical to those for 3′ processing. In fact, *Protocol 9* can be used to assay for DNA strand transfer as well as 3′ processing. The 3′ processing products go on to insert their 3′ ends into other substrate molecules. The resulting DNA strand transfer products form a series of bands that migrate more slowly than the starting DNA substrate However, it is often convenient to use a DNA substrate (*Protocol 10*) that differs from the 3′ processing substrate as follows:

(a) The substrate DNA lacks the terminal two nucleotides that are removed by integrase in the 3′ processing reaction. Since this substrate bypasses the processing reaction, only DNA strand transfer is assayed.

(b) The radiolabel is placed at the 5′ end of the DNA strand that is covalently joined to target DNA in the strand transfer reaction. With this modification, all labelled substrate molecules that undergo strand transfer give rise to a labelled product; with the 3′ processing substrate, strand transfer products resulting from insertion into the unlabelled strand of the substrate DNA will not be detected.

Protocol 10. Preparation of DNA strand transfer substrate[a]

Equipment and reagents

- Oligo C (U5 plus-strand) 5′-GTGTGGAAAATCTCTAGCA 0.1 mg/ml
- Oligo B (U5 minus-strand) 5′-ACTGCTAGAGATTTTCCACAC 0.1 mg/ml
- 10 × kinase buffer: 0.7 M glycine–NaOH pH 9.5, 0.1 M $MgCl_2$, 50 mM DTT
- [γ-^{32}P]ATP, at a specific activity of 3000 Ci/mmol (New England Nuclear NEG-002A)
- T4 polynucleotide kinase (PNK, Pharmacia 27–0736–01)
- 0.25 M EDTA pH 8.0
- 1 M NaCl

Method

1. Mix the following in a microcentrifuge tube:

• H_2O	2.5 μl
• 10 × kinase buffer	1.5 μl
• oligo C (0.1 mg/ml)	1.0 μl
• [γ-^{32}P]ATP	10.0 μl
• PNK	1.5 μl

2. Incubate the mixture for 45 min at 37 °C.
3. Stop by adding 1.0 μl of 0.25 M EDTA pH 8.0.
4. Heat inactivate the PNK by incubation at 85 °C for 15 min.
5. Cool on ice.
6. Add the following:

• H_2O	20.0 μl
• 1 M NaCl	4.0 μl
• oligo B (0.1 mg/ml)	1.0 μl

7. Heat to 85 °C and then cool slowly to room temperature.
8. Adjust the volume to 50 μl with H_2O and separate the labelled substrate

from the unincorporated [γ-^{32}P]ATP as described in *Protocol 8*, step 8. Store the substrate at −20°C.

[a] DNA strand transfer reactions are carried out as described for 3′ processing in *Protocol 9*, except that 1.75 μl of strand transfer substrate is substituted for the 1.75 μl of 3′ processing substrate.

4.4 Assay for disintegration

Reaction conditions for disintegration are the same as for 3′ processing and DNA strand transfer. The branched DNA substrate for the disintegration reaction is made as described in *Protocol 11*.

Protocol 11. Preparation of DNA substrate for the disintegration reaction[a]

Equipment and reagents

- Oligo D: 5′-GAAAGCGACCGCGCC at 0.1 mg/ml
- Oligo E: 5′-GGACGCCATAGCCCCGGCGCGGTCGCTTTC at 0.2 mg/ml
- Oligo B: 5′-ACTGCTAGAGATTTTCCACAC at 0.14 mg/ml
- Oligo F: 5′-GTGTGGAAAATCTCTAGCAGGGGCTATGGCGTCC at 0.23 mg/ml
- 10 × kinase buffer, [γ-^{32}P]ATP, PNK, and 0.25 M EDTA pH 8.0 (see *Protocol 10*)
- 1 M NaCl

Method

1. Label 1.0 μl of oligonucleotide D as described in *Protocol 10*, steps 1 and 2.
2. Stop the reaction and heat inactivate the PNK (*Protocol 10*, steps 3 and 4).
3. Cool on ice.
4. Add the following:
   - H_2O 18.0 μl
   - 1 M NaCl 4.0 μl
   - oligo B (0.14 mg/ml) 1.0 μl
   - oligo E (0.20 mg/ml) 1.0 μl
   - oligo F (0.23 mg/ml) 1.0 μl
5. Heat to 85°C and then cool slowly to room temperature.
6. Adjust the volume to 50 μl with H_2O and separate the labelled substrate from the unincorporated [γ-^{32}P]ATP (*Protocol 8*, step 8). Store the substrate at −20°C.

[a] Disintegration reactions are carried out as described for 3′ processing in *Protocol 9*, except that 1.75 μl of disintegration substrate is substituted for the 1.75 μl of 3′ processing substrate.

4.5 Modification of the integrase assays

There are many possible variations on the assays described above. Some possibilities are noted below:

(a) Essentially any DNA molecule, such as a different oligonucleotide or a plasmid DNA, can be included in the reaction mixture as a target for DNA strand transfer. In the latter case the reaction products are detected by electrophoresis in agarose followed by autoradiography. If the DNA concentration in the reaction mixture is changed, integrase should be titrated to determine the optimum concentration.

(b) Different oligonucleotides can be used as the DNA substrate. For example, sequences corresponding to the U3 end of HIV DNA are suitable, although they work somewhat less efficiently than U5 ends. Many variations of the disintegration substrate have been described (6, 7).

(c) Labels other than ^{32}P can be used.

(d) By using substrate DNA with biotin on one DNA strand and a label, such as ^{32}P, on the other DNA strand it is possible to separate labelled strand transfer products from unreacted substrate without the need for gel electrophoresis. After the reaction, the DNA is bound to avidin that is coupled to a solid support. After washing with alkali, the DNA is released by disrupting the avidin–biotin interaction. The only labelled DNA strands that are recovered after this procedure are DNA strand transfer products. This assay can be carried out in the wells of microtitre plates and thus can be easily adapted for high throughput screening of possible inhibitors (8).

(e) The DNA substrate may be coupled to a solid support.

5. Utility of *in vitro* integration assays in screening for inhibitors of HIV integrase

Current thinking favours the notion that each of the reactions catalysed by integrase uses a common catalytic site. If this is correct, all three assays described above may be equally well suited to identify compounds that inhibit at the catalytic step. However, the requirements for the 3′ processing and DNA strand transfer reaction are more stringent than for disintegration—a fragment of integrase that contains only the central region of the protein can carry out disintegration, whereas the entire protein is required for 3′ processing and DNA strand transfer (9, 10). An assay for DNA strand transfer that requires integrase to first process the ends of the DNA substrate may therefore present more targets for which inhibitors may be found. The feasibility of such an assay in microtitre well format has been demonstrated and thus high throughput screening for inhibitors of integrase is technically feasible (8).

Although many integrase molecules enter an infected cell with the viral RNA, they are needed to perform only a single integration event. It is therefore hardly surprising that *in vitro* studies reveal integrase to be a very sluggish enzyme. Integrase can also perform a variety of reactions *in vitro*, such as nicking of closed circular DNA, that do not strictly reflect its biological role. These observations raise the possibility that searching for 'stimulators' may identify compounds that cause integrase to destroy the viral DNA with which it is associated. Such hypothetical stimulators could work by causing the viral DNA to integrate into itself before it has a chance to find its intended target: host DNA. Alternatively, they could induce integrase to act as a nuclease that destroys the viral DNA.

It is hoped that the methodologies outlined in this chapter will stimulate the search for possible therapeutic agents that block the DNA integration step in the HIV replication cycle.

References

1. Studier, F. W. and Moffatt, B. A. (1986). *J. Mol. Biol.*, **189**, 113.
2. Rosenberg, A. H., Lade, B. N., Chui, D., Lin, S.-W., Dunn, J. J., and Studier, F. W. (1987). *Gene*, **56**, 125.
3. Engelman, A. and Craigie, R. (1992). *J. Virol.*, **66**, 6361.
4. Sambrook, J., Fritsch, E. F., and Maniatis, T. (ed.) (1989). *Molecular cloning, a laboratory manual.* Cold Spring Harbor Press, Cold Spring Harbor, NY.
5. Sherman, P. A. and Fyfe, J. A. (1990). *Proc. Natl Acad. Sci. USA*, **87**, 5119.
6. Chow, S. A., Vincent, K. A., Ellison, V., and Brown, P. O. (1992). *Science*, **255**, 723.
7. Vincent, K. A., Ellison, V., Chow, S. A., and Brown, P. O. (1993). *J. Virol.*, **67**, 425.
8. Craigie, R., Mizuuchi, K., Bushman, F. D., and Engelman, A. (1991). *Nucleic Acids Res.*, **19**, 2729.
9. Bushman, F. D., Engelman, A., Palmer, I., Wingfield, P., and Craigie, R. (1993). *Proc. Natl Acad. Sci. USA*, **90**, 3428.
10. Vink, C., Oude Groeneger, A. A. M., and Plasterk, R. H. A. (1993). *Nucleic Acids Res.*, **21**, 1419.
11. Engelman, A., Mizuuchi. K., and Craigie, R. (1991). *Cell*, **67**, 1211.
12. Vink, C., Yeheskiely, E., van der Marel, G. A., van Boom, J. H., and Plasterk, R. H. A. (1991). *Nucleic Acids Res.*, **19**, 6691.

5

HIV protease

C. DEBOUCK, T. A. TOMASZEK JR, L. A. IVANOFF, and J. CULP

1. Introduction

The retroviral protease, which plays an essential role in virion maturation (1), has been subjected to intensive studies and drug discovery efforts. The HIV-1 protease has a unique cleavage specificity and belongs to the family of aspartyl proteases. It is required for the proteolytic processing of the large Gag and Gag–Pol viral polyprotein precursors into the mature virion structural proteins (matrix, capsid, and nucleocapsid) as well as the virion enzymes (protease, reverse transcriptase, and integrase) (*Figure 1*). When the HIV-1 protease is inactivated by mutation or by a specific inhibitor, non-infectious virions with immature morphology are formed and the life cycle is halted. It is the essential nature and the unique specificity of the HIV-1 protease that called for its consideration as a therapeutic target.

Two technical advances have aided the effort to discover HIV-1 protease inhibitors by rational drug design and/or random compound screening. First, a sufficient and safe source of pure, authentic HIV-1 protease was developed by recombinant DNA technology and, secondly, convenient *in vitro* proteolytic assays were developed. In this chapter, we describe an expression system for the HIV-1 protease as well as methods used to produce recombinant HIV-1 protease in *Escherichia coli*, to purify and assay the enzyme *in vitro*, and to formulate HIV-1 protease–inhibitor complexes for structural studies.

2. Bacterial expression of HIV-1 protease

Virions have been used as a natural, but scarce, source of the HIV-1 protease. The 99 amino acid long protease has also been produced by total chemical synthesis. However, it was the construction of recombinant expression systems for HIV-1 protease that was the key in providing the first abundant and safe source of authentic and active enzyme. This resulted in the rapid development of *in vitro* peptidolytic assays for the enzyme and the solution of its three-dimensional structure by X-ray diffraction crystallography. It also

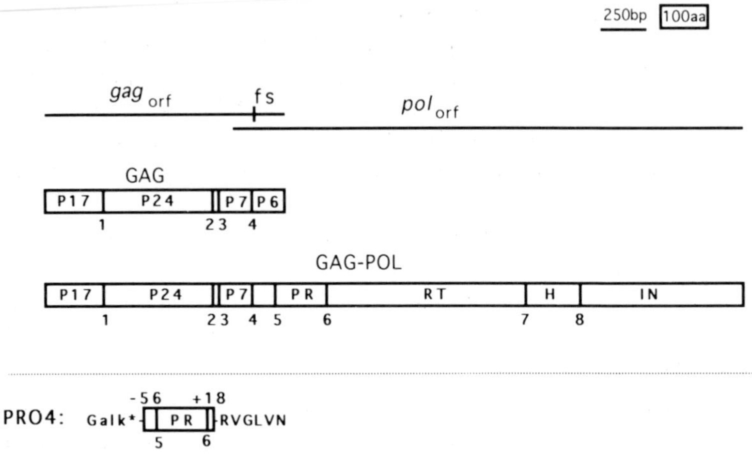

Figure 1. Organization of the HIV-1 Gag and Gag–Pol polyproteins. The *gag* and *pol* open reading frames (orf) and the position of the translational frame shift (fs) between *gag* and *pol* are shown. The organization of $Pr55^{Gag}$ and $Pr160^{Gag-Pol}$ is represented schematically with the position of eight cleavage sites for HIV-1 protease (1–8) and the mature proteins (p17 Gag, p24 Gag, p7 Gag, p6 Gag, protease (PR), reverse transcriptase (RT), ribonuclease H (H), and integrase (IN). The structure of the PRO4 construct described in the text is also depicted. Galk* indicates the translational fusion to the first 56 codons of galactokinase; −56 and +18 indicate the number of residues from Pol preceding and following the mature protease sequence, respectively; RVGLVN represent the last residues of the PRO4 construct using the one letter code. The scale is shown in base pairs (bp) and amino acids (aa).

allowed the thorough characterization of the structure and function of this protease by site-specific mutagenesis.

To date, a number of laboratories have reported recombinant expression of HIV-1 protease using various expression constructs in bacteria or yeast (see ref. 2 for a recent review). In our opinion, the expression systems that provide the most reliable source of active HIV-1 protease are those in which the enzyme is produced in *E. coli* by autoprocessing from a precursor species that includes sequences from the HIV-1 *pol* open reading frame on either side of the region encoding the 99 amino acid long mature protease. Several groups have over-produced HIV-1 protease by bacterial expression of its mature sequence (residues 1–99), but the enzyme is then found primarily in inclusion bodies in the insoluble fraction of bacterial extracts. These inclusion bodies can be solubilized in buffers containing chaotropic salts and refolded into active protease, but this often leads to the improper folding and insolubility of the protease. Perhaps this is not surprising, since the genuine enzyme is naturally translated and initially folded as part of a polyprotein precursor that undergoes proteolytic processing to yield the mature, properly folded, active enzyme.

In the PRO4 construct, which is described in more detail elsewhere (2, 3), the sequence for mature HIV-1 protease (PR, see *Figure 1*) is flanked by amino acid residues derived from the HIV-1 *pol* open reading frame. To maximize the level of expression of the precursor protein, the protease coding fragment is inserted downstream of the first 56 codons of galactokinase. Since intracellular production of the HIV-1 protease is highly cytotoxic, we used for its expression the strong and tightly regulated lambda P_L promoter on a high copy number pBR322-derived expression vector, pOTSKF33. The P_L promoter is typically induced by addition of nalidixic acid to the culture medium or by temperature upshift (when a thermolabile repressor is present). *Protocol 1* describes a one litre scale nalidixic acid induction of HIV-1 protease from the PRO4 construct in *E. coli*.

Protocol 1. Production of HIV-1 protease in *E. coli* by nalidixic acid induction of the P_L promoter in the PRO4 construct

Equipment and reagents

- 1 litre of LB broth (10 g Difco bacto tryptone, 5 g Difco yeast extract, 10 g NaCl) in a 2 litre shake flask sterilized by autoclaving
- 200 × stock solution of ampicillin at 10 mg/ml, filter sterilized
- 40 ml overnight culture of *E. coli* strain containing the PRO4 plasmid in LB broth + ampicillin (at 50 μg/ml)
- Controlled environment incubator shaker for culturing bacteria set at 37°C
- Spectrophotometer to monitor bacterial growth by optical density set at 650 nm
- 1000 × solution of nalidixic acid at 60 mg/ml in 1 M NaOH (nalidixic acid is not soluble in plain water)
- SDS–PAGE loading buffer: 125 mM Tris–HCl pH 6.8, 20% glycerol, 10% β-mercaptoethanol, 4% SDS, pinch of bromophenol blue
- Low-speed centrifuge and centrifuge bottles

Method

1. Add 5 ml of ampicillin stock solution to the 2 litre shake flask containing the LB broth. The presence of the antibiotic in the culture medium ensures the maintenance of the expression plasmid in all cells.
2. Inoculate the flask with the 40 ml overnight culture and aerate vigorously in the shaker at 37°C.
3. Monitor the absorbance at 650 nm (A_{650}) of the culture periodically starting 1 h after inoculation. After 1–2 h, the A_{650} will reach 0.6–0.8, which is the ideal density for induction with nalidixic acid.
4. Add 1 ml (from 1000 × stock solution) nalidixic acid to the culture.
5. Aerate vigorously in the shaker at 37°C for 5 h.
6. Analyse the induced proteins by SDS–PAGE and Western blot. Take a 500 μl aliquot of induced cells, microcentrifuge for 1 min, resuspend the cell pellet in 50 μl of SDS–PAGE loading buffer, and apply 1–10 μl of sample for SDS–PAGE.
7. Collect the induced bacterial cells by low speed centrifugation at 4°C.

Protocol 1. *Continued*

8. Discard the cleared supernatant and freeze the cell paste at −20°C until further use. This procedure should yield 2–4 g of wet cell paste per litre of induced culture.

The expression of HIV-1 protease achieved using *Protocol 1* can be readily analysed by Western blot analysis of the induced bacterial extract using an antibody specific for HIV-1 protease. As shown in *Figure 2*, a protein of about 11 kilodalton (kDa) in size is detected in the induced PRO4 extract (lane 2) but not in the induced parent vector without insert (lane 1). This 11 kDa protein is the authentic HIV-1 protease produced by autoprocessing of the PRO4 protease precursor at two Phe–Pro cleavage sites (sites 5 and 6 in *Figure 1*). This processing is eliminated if a catalytic site mutation is introduced in the PRO4 construct resulting in the detection of a protein of 25 kDa, the size expected for the product of the PRO4 precursor protein (lane 3).

2.1 Purification of recombinant HIV-1 protease

Recombinant HIV-1 protease produced by induction of the PRO4 construct is authentic, active, and soluble after lysis of the bacterial cells. The procedure for purifying HIV-1 protease using this expression system has been published (4) and is described in detail in *Protocol 2*. The Manton-Gaulin cell homogenizer is preferable to sonication or detergents for lysis of large

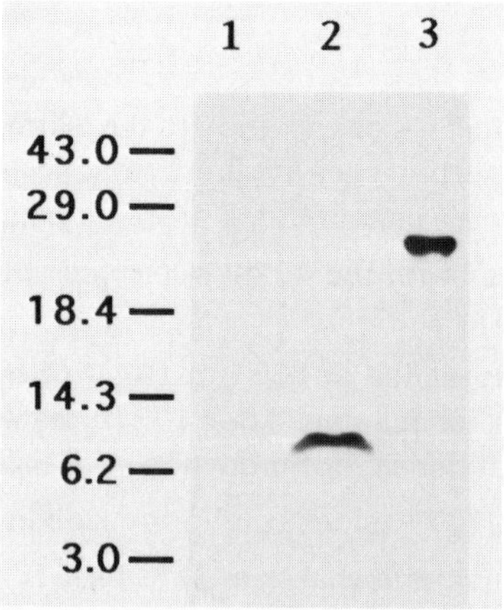

Figure 2. Expression of HIV-1 protease in *E. coli.* After induction of the P_L promoter by nalidixic acid, the bacterial extracts were subjected to 15% SDS–PAGE followed by immunoblot analysis with an HIV-1 protease-specific antibody. Lane 1: pOTSKF33 parent vector without gene insert; lane 2: PRO4 wild-type; lane 3: PRO4 inactivated by an active site mutation D25N. The position and size of molecular weight markers is shown.

quantities of frozen cells, because the temperature remains constant during lysis, membrane proteins are not solubilized, and lysis is efficient (> 80% lysis by microscopic observation). After lysis and centrifugation, the protease activity is concentrated by ammonium sulfate precipitation and efficiently recovered from the precipitate using dialysis.

The initial chromatography step on Superose-12 removes the majority of contaminating proteins resulting in highly pure protease (*Figures 3* and *4*).

(a) 10 mM dithiothreitol (DTT) is absolutely required for this step to prevent aggregation and elution from the column in a broad peak.

(b) HIV-1 protease interacts with the Superose-12 support and elutes after the bulk of the proteins. This phenomenon was originally described for a TSK-3000 column (5).

(c) Fractions from the Superose-12 column are pooled according to protease activity and purity.

For kinetic studies, the pooled protease fractions are made 40% in glycerol and they can be stored for more than one year at −20°C without loss of activity. For crystallization purposes, the purification is continued by applying the pooled Superose fractions to Q–Sepharose. Under these conditions, virtually all of the protease flows through the column. Without Q–Sepharose chromatography, HIV-1 protease does not bind efficiently to the subsequent S–Sepharose column, probably due to contaminating nucleic acids. These impurities also prevent crystallization if they are not removed. The fractions containing HIV-1 protease are pooled, applied to S–Sepharose, and eluted

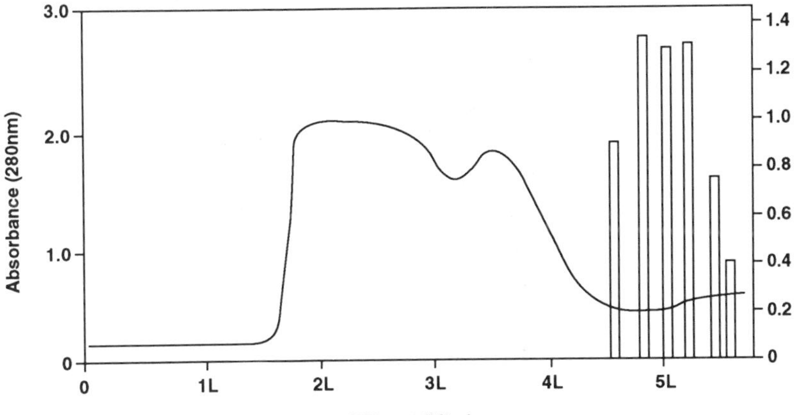

Figure 3. Profile of the Superose-12 chromatography of HIV-1 protease. The chromatography was carried out according to *Protocol 2*. Absorbance of the effluent was monitored at 280 nm. Protease activity was measured using the HPLC quantitation assay described in *Protocol 4* and reported as milligrams of protease per fraction (open bars).

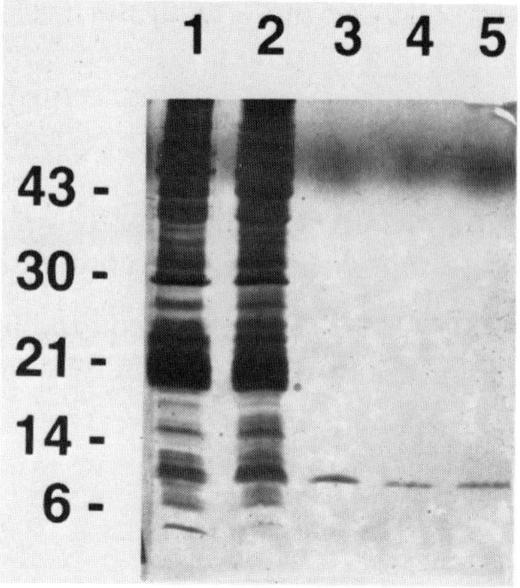

Figure 4. Analysis of purified recombinant HIV-1 protease by SDS–PAGE. Proteins present at various steps of *Protocol 2* were visualized by silver staining after 20% SDS–PAGE. Lane 1: total cell lysate; lane 2: dissolved ammonium sulfate pellet; lane 3: pooled fraction from Superose-12 chromatography; lane 4: pooled fraction from QSFF chromatography; lane 5: pooled fraction from SSFF chromatography. The position and size of molecular weight markers is shown.

in a concentrated form. Although the protease is purified to apparent homogeneity after elution from S–Sepharose (*Figure 4*), the enzyme is susceptible to rapid autodegradation even at 4°C.

(a) Add HIV-1 protease inhibitor as soon as possible after the S–Sepharose step.
(b) Concentrate the complex via ultrafiltration for crystallography.
(c) Alternatively, add 40% (v/v) glycerol to the S–Sepharose protease pool. This material is suitable for kinetic studies and can be stored for more than one year at −20°C without loss of activity.

Protocol 2. Purification of recombinant HIV-1 protease (PRO4) from induced bacterial cells

Equipment and reagents

- Frozen (−70°C) cell paste of induced *E. coli* cells containing recombinant HIV-1 protease (PRO4 construct)
- 2 litre lysis buffer: 50 mM Tris base pH 8.0, 5 mM EDTA, 10 mM DTT, adjust to pH 8.0 at 4°C with 6 M HCl—immediately before use, add 1 ml of 1 M phenylmethylsulfonyl fluoride (PMSF) freshly made in methanol
- 15 litres superose buffer: 50 mM Tris–HCl pH 8.0, 5 mM EDTA, 10 mM DTT, 200 mM NaCl, 10% glycerol
- Superose-12 column (11 × 46 cm, ~ 5 litre bed volume) equilibrated with superose buffer at 25 ml/min
- 4 litres Q buffer: 50 mM Tris–HCl pH 8.0, 5 mM EDTA, 10 mM DTT, 10% glycerol
- 1 litre Q elution buffer: 50 mM Tris–HCl pH 8.0, 5 mM EDTA, 10 mM DTT, 1 M NaCl, 10% glycerol
- Q–Sepharose Fast Flow (QSFF) column (5 × 15 cm, ~ 300 ml) equilibrated in Q buffer at 20 ml/min

- 2 litres S buffer: 50 mM Na acetate pH 5.0, 2 mM EDTA, 5 mM DTT
- 1 litre S elution buffer: 50 mM Na acetate pH 5.0, 2 mM EDTA, 5 mM DTT, 1 M NaCl
- S-Sepharose Fast Flow (SSFF) column (2.2 × 10 cm) equilibrated at 9 ml/min in S buffer
- 15% or gradient sodium dodecyl sulfate (SDS) polyacrylamide gels for analysis using SDS–polyacrylamide gel electrophoresis (SDS–PAGE)
- 0.2 μm Millipak filter (Millipore)
- Dialysis tubing with a < 6000 molecular weight cut-off (Spectrum)
- Tekmar Tissuemizer
- Manton-Gaulin cell homogenizer or equivalent mechanical cell disrupter
- Q–Sepharose (5 × 15 cm, ~ 300 ml), S–Sepharose (2.2 × 10 cm, ~ 40 ml), Superose-12 (11 × 46 cm, ~ 5 litre bed volume), SDS–PAGE standards, and column hardware (Pharmacia Fine Chemicals)—all columns are prepared according to manufacturer's recommendations
- FPLC system (Pharmacia)
- System Gold HPLC (Beckman Instruments) equipped with C18 columns (Vydac)
- BCA Protein Assay Kit (Pierce Chemical Co.) for total protein determinations (assays are performed according to manufacturer's instructions using BSA as a standard)

Method[a]

1. Resuspend 500 g frozen *E. coli* (PRO4) cells in 1 litre lysis buffer using a Tekmar Tissuemizer.
2. Lyse the cells by passing the suspension through a Manton-Gaulin cell homogenizer twice at 6–8000 p.s.i.
3. Centrifuge at 30 000 *g* for 45 min, decant the supernatant into a large beaker, and discard the pellet.
4. Add 209 g ammonium sulfate per litre of supernatant (35% saturation) gradually and stir for 1–2 h.
5. Centrifuge as in step 3, discard the supernatant, and resuspend the pellet in a minimal volume of lysis buffer (50–100 ml).
6. Dialyse resuspended ammonium sulfate pellet overnight against 5 litres superose buffer.
7. Centrifuge as in step 3 to remove any insoluble material from the dialysed ammonium sulfate sample and filter the supernatant through a 0.2 μm Millipak filter.
8. Load the filtrate on to a Superose-12 column at 25 ml/min and elute the column with superose buffer. HIV-1 protease elutes after 3 litres or 60% of the total column volume.
9. Analyse fractions using SDS–PAGE, the protease activity assay (see *Protocol 4*), and/or the HPLC quantitation assay: inject samples containing unknown amounts of protease on to a Vydac C18 column; elute with a gradient of 25–75% acetonitrile in 0.1% TFA; integrate peak areas at 220 nm using a Beckman 166 detector and System Gold software; compare peak areas to a standard curve of peak areas obtained using a purified HIV-1 protease standard quantitated by amino acid analysis. Pool the fractions containing the majority of the purified protease based on SDS–PAGE, activity, and/or HPLC quantitation analysis.

Protocol 2. *Continued*

10. Dilute Superose pool with equal volume Q buffer and load the Superose pool on to the Q–Sepharose Fast Flow column at 20 ml/min.

11. Collect the entire non-bound protein fraction as the QSFF protease pool and remove any protein or DNA bound to the column with Q elution buffer and discard.

12. Adjust the pH of the QSFF flow-through pool to 5.0 with 10% acetic acid (final concentration 0.1%) and dilute with equal volume S–Sepharose buffer.

13. Load the pH 5 QSFF flow-through pool on to an SSFF column at 9 ml/min. Wash the column with SSFF buffer; elute HIV-1 protease with a ten column volume linear gradient from 0–1 M NaCl.

14. If the enzyme is to be used for kinetic studies, glycerol may be added to a final concentration of 40% (v/v) and the solution will be stable at −20°C. Alternatively, protease can be prepared for crystallography by the addition of peptide inhibitors immediately after elution (see *Protocol 3*).

[a] Perform all procedures at 4°C unless otherwise indicated.

Using the method described, active HIV-1 protease can be purified to homogeneity in a few days. A yield of 2–3 mg of pure protease can be obtained from 50 g of induced cells. In addition to HIV-1 protease, SIV_{mac} protease can be purified by the same method with similar yield (6). This procedure has also been used successfully with low levels of protease expression.

2.2 Formulation of HIV-1 protease–inhibitor complexes for crystallography

The rational design of HIV-1 protease inhibitors can best be achieved with the knowledge of the three-dimensional structure of known protease inhibitors complexed with the protease. For X-ray crystallography, relatively high concentrations of HIV-1 protease are required in addition to an excess of inhibitor. The preparation of enzyme for crystallography is described in detail in *Protocol 2* and its formulation with inhibitors for structural studies is given in *Protocol 3* below.

Protocol 3. Formulation of purified HIV-1 protease with inhibitors for structural studies

Equipment and reagents

- Dimethyl sulfoxide (DMSO, Pierce)
- 0.2 μm Millex GV (Millipore)
- 10 ml ultrafiltration stirred cell and YM-10 ultrafiltration membrane (Amicon)
- HIV-1 protease inhibitor
- HIV-1 protease (~ 0.3 mg/ml, or 15 μM)

A. *Complexes with water soluble inhibitors*

1. Prepare a concentrated stock solution of the inhibitor in water (~ 10 mM).
2. Add pooled protein (~ 0.3 mg/ml, or 15 μM) from the S–Sepharose column (see *Protocol 2*) to a final inhibitor concentration of 15–150 μM.
3. Dilute the protease with an equal volume of fresh S buffer (see *Protocol 2*) in order to prevent precipitation of the protein which occurs in concentrations of NaCl greater than 200 mM.
4. Go to steps 3–5 in part B.

B. *Complexes with hydrophobic inhibitors*

1. Dilute the protease 1:1 in S buffer (see *Protocol 2*) with DMSO added to a final concentration of 5%.
2. Add a concentrated solution of inhibitor (~ 10 mM) in DMSO to a final inhibitor concentration of 10–100 μM. At this stage, the protease can be stored for several weeks at 4°C without significant degradation.
3. Filter using 0.2 μm Millex GV to remove any precipitated protein or insoluble peptides.
4. Concentrate the protease–inhibitor complex at room temperature with an Amicon YM-10 ultrafiltration membrane and concentrate to a final protein concentration ~ 5 mg/ml (0.25 mM). Tightly bound inhibitors (K_i ~ nM) co-concentrate with the protease. Weakly bound inhibitors (K_i ~ mM) must be added before, during, and after ultrafiltration in order to maintain their concentration levels.
5. Filter as in step 3 before storage at room temperature. In the presence of equimolar concentrations of a number of inhibitors, HIV-1 protease was stable for several weeks at room temperature.

3. Biochemical assays for HIV-1 protease

Peptide substrates closely resembling in sequence the processing sites found within the HIV-1 polyproteins (see *Figure 1*) have been used to develop peptidolytic assays. A variety of peptidolytic assays have been reported for HIV-1 protease:

(a) Separation of products from unreacted substrates by high-performance liquid chromatography (HPLC) (4, 7).

(b) Continuous spectrophotometric analysis of a peptide substrate containing an appropriate chromophoric residue, such as *p*-nitrophenylalanine, at one of its scissile positions (8–11).

(c) Ion-exchange separation of the radiolabelled carboxylic product from its uncharged labelled substrate (12).

(d) Continuous fluorometric analysis of a peptide substrate analogue designed to be fluorogenic upon cleavage (13, 14).

The most widely used peptidolytic assays for HIV-1 protease are the HPLC and spectrophotometric assays described in detail in *Protocols 4* and *5*, respectively. In the HPLC method, which is a stopped time assay, various concentrations of the peptide substrate are incubated with the protease in small volumes (< 0.1 ml). After quenching with acid at selected times, the samples are subjected to reverse phase HPLC on octyldecylsilane columns. The resulting peptidolytic products are thereby separated from the remaining substrate with linear gradients of acetonitrile in 0.05% trifluoroacetic acid with spectrophotometric detection of the peptides at 220 nm. The advantages of the HPLC assay are:

- versatility (any soluble peptide substrate can be assayed by this method)
- sensitivity (reaction mixtures of < 100 μl containing micromolar concentrations of peptides can be analysed)

A disadvantage, however, lies in the inherent low precision of the measurement of enzymatic initial rate by stopped time assay. The appropriate quantitation of initial rates in an HPLC assay requires that the time courses for each rate measurement be established to ensure linearity so that initial rate conditions prevail. Additionally, the integration areas of the product peaks need to be quantified via calibration of the actual molar amounts of product formed. In practice, the time courses of peptidolysis for a variety of peptide substrates as measured by HPLC are found to be apparently linear at fractions of product formation of 20% or less when obtained over a period of 0 to 60 minutes.

Protocol 4. HPLC assay for the HIV-1 protease

Equipment and reagents

- HIV-1 protease oligopeptide substrates custom ordered from Bachem BioSciences or commercially available:
  Ac-SQNY*PVV-NH_2 (similar to Gag residues 129 to 135)
  Ac-RASQNY*PVV-NH_2 (similar to Gag residues 127 to 135)
  Ac-RKIL*FLDG-NH_2 (similar to Pol residues 724 to 731)
  The cleavage site within each substrate is shown by an asterisk
- Purified HIV-1 protease enzyme (see *Protocol 2*)
- HIV-1 protease inhibitors
- HPLC grade water and acetonitrile (CH_3CN)
- Reagent grade trifluoroacetic acid (TFA) (Pierce)
- Reagent grade 2-[*N*-morpholino]ethanesulfonic acid (MES), NaCl, dithiothreitol (DTT), ethylenediaminetetraacetic acid (EDTA), Triton X-100 (Sigma Chemical Corp.)
- DMSO gold label (Aldrich Chemical Co.)
- Hewlett Packard HPLC equipped with ternary solvent delivery system, autosampler, diode array spectrophotometer, and digital integrator
- 5 × MENDT buffer: 250 mM MES, 5 mM EDTA, 1.0 M NaCl, 5 mM DTT, 0.5% (v/v) Triton X-100, adjust to pH 6.0—store in 1.0 ml aliquots at −20°C
- Fresh aqueous solutions (10–50 mM) of each oligopeptide substrate from lyophil-

ized material: determine the concentration of the substrate solutions from their UV spectra assuming extinction coefficients of $\epsilon^{275\ nm} = 1420/M/cm$ and $\epsilon^{257\ nm} = 197/M/cm$ for tyrosyl and phenylalanyl containing peptides, respectively

- Inhibitor dissolved in 100% DMSO (when appropriate)

Method

1. Carry out the peptidolytic assays in 50 μl reaction volumes at 37 °C:
   - 5 × MENDT buffer 10 μl
   - substrate (final concentration, 0.1–10 mM) 10 μl
   - inhibitor or DMSO (where appropriate) 5 μl
   - water 15 μl

   Substrates are usually varied over a concentration range equivalent to 0.2–2.0 times the values of their Michaelis constants.
2. Pre-incubate for 10 min at 37 °C.
3. Initiate the reaction by adding 10 μl of pure HIV-1 protease (5–50 nM; see *Protocol 2*).
4. Quench the reactions at various times (< 60 min) by adding 50 μl of 1% TFA.
5. Use 50 μl of the 100 μl reaction samples to separate the products and substrates on an octyldecylsilane column (Beckman Ultrasphere ODS, 4.5 × 250 mM, 5 mM) with a mobile phase (flow rate = 1.5 ml/min) composed of 0.05% TFA and the following gradients of acetonitrile:
   gradient 1: 5–20% (7 min), 20% (5 min), for substrates 1 and 2 (see *Table 1*)
   gradient 2: 5–40% (20 min), 40% (5 min) for substrate 3 (see *Table 1*).
6. Calculate the fraction of reaction as the ratio of P/[P + S] from the digital integration of the pairs of substrate (S) and product (P) peaks at 220 nM shown in *Table 1*.

The kinetic data obtained from the HPLC protease assay are fitted to the appropriate rate equations by using the Fortran programs of Cleland (15). In the equations that follow, v is the initial velocity, V is the maximum velocity, K is the Michaelis constant, A is the concentration of the variable substrate, I is the inhibitor concentration, K_{is} and K_{ii} are slope and intercept inhibition constants, respectively. The nomenclature used in the following rate equations is that of Cleland (16). Values of K, V, and V/K (see *Table 1*) are obtained by fitting the initial velocity data at variable concentrations of the oligopeptide substrates to eqn 1.

$$v = \frac{VA}{K + A} \tag{1}$$

Patterns conforming to linear competitive inhibition, linear non-competitive inhibition, or linear uncompetitive inhibition are fitted to Eqs 2–4, respectively.

Table 1. HIV-1 protease substrates and their products in the HPLC assay

	Substrate	R.T.[a]	Product	R.T.
Substrate 1	Ac-SQNY*PVV-NH_2	9.8	Ac-SQNY	4.5
Substrate 2[b]	Ac-RASQNY*PVV-NH_2	9.9	Ac-RASQNY	5.5
Substrate 3	Ac-RKIL*FLDG-NH_2	17.8	FLDG-NH_2	10.5
			Ac-RKIL	12.1

[a] R.T. is retention time in minutes.
[b] An example of a HPLC chromatograph using substrate 2 in the enzyme assay is shown in *Figure 5*.

When the type of kinetic pattern is in question for an experiment, data are fitted to all of the appropriate equations and a comparison of the resulting *s* values (square root of the average residual least square) is used to determine the best fit. A representation of competitive inhibitors is shown in *Table 2*.

$$v = \frac{VA}{K[1 + I/K_{is}] + A} \tag{2}$$

$$v = \frac{VA}{K[1 + I/K_{is}] + A[1 + I/K_{ii}]} \tag{3}$$

$$v = \frac{VA}{K + A[1 + I/K_{ii}]} \tag{4}$$

For the spectrophotometric assay of HIV-1 protease (see *Protocol 5*), the catalysed cleavage of peptide substrates bearing the chromogenic *p*-nitrophenylalanine is analysed. Cleavage of a peptide substrate containing a *p*-nitrophenylalanyl group at either the P1 or P1′ position results in small shifts in the ultraviolet absorbance spectrum, such that a time course of absorbance can be generated by monitoring wavelengths between 300–330 nm. The advantage of the spectrophotometric assay is that one obtains a continuous monitoring of the initial rate, which upon extrapolation of the initial linear region of the time course to zero time allows an accurate measure of the enzymatic initial rate. The use of an enzymatic time course also helps in evaluating the kinetics of inhibitors which are slow and/or tight binding. Compared to the HPLC assay, the spectrophotometric assay has a few disadvantages:

(a) Limited to use of those peptide substrates that contain a *p*-nitrophenylalanine residue at one of the scissile positions.
(b) Larger reaction volumes.
(c) Small values of $\Delta\epsilon^{320\ \mathrm{nm}}$ (*c.* 1500/M/cm) against a high absorbance background (λ_{max} 280 nm; *c.* 7000/M/cm).

Table 2. HPLC assay values for peptide substrates and inhibitors of HIV-1 protease

Substrate	**K (mM)**	**V/E_t (sec−1)a**	**V/KE_t (mM^{-1} sec^{-1})**
Ac-SQNY*PVV-NH_2	5.5	29	5.7
Ac-RASQNY*PVV-NH_2	3.9	29	7.5
Ac-RKIL*FLDH-NH_2	2.1	7.7	3.6

Inhibitor

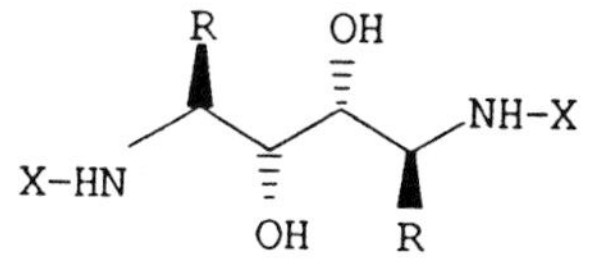

X	**R**	**K_i (nM)**
Cbz–Ala–Ala	*i*-Bu	580
Ala–Ala	*i*-Bu	800
Cbz–Ala	*i*-Bu	2400
Cbz–Val	*i*-Bu	11
Pepstatin A		3000

[a] V/E_t = the maximum velocity (V) per total amount of HIV protease (E_t); Cbz, benzyloxycarbony; *i*-Bu, isobutyl.

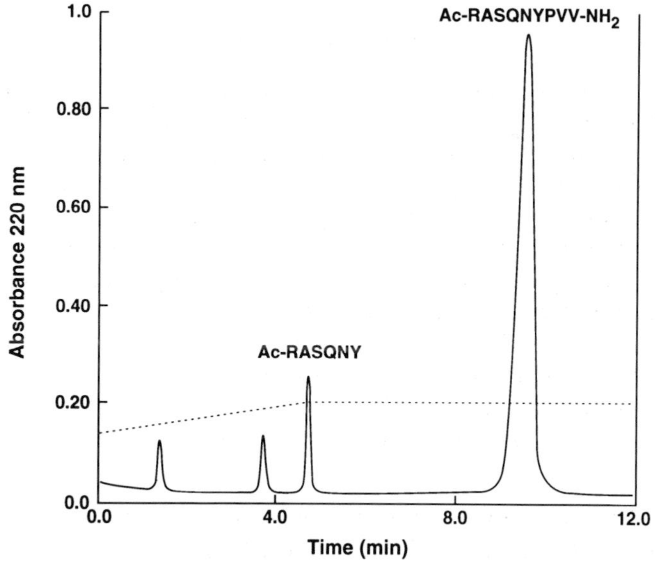

Figure 5. HPLC assay for HIV-1 protease. Analysis of Ac-RASQNY* PVV-NH_2 incubated with purified HIV protease by reverse phase HPLC (Beckman Ultrasphere C18, 4.5 mm × 25 cm); mobile phase: 5–20% acetonitrile (7 min), 20% acetonitrile (5 min) in 0.05% trifluoroacetic acid at 1.5 ml/min. Detection at 220 nm.

Protocol 5. Spectrophotometric assay of HIV-1 protease

Equipment and reagents

- HIV-1 protease oligopeptide substrate containing *p*-nitrophenylalanine custom ordered from Bachem BioSciences: Ac-RKIL*F(pNO$_2$)LDG-NH$_2$ (similar to Pol residues 724 to 731)
- Reagent grade sodium acetate
- Inhibitors and all other chemicals are as in *Protocol 4*
- Inhibitor dissolved in 100% DMSO (when appropriate)
- Perkin Elmer Lambda 4C spectrophotometer equipped with a constant temperature bath (37°C) which circulates water through jacketted 1 cm (0.5 ml) cuvettes
- 2 × AEND buffer: 160 mM sodium acetate, 2 mM EDTA, 2 mM DTT, 1.6 M NaCl, adjust to pH 4.7
- Fresh aqueous solutions of Ac-RKILF(pNO$_2$)LDG-NH$_2$ (1–10 mM) from lyophilized material

Method

1. Carry out assays in 400 μl reactions volumes:

• 2 × AEND buffer	200 μl
• substrate (0.1–2.0 mM, final concentration)	80 μl
• inhibitor (DMSO or water as controls)	40 μl
• water	70 μl

2. Pre-incubate the mixture for 5 min at 37°C.
3. Initiate the reaction by adding 10 μl of HIV-1 protease (5–50 nM; see *Protocol 2*).
4. Record the absorbance at 310 nm.

For kinetic analysis of the data:

(a) Determine initial rates from the slope of the observed linear decrease in absorbance at 310 nm (see *Figure 6*).

(b) Determine extinction coefficients from the total absorbance change at 310 nm resulting from the complete consumption of the peptide substrate in reaction mixtures initiated by the addition of a small concentration of substrate (in our laboratory, this value is $\Delta\epsilon^{310\ \mathrm{nm}} = 1200/\mathrm{M/cm}$).

(c) Determine the initial velocity (v) using eqn 5:

$$v = \frac{y}{\Delta\varepsilon(l(x)}(\Delta A/\mathrm{min}) \qquad (5)$$

where, in our assay, $\Delta\epsilon = 1200/\mathrm{M/cm}$; l = the cuvette width (1 cm); x = the volume of enzyme in the reaction (10 μl); y = the total reaction volume (400 μl); $\Delta A/\mathrm{min}$ = the change in absorbance per minute (as determined from the slope of the A versus time graph shown in *Figure 6*).

(d) Subsequently, fit the kinetic data as described above for the HPLC assay.

Representative results are shown in *Table 3*.

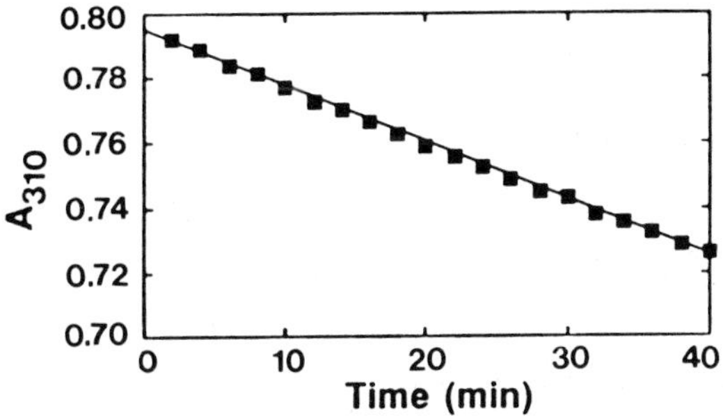

Figure 6. Spectrophotometric assay for HIV-1 protease. The absorbance (*A*) versus time profile illustrates the decrease (0–40 min) in absorbance at 310 nm upon incubation of Ac-RKIL*F(pNO$_2$)LDG-NH$_2$ with HIV-1 protease. The slope was calculated to be -1.9×10^{-3}/min.

Table 3. Spectrophotomeric assay values for Ac-RKIL*F(pNO$_2$)LDG-NH$_2$ and inhibitors of HIV-1 protease

Substrate	***K*** **(mM)**	***V/E***$_t$ **(sec−1)*a***	***V/KE***$_t$ **(mM^{-1} sec^{-1})**
Ac-RKIL*F(pNO$_2$)LDG-NH$_2$	0.28	7.7	28
Inhibitors	***K***$_i$ **(μM)**		
Pepstatin A	0.14		
Ac-SQNYΨ[CH$_2$NH]PVV-NH$_2$	7.1		

4. Conclusion

Recombinant expression of HIV-1 protease in bacteria has provided a safe and abundant source of authentic, active HIV-1 protease. The fact that the protease is fully active within bacteria has greatly facilitated its purification and the development of biochemical assays. This work then formed the foundation for more detailed enzymological and structural studies on the HIV-1 protease and for the identification of potent protease inhibitors with therapeutic potential.

References

1. Debouck, C. (1992). *AIDS Res. Hum. Retroviruses*, **8**, 153.
2. Stebbins, J. and Debouck, C. (1994). In *Methods in enzymology*, (ed. L. C. Kuo and J. A. Shafer), Vol. 241, pp. 3–16. Academic Press, San Diego.
3. Debouck, C., Gorniak, J. G., Strickler, J. E., Meek, T. D., Metcalf, B. W., and Rosenberg, M. (1987). *Proc. Natl Acad. Sci. USA*, **85**, 2449.
4. Hyland, L. J., Tomaszek, T. A. Jr., Roberts, G. D., Carr, S. A., Magaard, V. W., Bryan, H. L., *et al.* (1991). *Biochemistry*, **30**, 8441.

5. Strickler, J. E., Gorniak, J., Dayton, B., Meek, T., Moore, M., Magaard, V., *et al.* (1989). *Proteins*, **6**, 139.
6. Grant, S. K., Deckman, I. C., Minnich, M. D., Culp, J., Franklin, S., Dreyer, G. B., *et al.* (1991). *Biochemistry*, **30**, 8424.
7. Darke, P. L., Nutt, R. F., Brady, S. F., Gasky, V. M., Ciccarone, T. M., Leu, C. T., *et al.* (1988). *Biochem. Biophys. Res. Commun.*, **156**, 297.
8. Nashed, N. T., Louis, J. M., Sayer, E. M., Wondrak, P. T., Oroszlan, S., and Jerina, D. M. (1989). *Biochem. Biophys. Res. Commun.*, **163**, 1079.
9. Tomaszek, T. A. Jr., Magaard, V. W., Bryan, H. G., Moore, M. L., and Meek, T. D. (1990). *Biochem. Biophys. Res. Commun.*, **168**, 274.
10. Richards, A. D., Phylip, L. H., Farmerie, W. G., Scarborough, P. E., Alvarez, A., Dunn, B. M., *et al.* (1990). *J. Biol. Chem.*, **265**, 7733.
11. Phylip, L. H., Richards, A. D., Kay, J., Konvalinka, J., Strop, P., Blaha, I., *et al.* (1990). *Biochem. Biophys. Res. Commun.*, **171**, 493.
12. Hyland, L. J., Bayton, B. D., Moore, M. L., Shu, A. Y. L., Heys, J. R., and Meek, T. D. (1990). *Anal. Biochem.*, **188**, 408.
13. Matayoshi, E. D., Wang, G. T., Krafft, G. A., and Erickson, J. (1990). *Science*, **247**, 954.
14. Tyagi, S. G. and Carter, C. A. (1992). *Anal. Biochem.*, **200**, 143.
15. Cleland, W. W. (1979). In *Methods in enzymology* (ed. D. L. Purich), Vol. 63, pp. 103–37. Academic Press, San Diego.
16. Cleland, W. W. (1963). *Biochim. Biophys. Acta*, **67**, 104.

6

Expression of HIV proteins using recombinant baculoviruses

IAN JONES

1. Introduction

Why express viral proteins in foreign systems? After all, the expression of a viral product in isolation may overlook an essential factor in the host cell environment. A variety of reasons are commonly proffered:

(a) The level of expression of any given gene product may be improved to allow antibody generation or commercial development. This is particularly true of non-structural proteins involved in control rather than virion assembly.
(b) The expressed viral product can be manipulated by mutagenesis to determine the interplay of structure and function.
(c) The function of individual viral proteins can be distinguished from those of the virion background.
(d) Experimental work with the 'real' virus is difficult or dangerous.

These factors provide the driving force behind the sometimes difficult search for suitable expression systems for HIV genes. However, it is important from the outset to realize the limitations of what can be achieved with expression systems alone. The study of one particular gene product can only partially represent its complete role in the viral life cycle since it is the **interaction** of viral proteins together that results in virus replication and associated pathogenesis. Reservations aside, however, expression of a single viral product can provide much basic information on the way in which a protein functions. If expression is adopted by a project, a number of factors must be considered prior to the choice of a suitable expression system:

(a) For the expression of large amounts of a relatively short protein sequence (e.g. to raise a serum), use of an *E. coli* fusion system (e.g. the GST, MalB, or Hex-a-His systems) would be the natural first step. These combine ease of use with, potentially, large yields, and simple purification.

(b) For more complex proteins, eukaryotic expression systems offer the greatest opportunity for a functional product. Glycosylation, phosphorylation, myristoylation, and propeptide processing are some of the more common post-translational modifications that essentially **demand** the use of eukaryotic expression systems and a choice must then be made between systems that express the product transiently (e.g. during a viral infection cycle) or in stable cell lines.

In general, transient expression systems are faster and fairly sure of success, although the yield of any particular product is difficult to predict. Stable cell lines are more time-consuming to produce but often allow for an improvement of yield by gene amplification or manipulation of the growth conditions (e.g. fermentation). The expression of proteins in cell lines under the control of inducible promoters offers a compromise between these systems and allows for the generation of permanent cell lines producing cytotoxic products.

Recombinant virus systems are a relatively popular method of transient expression and a number of viruses have been developed for this purpose:

(a) Recombinant vaccinia viruses have found wide use for the direct immunization of animals and can express foreign viral components in a variety of mammalian cell types.
(b) Recombinant adenoviruses allow targeted expression in restricted cell types.
(c) Recombinant baculoviruses offer a method for the generation of sufficient product to enable functional and structural studies to be easily undertaken.

The method of recombinant baculovirus production is not difficult and the level of expression achieved is often very high. Frequently the product can be seen as a major band in crude cell extracts fractionated by polyacrylamide gel electrophoresis, helping to ensure a healthy level of confidence in the experimenter and supervisor alike!

In this laboratory, many HIV and HIV-related proteins have been successfully expressed using recombinant baculovirus technology (*Table 1*) and the resulting recombinants have often shown inherent properties beyond the simple expression of the desired HIV antigen. In this chapter the application of recombinant baculoviruses to the study of two of the structural proteins of HIV, the p55 Gag polyprotein and the *env* surface glycoprotein, are outlined to illustrate the particular use of this system for the expression of HIV proteins.

1.1 Recombinant baculoviruses

Baculoviruses have been exploited as expression vectors for almost ten years (1). During this time, the methods for preparing recombinant virus have

Table 1. HIV proteins expressed in recombinant baculoviruses

Protein	Comments
HIV-1 gp120	Secreted and functional in CD4 binding
HIV-2 gp120	Secreted and functional in CD4 binding
SIV gp120	Secreted and functional in CD4 binding
HIV-1 gp160	Cell bound, functional, and immunogenic
SIV gp160	Cell bound, functional, and immunogenic
HIV-1 p55 Gag	Forms immature virus-like particles
HIV-1 p24	High yield and evidence of multimer formation
Human CD4	Binds to gp120
Human CD26	Cell surface expression
Mouse furin	Cleaves gp160

improved dramatically but the basic concept remains unchanged: a non-essential, highly expressed baculovirus gene is replaced by a foreign gene. Most recombinant viruses, including the HIV recombinants described here, make use of the **polyhedrin** gene transcription unit of *Autographica californica* nuclear polyhedrosis virus (AcMNPV) for the expression of foreign proteins. The polyhedrin protein is wholly dispensable for virus growth in tissue culture and can be substituted by *in vivo* recombination for a variety of genes. Moreover, the genome of AcMNPV can accommodate large inserts (at least 10 kb) with no apparent effect on packaging efficiency or virus stability. HIV gene fragments are incorporated into the virus genome such that they are transcribed, late in the virus life cycle, by the polyhedrin promoter.

Methods for the direct incorporation of genes into the genome of AcMNPV by cloning have been described (2, 3). However, the great majority of recombinants are generated by *in vivo* recombination within insect cells between two DNA components:

- **transfer vector DNA**, which contains the HIV gene flanked by baculovirus control sequences
- **baculovirus genomic DNA**, which may be linearized to enhance the recovery of recombinant genomes (4)

Several companies (Clonetech, Invitrogen, PharMingen) now market the components necessary for generating recombinant baculoviruses and the purchase of linearized virus DNA for transfections is highly recommended for laboratories new to this technology. As only circular viral DNA is infectious, DNA cleaved at the polyhedrin locus can not initiate an infection unless it is rescued by re-circularization via recombination with the transfer vector (*Figure 1*). In practice, although the recombination rate remains the same when cells are transfected with linear or circular DNA, the wild-type background is reduced substantially when linearized DNA is used. When viral DNAs such as pAcBAK6, engineered to contain three cleavage sites for the

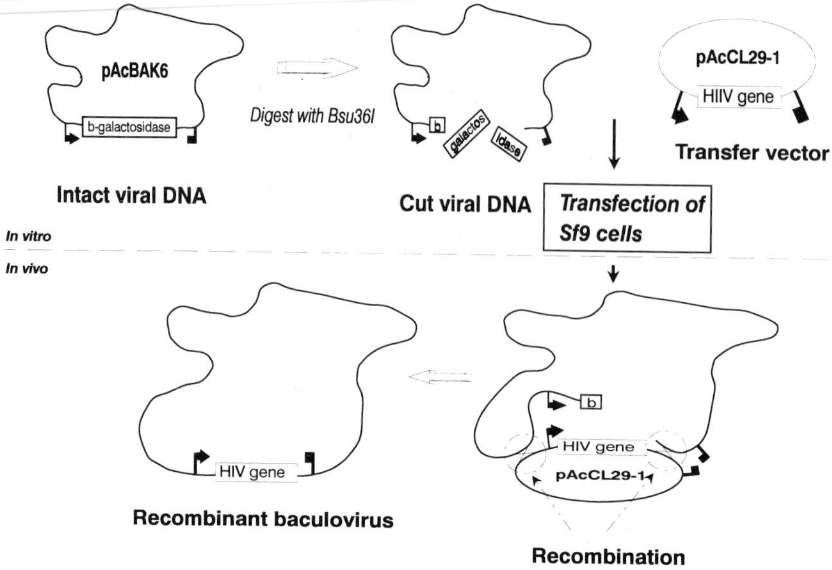

Figure 1. Cartoon representation of the process leading the production of a recombinant baculovirus.

enzyme *Bsu*36I (as sold in the Clonetech and PharMingen Kits) is used, almost no background virus is produced and recombinant virus selection is dependent solely on successful genetic modification of the transfer vector and good insect cell tissue culture.

2. Experimental considerations

2.1 Baculovirus vectors

For the HIV gene products of *gag* and *env*, maximum yields of the complete proteins is the desired goal. The transfer vector should be:

- optimized for the expression of full-length proteins
- easily manipulated with many cloning sites for the insertion of foreign DNA
- useful for mutation analysis without recourse to subcloning

Many transfer vectors have already been constructed and characterized and, in general, those in current use have already been optimized for a high level of transcription and translation of the insert. A typical transfer vector in many respects and the one used for the expression of both *gag* and *env* genes is pAcCL29–1 (*Figure 2*) (5) in which the key points are:

(a) The vector backbone (pUC118) allows good yields of plasmid DNA and the opportunity of producing positive sense (with respect to the inserted ORF) single-strand DNA for the construction of site-directed mutants.

(b) About 5 kb of baculovirus DNA derived from the genomic *Eco*RI (I)

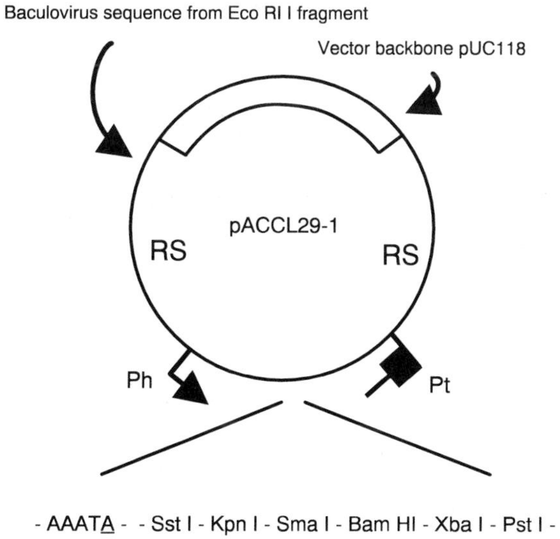

Figure 2. The baculovirus transfer vector pAcCL29–1.

fragment provide the control sequences for polyhedrin promoter driven transcription.

(c) The ill-defined recombination sequences (RS) either side of the transcription unit provide for recombination with the baculovirus genome.

(d) A polylinker cloning site suitable for the insertion of a variety of restriction fragments has been inserted at an optimized distance downstream of the polyhedrin promoter.

The **A** underlined in *Figure 2* is the first nucleotide of the methionine codon (ATG) for the polyhedrin open reading frame. ORFs inserted at the polylinker site therefore have the full-length polyhedrin 5′ non-translated region appended to them and this ensures a high level of expression. Minimizing the length of the 5′ non-translated region of the incoming HIV fragment is a worthwhile step that, for many original constructions, required elaborate subcloning procedures. However, the advent of PCR has meant that any ORF-containing fragment can be tailored to juxtapose the ATG to the restriction site used for the cloning. In general therefore, expression levels of many HIV products are near their possible maximum and there is, today, relatively little to be gained from vector manipulation. *Protocol 1* gives a typical working plan for the generation of a recombinant transfer vector encoding the full-length *env* product gp160.

Vectors also exist in which an ATG (sometimes the polyhedrin ATG) is provided 5′ to the cloning site. These can be used successfully for the expression of fusion proteins using recombinant baculoviruses. Recently, a set of

Protocol 1. Production of transfer vector for expression of HIV protein

Equipment and reagents

- Template for PCR, e.g. linearized HIV genome or infected cell lysate
- Plasmid preparation of transfer vector (pAcCL29–1)
- Kit for agarose gel electrophoresis
- Thermal cycler
- Oligonucleotides for PCR incorporating restriction sites for cloning into transfer vector that are not present in target. The following oligos were used to amplify gp160 from a full-length clone of HIV-I_{LAI}:
Forward primer 5′-GCGC*GAGCTC***ATGAGAGTGAAGGAGAAATAT**-3′
Reverse primer 5′-GCGC*TCTAGA***TTATAGCAAAATCCTTTCCAA**-3′
The design of our oligonucleotides generally follows the plan shown here; a four base CG 'clamp' at the 5′ end (underlined), the restriction sites used for cloning (here *Sac*I in the forward oligo and *Xba*I in the reverse oligo—italics), and 21 bases complementary to the template (bold).

Method

1. Amplify insert by PCR. Titrate both template amount and annealing temperature for maximum product yield and quality. 10 ng of template and an annealing temperature of 50°C would be commonly used values. Use only 25 cycles and choose the higher end of the template concentrations that give maximum product if using *Taq* polymerase to minimize errors. Alternatively use a low error rate *Taq* mimic such as Pfu, Vent, Ultma, or BioExact.
2. Digest pAcCL29–1 with (in this case) *Sac*I and *Xba*I. It is a good idea to check the digestion with each enzyme separately as well as in a double digest, particularly if you choose to use a 'universal' restriction buffer.
3. Gel elute both insert and vector to allow the efficient cloning of the PCR product. This laboratory uses low gelling temperature agarose and Agarase as the standard method for the elution of DNA from agarose gels. Solubilized bands should be phenol/chloroform extracted and precipitated before further work.
4. Restriction of PCR product after gel elution and ligation to double cut vector is done by standard procedures. PCR screening and restriction analysis confirm the vector construction. Miniprep DNA is satisfactory for the transfection process but must be clean and free of RNA.

fusion vectors that contain affinity tags for the one-step purification of baculovirus expressed proteins have also been described (6). A range of the current vectors available for AcMNPV can be found in recent reviews of the baculovirus expression system (7, 8).

2.2 Cell culture

The key to obtaining good virus growth and expression when working with recombinant baculoviruses is good insect cell culture. Two *Spodoptera frugiperda (Sf)* cell lines are used widely for baculovirus propagation, *Sf*9s and *Sf*21s. Laboratories vary in their belief that one line is better than another, but all the HIV gene expression done in this laboratory has used *Sf*9 cells.

(a) *Sf*9s will grow in suspension culture in a simple stirred vessel (up to about 1 litre culture in a 3 litre flask) or as adherent cells and can be switched between the two without any necessary adaptation.

(b) The temperature of growth should not exceed 28°C and cells can be grown satisfactorily on the bench if a dedicated incubator can not be found although growth rate and performance will parallel the growth temperature.

TC100 media (Gibco, Imperial, SeraLab) is a widely used base mineral salts medium for insect cells. Media is supplemented with 5% (v/v) fetal calf serum to make a complete medium for insect cell growth. It is essential to maintain the culture in a growing state. Once *Sf*9 cells have reached saturation the culture quickly loses viability. Cells with a viability of less than 95% should not be used for infections as they are unlikely to give the best yields. Recently, serum-free media have become available from a number of companies which support a higher cell density than TC100 plus FCS. Cell densities that approach 2×10^7 cells/ml at saturation in simple stirred or shaken flasks are achievable. These media are expensive and cells often need to be adapted to them before optimum growth is achieved. Thus, cells grown and adapted to serum-free media can not be readily used in TC100 media and vice versa. TC100-based media supplemented with FCS are, therefore, likely to remain the most commonly used media for most applications, particularly for routine baculovirus growth and recombinant virus isolation.

Cell lines derived from *Mamestra brassicae* have also been shown capable of supporting AcMNPV growth with the added bonus of improved expression on a volume for volume basis with *Sf*9 cells (9). As with serum-free media however, experimentation with other cell lines is best considered as part of an optimization program **after** the initial isolation and characterization of the recombinant which, for many experimental purposes, may already produce sufficient material for the desired research aims.

The following tips are useful when growing insect cells for recombinant baculovirus work:

(a) Keep 100–200 ml in suspension culture as the routine source of cells. At each passage set aside one 75 cm^2 flask of cells as a back-up. Discard the last but one back-up flask at each split.

(b) Keep the culture density below $1–2 \times 10^6$ cells/ml by regular passage.

Cells will double every 24 h at 28°C in TC100 + 5% FCS. If cells are not needed routinely, make large dilutions (1 in 50 is OK). If cells are required regularly passage one in four every other day. DON'T take an overgrown cell culture (2–3 × 10^6 cells/ml) and dilute it two- or three-fold prior to an infection, this will only result in a waste of time and viral stocks.

(c) Maintain the incubation temperature at or below 28°C. Cells will die very quickly above this temperature.

(d) Do not be tempted to cut corners with tissue culture. Use of 'old' cells will not give reproducible results and is sure to give less than optimal virus growth and expression level.

2.3 Transfection, plaque assay, and virus isolation

The methods for transfection of *Sf*9 cells with a mixture of transfer vector and linearized baculovirus DNA is well reviewed in a number of manuals dealing with the baculovirus expression system (7, 8). If the viral DNA to be used is purchased it will come with a recommended protocol which should be adhered to. The transfer vector DNA must be clean for the transfection to be efficient and DNA made by one of the currently available miniprep kits is very suitable for this purpose. A typical protocol would be *Protocol 2*.

Protocol 2. Transfection of *Sf*9 cells

Equipment and reagents

- 500 ng of transfer vector (Wizard, Qiagen, or equivalent quality)
- 100 ng of linearized baculovirus DNA (as supplied)
- One 3.5 cm^2 dish of *Sf*9 cells seeded at about 50% confluence (1 × 10^6 cells)
- Lipofectin (as supplied)

Method

1. Mix the transfer vector DNA and linearized viral DNA in a total of 16 μl of sterile water.
2. Add 8 μl of Lipofectin and allow to stand at room temperature for 15 min.
3. Wash cell monolayer twice with 1 ml TC100 medium without FCS. Overlay monolayer with 1 ml of TC100 (without FCS).
4. Add the Lipofectin/DNA mix to the TC100 medium overlay dropwise and swirl over the cells during the addition.
5. Incubate at 28°C for at least 6 h (overnight is OK).
6. Add a further 1 ml of complete medium and continue incubation at 28°C for a further two to three days.
7. Harvest the transfection supernatant and store at 4°C whilst the plaque assay is done.

The typical yield of virus from such a transfection is about 1×10^4 p.f.u./ml at three days post-infection. Longer incubations give higher virus yield but the proportion of recombinants does not change. As stated earlier, use of the *Bsu*36I digested pAcBAK6 DNA as the source of viral DNA for transfections leads to an apparent recombination frequency of near 100%.

A plaque assay is recommended to screen for recombinants by one of the following ways:

(a) Plaques can be easily checked for expression by direct inoculation into a 24-well plate of *Sf*9 cells followed by a Western blot of the infected cell lysate at two days post-infection. An antibody-based screen is preferable in the early phase of virus isolation as it allows for the detection of a protein that is only poorly expressed.

(b) If an antibody is not available, a PCR screen using the original cloning primers and 1 μl of test supernatants as template can be used to confirm a virus recombinant.

It is important to note that a stained band on a protein gel represents a very highly expressed product and is not a suitable way to screen first round plaques. In the case of gp160 for example, the expression level is barely sufficient to identify a unique new band on a gel even after synchronous infection using a high titre stock. Once a recombinant is shown to be positive for the expression of the desired HIV product it should be grown in moderate volume (10–50 ml) to provide a high titre stock of virus for subsequent use. Store a small seed stock of virus at −70 °C. Store the bulk at 4 °C in the dark. The titre should not drop more than 1 log/yr as long as the temperature does not increase substantially.

3. Analysis of HIV expression

3.1 HIV-1 Env expression

Infection of *Sf*9 cells with a high titre stock of a recombinant baculovirus ensures that all cells are synchronously infected. A multiplicity of infection of at least five will achieve this (in practice 1 ml of virus stock ($> 10^7$ p.f.u./ml) per 35 mm dish of cells). The polyhedrin promoter is a late viral promoter and, for most expressed proteins, the time course of expression follows a well observed pattern:

- slight expression of product at one day post-infection (pi)
- abundant expression at two days pi which is often maintained for a third day pi
- cytolysis may begin by four days pi and the infection is best harvested before this time

Despite this common pattern of expression, a time course of expression should be done for all new recombinants. A case in point would be the

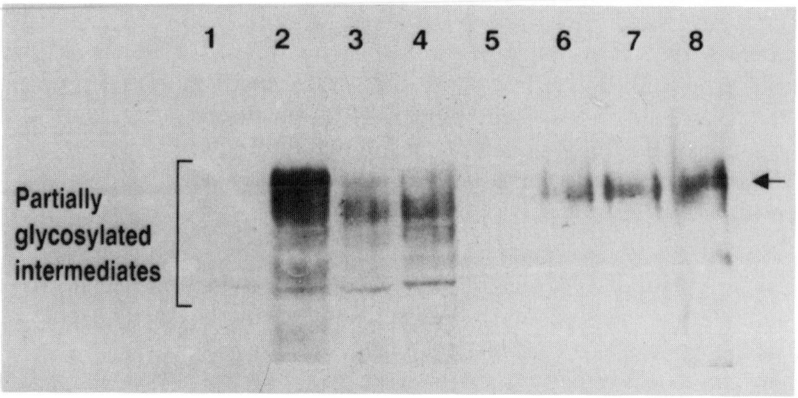

Figure 3. Western blot analysis of HIV-1 gp120 expression time course by recombinant baculoviruses. Tracks 1–4 are cell lysates and tracks 5–8 are supernatants. The days post-infection are: tracks 1 and 5—day one; tracks 2 and 6—day two; tracks 3 and 7—day three; tracks 4 and 8—day four.

analysis of HIV-1 gp120 expression in this system (*Figure 3*). The highly glycosylated nature of gp120 means there is an unusually long folding and secretion time. As a result of this, although the level of intracellular antigen peaks at two days pi as expected (*Figure 3*, track 2) the level of extracellular antigen is higher on day three (*Figure 3*, track 7) reflecting a lag in the secretion of this complex glycoprotein from cells. The expression of gp120 also illustrates that downstream bottlenecks of expression can occur subsequent to translation. The intracellular contents of infected cells reveal many partially glycosylated forms of gp120 that are never released from cells (*Figure 3*, tracks 1–4). In this case therefore, improved vector efficiency is unlikely to result in higher levels of mature glycoprotein expression. The secreted nature of gp120 has enabled simple capture ELISA methods to be developed for the analysis of immune reactivity and function using the supernatants of infected cells directly (see Chapter 12, Volume 1).

3.2 Taking advantage of insect cell properties

The glycosylation of proteins during their passage through the secretory pathway is the one secondary modification that is markedly different in insect cells when compared to mammalian cells. The early steps of the glycosylation pathway are shared, but after the trimming of the precursor to the trimannose core insect cells fail to assemble many complex carbohydrate structures (10). As a result, the total carbohydrate load added to insect cell expressed glycoproteins is marginally less than that added during secretion from a mammalian cell and baculovirus expressed glycoproteins are correspondingly smaller than

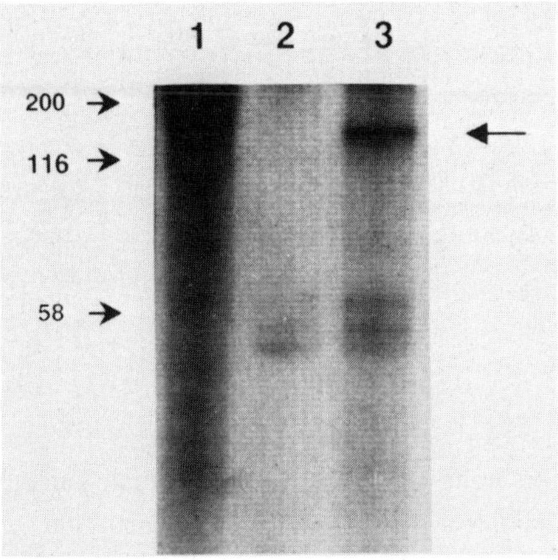

Figure 4. [^{3}H]mannose labelling of *Sf9* cells following infection by wild-type baculovirus (track 2); HIV-1 gp160 recombinant baculovirus (track 3); no virus mock infection (track 1). The sizes on the left of the panel are in kilodaltons.

their mammalian counterparts (e.g. gp120 has an apparent mobility of about 105 kDa following expression using recombinant baculoviruses). The carbohydrate that does remain on proteins secreted from insect cells is high in mannose content and this can be used to advantage for detection and purification. In the case of HIV-1 gp160, a pulse labelling of recombinant baculovirus infected cells at two days post-infection using [^{3}H]mannose reveals only one predominant glycoprotein at around 160 kDa (*Figure 4*, track 3) that is absent from the mock infected and wild-type virus controls (*Figure 4*, tracks 1 and 2 respectively). Trace amounts of the major baculovirus structural glycoprotein gp64 can be seen in both the wild-type and recombinant infections (*Figure 4*, tracks 2 and 3) but background labelling is otherwise very low. The high mannose content of gp160 has, therefore, made its detection straightforward and this property can be usefully used when looking for the expression of a poorly expressed product for which no antibody is available.

The high mannose content of insect cell expressed glycoproteins can also be used as a general method of purification. *Protocol 3* shows the procedure adopted in this laboratory to purify large quantities of SIV gp160 for trial vaccine usage. A monoclonal antibody column that was successfully used to purify SIV gp120 proved not to be efficient for the uncleaved precursor and Lentil lectin chromatography was substituted in its place. The final level of SIV gp160 was about 10% of the total glycoproteins fractionated from infected insect cells and proved sufficient to generate substantial levels of anti-SIV Env antibody.

Protocol 3. Purification of SIV gp160 from baculovirus-infected cells

Equipment and reagents

- 1 litre *Sf*9 cells grown in TC100 media +5% FCS and infected at 5×10^5 cells/ml with recombinant baculoviruses expressing SIV gp160 at an m.o.i. of 5
- Lentil lectin Sepharose column (Sigma)
- Wash buffer: 50 mM Tris pH 8.0, 50 mM NaCl
- Lysis buffer: 1% deoxycholate in wash buffer
- Elution buffer: 1 M methyl-α-D-glucopyranoside in wash buffer
- Protease inhibition cocktail (Boehringer Mannheim)

Method

1. Harvest infected cells at three days pi Wash once in wash buffer. Resuspend in 1/20 vol. of wash buffer. Place on ice.
2. Add equal volume of lysis buffer and protease inhibitor cocktail to recommended levels. Incubate on ice for 20 min.
3. Clarify lysate by centrifugation at 10 000 r.p.m. for 30 min at 4°C using a Beckman 8×50 rotor in a J2-21 centrifuge.
4. Load at approx. 0.5 ml/min to Lentil lectin column (20 ml) previously pre-eluted with 1 M elution buffer. Wash with five column volumes of wash buffer. Wash with wash buffer + 0.5% deoxycholate until OD_{280} is stable and near zero (several column volumes). Elute at 0.5 ml/min with elution buffer collecting 1 ml fractions for about 30 ml. Detect eluted protein by A_{280} of eluted fractions. Pool positive fractions. Concentrate by Centriprep (Amicon) and desalt to PBS (+/− detergent) or desired final buffer using prewashed PD-10 column (Pharmacia).
5. Western blot next to standard to confirm gp160 recovery, assess breakdown, and give approximate concentration. Total gel stain to estimate per cent purity.

This protocol enriches all the cellular glycoproteins as well as those that are virally encoded. Thus, the relative levels of recombinant expressed glycoprotein are directly related to the efficiency of the infection and stress the need for good viral stocks and good tissue culture to ensure maximum yields. It is also important to note the following to ensure high yields:

(a) Avoid dialysis tubing wherever possible as this leads to losses of product.

(b) Try to work through from lysis to final fraction in one go, this minimizes gp160 breakdown.

(c) It is better to store whole infected cells frozen and start on a clear day rather than begin the lysis and hold the procedure before the final fractions are collected.

3.3 Expression of Gag proteins

One feature of the recombinant baculovirus expression system that is frequently stressed is the high yield of expressed protein that can often be achieved. This is by no means guaranteed, but when the expression level is very high, inherent properties of the expressed product that require local high protein concentrations become apparent.

For many viral structural proteins, expression at high level allows protein self-assembly and the formation of virus-like particles and this is particularly true for a number of retroviral core antigens (listed in ref. 11). Following infection by a recombinant baculovirus expressing the HIV-1 *gag* gene the insect cell shows large numbers of HIV-like particles (VLPs) budding from the cell surface (*Figure 5*). The dimensions of the VLPs are essentially the same as those found in HIV-infected lymphocytes. Because no other HIV protein is present, these results clearly indicate that the Gag antigen encodes all the information necessary for particle assembly. A scanning micrograph of the infected cell surface shows the abundance of the budding forms which all but cover the otherwise smooth cell surface (*Figure 5*, inset). Note that the VLP stays as the 'doughnut' immature shape as there is no protease present within the particle to effect maturation. Co-expression of the HIV protease with the Gag precursor leads to Gag processing within the infected

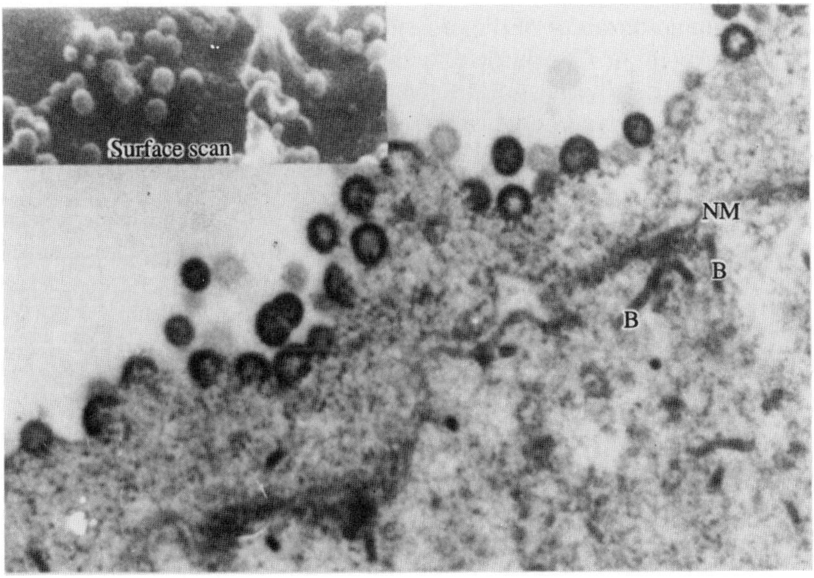

Figure 5. Electron micrographs of *Sf9* cells infected with a recombinant baculovirus expressing HIV-1 p55 Gag antigen. The nuclear membrane is indicated (NM) as are some baculoviruses (B).

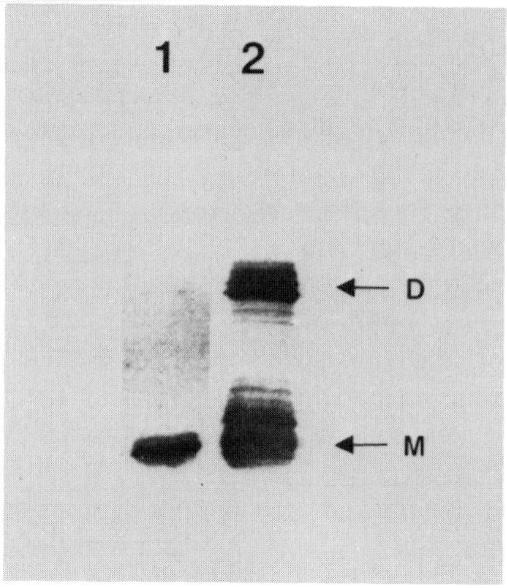

Figure 6. Western blot analysis of HIV-1 p24 Gag antigen following expression using recombinant baculoviruses and sample preparation as described in the text.

cell and an observed reduction in particle formation. The use of this and a similar vaccinia expression system for the analysis of Gag particle assembly has led to the delineation of a number of important domains within the p55 ORF. The amino-terminus of the protein directs membrane localization and allows secretion of the Gag particle. In the absence of membrane localization particles form intracellularly (12, 13). Mutations within the *gag* reading frame can also prevent surface budding, presumably by inducing a conformation incompatible with transport to the surface (14). The expression of deleted forms of *gag* have allowed some of the sequences involved in Gag particle formation to be mapped but the location of all the interacting surfaces is still the subject of ongoing research efforts. Cleaved Gag antigen does not assemble to form particles but evidence is accumulating that individual cleaved Gag antigens have assembly capabilities none the less. Following the baculovirus expression of the major core antigen of *gag*, p24, evidence can be seen of dimer formation (*Figure 6*). When the p24 antigen is treated with standard protein gel sample buffer (3% SDS, 5% β-mercaptoethanol) and heated prior to gel analysis and Western blot only the monomeric form of the protein (M) is revealed (*Figure 6*, track 1). But, when the sample is prepared in low SDS sample buffer (0.1% SDS) and not heated before gel analysis, a dimer of p24 (D) is clearly visible (*Figure 6*, track 2).

The level of expression of p24 from the baculovirus system is very high facilitating purification and an analysis of mutations within p24 for those that fail to dimerize should allow the functional domains involved in protein–protein interaction to be mapped.

4. Conclusions

Recombinant baculoviruses have proven useful for the production of a number of HIV proteins. Baculovirus expressed gp160 derived from HIV-1 forms the basis for the most advanced HIV-1 vaccine trial currently in progress (15) and baculovirus SIV gp160 was part of the successful protection experiments reported by Hu *et al.* (16). It remains to be seen if these studies, and those on Gag assembly, will lead to therapeutically beneficial products but the usefulness of the baculovirus expression system for the study of HIV protein structure and function does not seem in doubt.

Acknowledgements

I thank Dr Milan Nermut and Dr David Hockley for the electron micrographs shown in *Figure 5* and the MRC AIDS directed programme for support.

References

1. Smith, G. E., Summers, M. D., and Frazer, M. J. (1983). *Mol. Cell. Biol.*, **3**, 2156.
2. Patel, G., Nasmyth, K., and Jones, N. (1992). *Nucleic Acids Res.*, **20**, 97.
3. Luckow, V. A., Lee, S. C., Barry, G. F., and Olins, P. O. (1993). *J. Virol.*, **67**, 4566.
4. Kitts, P., Ayres, M., and Possee, R. (1991). *Nucleic Acids Res.*, **18**, 5667.
5. Livingstone, C. L. and Jones, I. M. (1989). *Nucleic Acids Res.*, **17**, 2366.
6. Davies, A., Jowett, J. B. M., and Jones, I. M. (1993). *Biotechnology*, **11**, 933.
7. King, L. A. and Possee, R. D. (ed.) (1992). *The baculovirus expression system: a laboratory guide*, 1st edn. Chapman and Hall, London.
8. O'Reilly, D. R., Miller, L. K., and Luckow, V. A. (1992). *Baculovirus expression vectors: a laboratory manual*, 1st edn. W. H. Freeman and Co., New York.
9. King, L. A., Mann, S. G., Lawrie, A. M., and Mulshaw, S. H. (1991). *Virus Res.*, **19**, 93.
10. Wells, D. E. and Compans, R. W. (1990). *Virology*, **176**, 575.
11. Jowett, J. B. M., Hockley, D. J., Nermut, M., and Jones, I. M. (1992). *J. Gen. Virol.*, **73**, 3079.
12. Overton, H. A., Fujii, Y., Price, I. R., and Jones, I. M. (1989). *Virology*, **170**, 107.
13. Gheysen, D., Jacobs, E., de Foresta, F., Thiriart, C., Francotte, M., Thines, D., *et al.* (1989). *Cell*, **59**, 103.
14. Hong, S. S. and Boulanger, P. (1993). *J. Virol.*, **67**, 2787.
15. Watson, T. (1992). *Nature*, **360**, 94.
16. Hu, S. L., Abrams, K., Barber, G. N., Moran, P., Zarling, J. M., Langlois, A. J., *et al.* (1992). *Science*, **255**, 456.

7

Envelope expression and purification

O. VAN OPSTAL, L. FABRY, M. FRANCOTTE, C. THIRIART, and C. BRUCK

1. Introduction

The envelope glycoproteins of human immunodeficiency viruses (HIV types 1 and 2) play a central role in the transmission of acquired immunodeficiency syndrome (AIDS) (1). The viral envelope mediates attachment of the virus to the CD4 receptor present on target cells (2–4) and the spread of virus by cell-to-cell fusion (5, 6). In the infected cells, the viral envelope of HIV-1 is synthesized as a precursor glycoprotein (gp160) which is subsequently cleaved into an external surface protein gp120 and a transmembrane protein gp41, upon maturation in the Golgi (7–9). After cleavage, the two glycoproteins (held together by non-covalent interactions) are transported to the cell surface.

The HIV-1 exterior envelope protein is heavily glycosylated: roughly one-half of the apparent molecular weight of the gp120 protein on SDS–PAGE can be attributed to asparagine-linked carbohydrate (7, 9). The gp41 transmembrane protein is also glycosylated (8). Glycosylation of the molecule (10–12) is required for HIV-1 infectivity and interaction of envelope antigen with CD4. Similarly, ability of envelope protein to bind to type-common neutralizing antibodies (13) is dependent on assembly of discontinuous protein segments.

In spite of an extensive variability mainly concentrated in five hypervariable regions (see Chapter 11, Volume 1), envelope proteins from various HIV-1 isolates maintain the ability to bind to the human CD4 receptor (14).

The viral glycoproteins are organized into surface spikes (15) of oligomeric nature. The exact degree of oligomer formation for HIV-1 envelope glycoprotein is still somewhat controversial: trimer formation has been shown for other retroviral glycoproteins (16), stable gp160 dimers are formed during processing of HIV-2 envelope (17), and evidence for tetramer formation has been described for HIV-1 envelope glycoprotein (18). The rapid loss of most of the visible spikes after virus maturation can be correlated with a release of the external glycoprotein gp120 resulting from loose association between the two glycoproteins within the surface spike (19).

The formation of oligomeric structures may also affect biochemical activity and antigenic composition of the envelope surface glycoprotein spike. Our work has focused on the expression of the gp160 precursor molecule in mammalian cells and its purification in non-denaturing conditions.

In this chapter we describe techniques for the preparation of milligram amounts of oligomeric HIV-1 gp160, which has retained the ability to bind immobilized sCD4. We have introduced a series of mutations in the 3′ end of the gp120 gene, affecting the proteolytic site between gp120 and gp41, to avoid separation of envelope subunits during culture and purification steps.

2. Vaccinia recombinants directing the expression of gp160

Expression of HIV envelope in mammalian cells using vaccinia has been described previously by several groups (20–22), and will not be extensively discussed in this chapter (see Chapter 8).

Genetically modified versions of the HIV_{BH10} (23) and HIV_{W61D} envelope genes are used as source of envelope sequences. The molecular clone W61D is derived from an isolate originating from a Dutch patient. This isolate is 'MN-like', since it includes the nonapeptide sequence IHIGPGRAF at the tip of the V3 loop region.

(a) The target sites for specific cleavage of gp160 into gp120 and gp41 within the envelope genes are modified by site-directed mutagenesis (24) to generate peptide structures not recognized by cellular furins. Specifically the gp120 **K**RRVVQRE**K**R carboxy-terminal sequence is modified to **E**RRVVQRE**E**R.

(b) The complete coding sequences of the modified BH10 and W61D *env* genes are cloned into vaccinia plasmid transfer vector pGS20 or pSC11 downstream of the vaccinia 7.5 promoter (25, 26) or into plasmid pMJ601 downstream of a synthetic late vaccinia promoter (27) using classical cloning techniques (24) and polymerase chain reaction.

Recombinant plasmids are transferred into the vaccinia virus (Elstree or WR strain) thymidine kinase (*TK*) gene by homologous recombination, as described elsewhere (25, 26).

(a) For recombinants deriving from vector pGS20, 100 vaccinia plaques are recovered with a Pasteur pipette, individually seeded on to CV-1 cells in a 24-well plate in a total volume of 1 ml of culture medium per well, for identification of recombinants expressing gp160.

(b) For recombinants deriving from vectors pSC11 (26) or pMJ601 (27), blue plaque selection limits the screening to a few plaques, since the majority of recombinant blue plaques are found to express antigen.

Recombinant viruses are identified and harvested as follows:

(a) Cells in the individual wells are detached using a rubber policeman.

(b) The cells are lysed in 1 ml of culture medium by repeated freeze–thaw cycles and addition of 1% Triton X-100.

(c) 100 μl aliquots are removed for screening for *env* expression by capture EIA (see *Protocol 8*).

(d) Two recombinant plaques giving the highest ELISA signal are further amplified and virus seeds prepared in CV-1 cells (see *Protocol 1*). Recombinant vaccinia seeds routinely contain 2×10^7 to 2×10^8 plaque-forming units (p.f.u.)/ml.

Figure 1 shows the results of a comparative expression experiment using vaccinia recombinants vMJ601-*env*W61D and vSC11-*env*W61D (respectively resulting from recombination with plasmids pMJ601-*env*W61D and pSC11-*env*W61D) in different cell lines. Constructs in vector pMJ601 generally yield a three- to sevenfold higher expression level of gp160 than pGS20 or pSC11, depending on the cell line used. CV-1 cells routinely yield a three- to fivefold higher expression level than BHK-21 cells. The establishment of a standard titration curve using purified gp160 (see 5.1) permits the calculation of absolute expression levels. At laboratory scale (T-flasks), expression levels ranging from 2–5 $\mu g/10^6$ cells are routinely observed for gp160 produced by vaccinia recombinants vMJ601-*env*W61D and vMJ601-*env*BH10 grown in BHK-21 cell suspension culture. Antigen purification experiments described below have been performed with the lower expressing vaccinia recombinant vGS20-*env*BH10 (0.5–1 μg of gp 160/10^6 BHK-21 cells at laboratory scale) and with vMJ601-*env*W61D.

Figure 2 shows the relative gp160 expression level using recombinant vaccinia vMJ601-*env*W61D at different time points after inoculation of a BHK-21

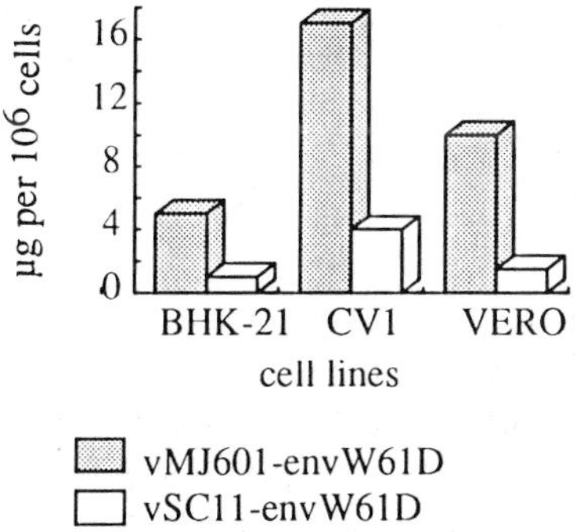

Figure 1. Expression of gp160 in BHK-21, CV1, and Vero cells infected with vaccinia recombinants vMJ601-*env*W61D and vSC11-*env*W61D at an m.o.i. of 0.15.

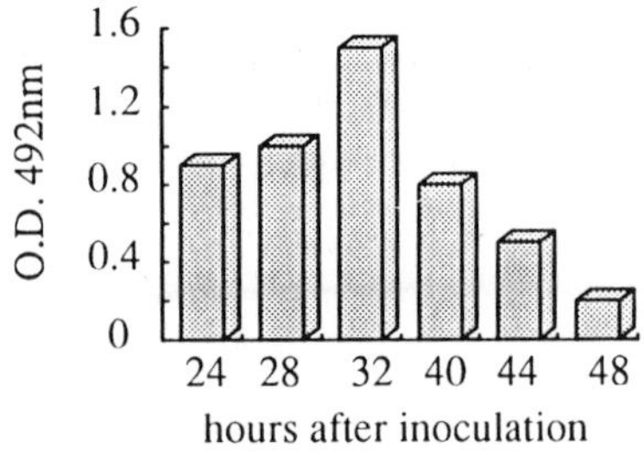

Figure 2. Expression of gp160 at different time points after infection of BHK-21 cells with vaccinia recombinant vMJ601-*env*W61D at an m.o.i. of 0.15.

culture. The optimal harvest time is found to be 32 hours after inoculation of BHK-21 cells at a multiplicity of infection (m.o.i.) of 0.15.

The production of uncleaved gp160 in these cell lines is verified by radio-immunoprecipitation, as described (28). All steps involving infectious vaccinia virus, including the lectin purification step (see *Protocols 4* and 5) are performed in a BL-2 laboratory (see Chapter 1, Volume 1).

3. Production of gp160 glycoprotein in cell culture

Our conditions are designed to produce milligram quantities of purified gp160. The example described here applies to the recombinant virus vMJ601-*env*W61D described above. The production procedure is developed essentially in two stages:

- generation of large virus stock
- virus incubation in cell culture for protein production

3.1 Preparation of virus stock

The production and amplification of virus recombinants is carried out in CV-1 cell culture. In addition, CV-1 cells are used to prepare large virus stock from the initial virus seed. It is advisable to initiate a master virus stock from which working stocks of second generation will be prepared—according to the seed lot principle—to provide enough virus material of the same origin and titre for consistent protein production. Virus stock is prepared by propagating CV-1 cells in roller bottles for incubation with the virus seed at low m.o.i., when the cells are in exponential growth phase. When cell lysis is observed, the cells are detached by a rubber policeman, concentrated by centrifugation, and the virus is released by a repeated freezing–thawing procedure as described in *Protocol 1*.

Protocol 1. Preparation of recombinant vaccinia virus stock

Equipment and reagents

- 900 cm^2 roller bottles (Costar Corporation, 3901)
- DMEM culture medium (Gibco Life Technologies, 074–02100)
- Scraper (Costar Corporation, 3010)

Method

1. Incubate CV-1 cells at 37 °C in 900 cm^2 roller bottle (Costar) with 300 ml of DMEM culture medium (Gibco) supplemented with 10% inactivated FBS.
2. Rinse exponentially growing cell culture with PBS (twice).
3. Infect cells at m.o.i. of 0.02–0.05 in 50 ml of DMEM culture medium without serum.
4. Incubate for 2 h at 37 °C.
5. Add 50 ml of fresh DMEM medium containing 10% FBS for a further 30 h incubation at 37 °C.
6. Collect cells by scraping them into a minimal volume of complete culture medium. Concentrate the cells by gentle centrifugation for 10 min at 160 *g*. Resuspend the pellet in 1–2 ml of medium per roller bottle.
7. Release the virus by two consecutive cycles of freezing–thawing.
8. Distribute the virus working stock in aliquots and store at −70 °C.

The virus stock is further titrated at regular intervals in CV-1 cell culture using a procedure adapted from Mackett *et al.* (25), see *Protocol 2*. P.f.u. titres observed range from 2×10^7 to 2×10^8/ml.

Protocol 2. Titration of virus stock

Equipment and reagents

- 6-well plates (Nunc 1–52795A).
- DMEM culture medium (see *Protocol 1*).
- Agarose type VII: low gelling temperature (Sigma, A-4018)
- Agarose type II: medium electroendosmosis (Sigma, A-6877)
- Neutral red stock solution prepared by dissolving 0.3% neutral red dye solution (Merck Ltd, 1369.0100) in water

Method

1. Seed CV-1 cells in 6-well plate (Nunc) with 4 ml of DMEM culture medium (Gibco), supplemented with 10% inactivated FBS, at 37 °C in

Protocol 2. *Continued*

a 5% CO_2 atmosphere so that near-confluent cell culture is obtained after 24 h.

2. Thaw the virus stock aliquot and prepare tenfold serial dilutions in DMEM medium—$10^{-6}/10^{-7}/10^{-8}$ dilutions are recommended to calculate the virus stock titre. Remove the culture medium from the 6-well culture. Add 1.5 ml of DMEM medium in triplicate culture—without serum—containing 0.1 ml of the virus dilutions. Incubate at 37 °C for 2 h.
3. Remove the medium and rinse once with PBS.
4. Pour 2 ml of type VII agarose, produced by mixing one part agarose (Sigma) and one part of 2 × complete DMEM medium (v/v), into each well and incubate at 37 °C.
5. Observe the plaque formation under the microscope after two or three days.
6. Pour 1 ml of agarose, prepared by one part type II agarose and one part 2 × PBS (v/v) containing 7.5% neutral red stock solution, into each well.
7. Allow the wells to dry for 3–4 h at room temperature.
8. Count plaques and calculate the titre of virus stock using plates displaying approximately 100 plaques.

3.2 Production of gp160 glycoprotein

In spite of the higher expression level observed in small scale experiments with CV-1 and Vero cell lines, BHK-21 cell culture were selected for gp160 production, since these cells can be cultured in suspension. This allows cells to be easily concentrated for extraction of cell-associated glycoprotein gp160. The procedure described in *Protocol 3* is adapted from the methods used in small scale studies. The most important change resides in dilution of the cell suspension as an alternative to cell concentration for virus inoculation, which results in similar expression levels. The basic procedure used for roller bottle culture or for 20–80 litre bioreactor development is essentially similar, although manipulation of bioreactor cultures requires specialized skills. BHK-21 cells are propagated for a few days at 37°C and exponentially growing cells are diluted with fresh medium for virus inoculation. After the incubation time (± 32 h), cells are collected, concentrated, and kept frozen prior to purification. In these conditions, gp160 glycoprotein production is usually around 2–3 mg/litre of culture suspension as assessed by ELISA (see *Protocol 8*).

Protocol 3. Production of gp160 glycoprotein in vaccinia virus infected cells

Equipment and reagents

- S-MEM medium (Gibco Life Technologies, 072–01400T)
- L-glutamine 200 mM stock solution (Gibco Life Technologies, 043–05030H)
- Tryptose phosphate broth (Difco Lab., 0060–01–6)
- Neomycin sulfate (Apothekerns Lab.)
- Penicillin/streptomycin solution (Gibco Life Technologies, 043–05070H)
- J2–21 Beckman centrifuge equipped with JA-10 angular rotor
- Pellicon cassette filter HVMP 000 C5 and Pellicon cassette stainless steel holder (Millipore)

Method

1. Initiate the culture of BHK-21 cells by seeding 0.25×10^6 cells/ml in S-MEM medium supplemented with inactivated FBS (5%), L-glutamine (4 mM), tryptose phosphate broth (1.5 g/litre) and antibiotics (neomycin sulfate 50 μg/ml, penicillin-streptomycin 1/500) in 900 cm^2 roller bottle at 37 °C.
2. Dilute the cell suspension three- to fourfold (in a 20 litre bioreactor) with fresh medium (without serum) when cells in roller bottle reach $3–4 \times 10^6$ cells/ml. This is usually after three days' growth.
3. Inoculate the appropriate amount of the virus stock at m.o.i. 0.1.
4. Incubate at 37 °C for 2 h.
5. Add FBS at a 5% final concentration and incubate for approximately 30 h.
6. Collect cells: centrifuge for 10 min at 160 *g* in angular rotor (Beckman) when collecting volumes under ± 5 litres.
7. Collect the cells by using cross-flow filtration Pellicon 0.45 μm membrane (Millipore) for large culture volumes. Adjust the flow rate and filtration rate to maintain transmembrane pressure near zero and concentrate the cell suspension six- to eightfold. Pellet the cells by low speed centrifugation for 10 min at 160 *g* in angular rotor (Beckman). Resuspend the pellet in a minimal volume of complete medium.
8. Keep the pelleted cells frozen (−20 °C) until extraction of gp160 glycoprotein by detergent treatment (see *Protocol 4*).

4. Purification of gp160

The full-length gp160, expressed as a cell-associated membrane protein, is purified according to a three-step scheme involving:

- detergent-mediated cell lysis and protein extraction
- affinity chromatography on immobilized Lentil lectin
- immunoaffinity chromatography

Table 1. Purification of recombinant gp160 BH10 from 60 litres vaccinia infected BHK-21 cells (6×10^{11} cells)

Preparation	Volume (ml)	Protein (mg)	gp160 (mg)	Purification (-fold)	Yield (%)
Lysate	4000	6770	46.8	1	100
Eluate Lentil lectin	700	203	37.4	26.7	80
Eluate immunoadsorbent	300	32.4	30	134	64
PEG concentrate	20	26	24	134	50

The procedure, essentially based on bio-affinity chromatography, permits purification of small amounts (mg) of non-clinical grade material (e.g. biochemical and immunological studies) to near homogeneity. Although less selective, classical chromatographic techniques (ion-exchange, size exclusion, hydrophobic interaction) are generally more advantageous for scale-up to industrial production scale and also more appropriate to meet regulatory requirements for preparation of clinical grade material. However, isolation of gp160 by these methods is not successful in our hands due to the micro-heterogeneity of the heavily glycosylated protein and its tendency to adsorb non-specifically to chromatographic supports.

The method described here has been developed at 1 litre culture scale and subsequently scaled-up linearly, permitting batches of 60 litre cultures to be processed in a single purification cycle (see *Protocols 4–7* and *Table 1*).

4.1 Extraction and solubilization of gp160 from vaccinia-infected BHK-21 cells

Membrane proteins are generally extracted from the harvested mammalian cells by detergent-mediated cell lysis and protein solubilization. Further purification steps also require the presence of a detergent to keep the protein in solution. The action of detergents and their use in biochemical research have been described (29). Key issues are:

(a) Detergent selection is often based on empirical research since various properties of the detergent such as critical micellar concentration (CMC), hydrophilic lipophilic balance (HLB), denaturing effects, purity, extraction efficacy, and interference with UV monitoring of proteins contribute to their efficiency when applied to different membrane proteins.

(b) Cost considerations also arise when used in larger scale production.

(c) The choice of a detergent is critical if the structural integrity of the protein should be preserved during purification.

In the case of HIV-1 gp160, it is important to maintain the biological activity of the recombinant protein (CD4 receptor binding) as well as the formation of oligomers which reflect the structure of viral gp160 arranged as spikes at the surface of the virion (30). Neutral detergents are often considered as being mild in their action and the non-ionic alkyl polyoxyethylene surfactant Thesit meets most of the above mentioned criteria for gp 160 solubilization (*Protocol 4*).

Protocol 4. Extraction of recombinant gp160 BH10 from vaccinia-infected BHK-21 cells

Equipment and reagents

- Beckman Model J2–21 centrifuge
- Lysis buffer: 30 mM Tris–HCl pH 8, 150 mM NaCl, 1% Thesit (Boehringer Mannheim 831620), 20 μg/ml aprotinin (Sigma A-6279)

Method

1. Centrifuge the thawed BHK-21 cell suspension from 60 litres (6×10^{10} cells) for 15 min at 4°C and 2000 *g* and discard the supernatant.
2. Resuspend the cell pellet in 3 litres of cold lysis buffer.
3. Lyse cells for 1 h on ice with occasional shaking by hand. Centrifuge the cell lysate for 15 min at 4°C and 11 300 *g*. Separate the supernatant.
4. Wash the pellet with 1 litre of lysis buffer. Centrifuge for 15 min at 4°C and 11 300 *g*.
5. Discard the pellet. Combine the supernatants for further purification (see *Protocol 5*).

The extraction results with Thesit are shown in *Table 1*.

Other neutral alkyl polyoxyethylene alcohols of the same type $CH_3(CH_2)_{x-1}$-O-$(CH_2CH_2O)_y$ (= C_xE_y), with x referring to the number of C atoms in the alkyl chain and y to the number of polyethylene glycol units, have been tested:

- $C_{10}E_6$, $C_{10}E_8$, $C_{10}E_{6/7}$, $C_{12}E_6$, $C_{12}E_8$, $C_{12}E_{10}$ (Genapol C-100)
- $C_{13}E_8$ (Genapol X-080)
- $C_{12}E_{23}$ (Brij 35)
- $C_{13}E_{10}$ (Emulphogen)

They show similar extraction efficacy as Thesit except for $C_{10}E_8$ and Brij 35 which tend to give 50% lower yields. Thesit ($C_{12}E_9$) has been chosen among the series of detergents for practical reasons (availability, purity, and cost).

4.2 Affinity chromatography

The centrifuged lysate is subjected to further chromatographic separation on immobilized Lentil lectin, having a sugar specificity for glucose and mannose (*Protocol 5*).

Protocol 5. Affinity chromatography of gp160 on Lentil lectin Sepharose 4B[a]

Equipment and reagents

- FPLC (Pharmacia) or equivalent equipment
- 90 ml Lentil lectin Sepharose 4B (Pharmacia 17–0444–01) in a XK26/40 column (Pharmacia 18–8768–01), equilibrated in 30 mM Tris–HCl pH 8, 150 mM NaCl 0.1% Thesit (Boehringer Mannheim 836630)
- Wash buffer: 30 mM Tris–HCl pH 8, 1 M NaCl, 0.1% Thesit.
- Elution buffer: equilibration buffer supplemented with 0.5 M methyl-α-D-mannopyranoside (α-MMP, Janssen Chimica 22.925.33)

Method

1. Load the clarified cell lysate overnight on the Lentil lectin column at a flow rate of 4 ml/min.
2. Wash the column with wash buffer at a flow rate of 5 ml/min until UV absorption reaches the baseline.
3. Desorb bound proteins at the same flow rate with elution buffer.
4. Pool gp160 positive fractions.

[a] Work in cold room (6–10°C).

The lectin column, which can be used for at least ten purification cycles without affecting the chromatographic performance, eliminates 95% of contaminating proteins (see *Table 1*). The antigen recovery based on the amount present in the lysate is 80% but occasionally a significant drop to 30–50% occurs. The occasional lower recovery is not due to poor gp160 binding on to the immobilized ligand but rather to inefficient elution. Alternative elution procedures such as increased displacer concentration (1 M α-MMP), detergent switching to other C_xE_y or 0.1 M borate buffer pH 6 do not improve the yields. The reduced elution efficiency is probably caused by batch-dependent minor glycosylation differences of the protein. For instance, an increased fucose content on the *N*-linked glycan antennae can dramatically increase the affinity of a glycoprotein for the lectin (31). Specific lectin blotting (see 5.1.3) has confirmed the presence of fucose on recombinant gp160. A similar heterogeneous elution behaviour due to glycan heterogeneity has been described recently for the purification of recombinant HIV gp120 on immobilized *Galanthus nivalis* agglutinin (32).

4.3 Immunoaffinity chromatography

The final purification step of recombinant gp160 BH10 is by immunoaffinity chromatography on immobilized monoclonal antibody 178.1. The selection of monoclonal antibody 178.1 as immunoaffinity ligand results from an

ELISA microtitre plate screening procedure, in which binding and desorption of gp160 from the different candidate monoclonal antibodies has been assessed. Monoclonal antibody 178.1 shows, in acid desorption buffer, the best reversible binding of gp160 among 20 monoclonal antibodies tested. The selected antibody recognizes the epitope KSIRI of the V3 loop of BH10 isolate and is therefore of very strict type specificity. As a consequence, the 178.1 antibody can not be used for envelope proteins from other HIV isolates. Nevertheless, it has been chosen as its use results in optimal elution characteristics and final product profile (see below). Possibly, V3 loop being one of the most exposed epitope on the surface of the spike structure, binding to the affinity column through this epitope has little effect on the oligomeric gp160 structure.

The generation of 178.1 IgG2a has been described (33) and the monoclonal antibody is isolated from ascites fluid by classical affinity chromatography on protein G Sepharose FF (Pharmacia 17–0618–01). The antibody is coupled on glutaraldehyde activated Trisacryl (Sepracor 261911), following manufacturer's instructions, at a density of 1.8 mg/ml resin. The Lentil lectin eluate is further purified according to *Protocol 6*.

Protocol 6. Immunoaffinity chromatography of gp160 BH10 on 178.1-Trisacryl[a]

Equipment and reagents

- FPLC (Pharmacia) or equivalent equipment
- 50 ml immunosorbent in a XK26/20 column (Pharmacia 18–8773–01), equilibrated in 30 mM Tris–HCl pH 8, 150 mM NaCl, 0.1% Thesit
- Stock solution 1 M Tris–HCl pH 8.8
- Wash buffer: 30 mM Tris–HCl pH 8, 1 M NaCl, 1% *n*-octyl-β-D-glucopyranoside (OGP) (Sigma O-8001).
- Elution buffer: 0.1 M citric acid buffer pH 3.3, 1% OGP

Method

1. Load the Lentil lectin eluate on to the immunoaffinity column by overnight recycling at 1 ml/min.
2. Wash the column at a flow rate of 3.3 ml/min with 20 column volumes wash buffer.
3. Invert the flow direction and elute gp160 at the same flow rate with elution buffer.
4. Immediately neutralize the elution fraction with the Tris stock solution and pool the antigen-positive fractions.

[a] Work in cold room (6–10°C).

The gp160 BH10 binds quantitatively on to the immunosorbent column which, in the absence of a specific regeneration step, can be used for over 30 purification cycles. The acid elution procedure allows a recovery of at least 80% of bound antigen (see *Table 1*). Please note:

(a) The detergent Thesit, having a low critical micellar concentration (CMC), cannot be removed easily for further adjuvantation of formulation of the protein for immunological studies. Therefore a detergent switch towards OGP (high CMC), which can be removed by simple dialysis, is included during the wash of the immunoaffinity column. The use of OGP at the lysis step is discouraged as gp160 extraction yields are very poor with this detergent.

(b) The monoclonal antibody 178.1 has also been coupled to carbonyl diimidazol activated Sepharose (34), to oxirane activated Eupergit C (Röhn Pharma) and to Affigel Hz (Bio-Rad 153–6060) but yields and/or purity of the recombinant gp160 are lower than for the Trisacryl-based column.

(c) As an alternative to monoclonal antibody 178.1, CHO-derived recombinant soluble CD4 (sCD4) receptor (35) has been immobilized on glutaraldehyde activated Trisacryl, with the hope of devising an affinity support of more general use. However, suitable elution conditions are difficult to find, presumably due to the high affinity of gp160 protein for the CD4 receptor.

4.4 Concentration of purified recombinant gp160

The concentration of the purified protein after immunoaffinity chromatography is rather low (50–100 μg/ml) and further concentration is achieved by precipitation with polyethylene glycol (PEG) (*Protocol 7*).

Protocol 7. Concentration of recombinant gp160 by PEG precipitation[a]

Equipment and reagents

- Stock solution of 40% PEG 6000 (Merck 12033)/ 2 M NaCl in MilliQ H_2O
- Beckman Model J2–21 centrifuge
- PBS supplemented with 1% OGP

Method

1. Add the PEG stock solution to the gp160 solution to a final concentration of 15% PEG 6000.
2. Allow the protein to precipitate overnight.
3. Centrifuge for 60 min at 4°C and 500 *g* and discard the supernatant.
4. Redissolve the pellet in an appropriate volume of PBS/OGP.

[a] Work in cold room (6–10°C).

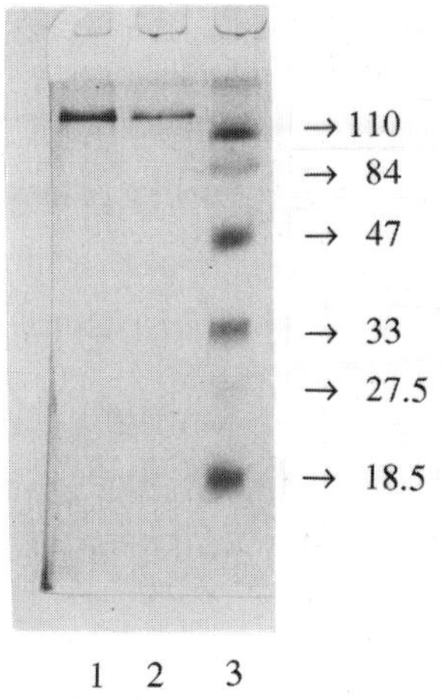

Figure 3. SDS–PAGE (10% acrylamide) and PAS staining of recombinant gp160 BH10: lane 1 = 600 ng gp160; lane 2 = 300 ng gp160; lane 3 = size markers (kDa).

Routinely, protein concentrations of about 1 mg gp160/ml are obtained after PEG precipitation. Protein recovery is 80% (see *Table 1*) but alternative concentration methods based on membrane technology showed much lower yields due to protein adsorption, even in the presence of detergent.

By the procedure outlined in *Protocols 4–7*, the gp160 BH10 being expressed at 0.7–1 mg/litre culture (1–1.5 × 10^9 cells) using vaccinia recombinant vGS20-*env*BH10 is purified to near homogeneity (see *Figure 3*) with a final yield of 0.4–0.5 mg/litre culture.

4.5 Purification of recombinant gp160 W61D

The protein derived from the W61D isolate is purified by the same procedure as outlined in section 4.3 except the antibody 178.1 has been replaced by monoclonal antibody 11H11G10 as the V3 loop sequence KSIRI of the BH10 isolate is mutated into KGIHI in the W61D isolate.

(a) The IgG1 has been prepared in BalbC mice primed with infectious vaccinia gp160 BH10, and boosted with purified gp160 BH10 in SAF-1 adjuvant.
(b) Monoclonal antibody 11H11G10 was selected on the basis of its reactivity with HIV envelope products of various isolates but its target epitope on gp160 is unknown.
(c) Recombinant gp160 of BH10 and W61D isolates behave in a similar way at the lysis and the Lentil lectin chromatography steps but recovery of W61D isolate protein from 11H11G10 coupled on to glutaraldehyde activated Trisacryl is lower. This illustrates the well-documented difficulty in finding optimal immunoaffinity ligands when gentle chromatographic conditions are required to preserve product characteristics.

5. Analytical methods

5.1 gp160 determination

Protein concentrations are determined by the Bradford method (36) using BSA as the standard.

The amount of gp160 is measured by a capture ELISA, outlined in *Protocol 8* (see also Chapter 12, Volume 1) for the BH10-derived envelope protein using purified gp160 as the reference. Absorbance readings are on a Multiskan MCC/340 (Titertek-Flow Laboratories) linked to a personal computer equipped with Titercalc (Hewlett Packard) software.

Protocol 8. HIV-1 gp160 BH10 ELISA

Equipment and reagents

- Capture antibody: resuspend 2 mg lyophilized sheep anti-gp41 monoclonal antibody (Biochrom Seromed D7323) in 2 ml MilliQ H_2O
- Indicator antibody: resuspend lyophilized anti-gp120 monoclonal antibody (Dupont 9284/7054) in MilliQ H_2O to a final concentration of 100 μg/ml; store stock aliquots at −20 °C but do not freeze–thaw more than twice
- Biotinylated anti-mouse antibody (Amersham RPN 1021) and biotin–streptavidin–peroxidase (Amersham RPN 1051) combined with H_2O_2/*ortho*-phenylenediamine (OPDA) according to manufacturer's instructions
- Use quantified (section 5.1) pure gp160 as the reference protein
- Saturation buffer: PBS supplemented with 0.1% Tween-20, 1% BSA, and 4% newborn calf serum (NCS); store at − 20 °C or not longer than four days at 4 °C
- Wash buffer: PBS supplemented with 0.1% Tween-20
- Dilution buffer: PBS containing 1% Triton X-100
- 96-well microtitre plates (96F-Nunc Immunoplate I 4–39454)

Method

1. Dilute capture antibody solution with PBS to 5 μg/ml and coat microtitre plate wells with 50 μl.
2. Incubate overnight at 4 °C.
3. Wash the wells six times with 300 μl of PBS, 0.1% Tween-20.
4. Saturate wells with 100 μl saturation buffer for 1 h at 37 °C and wash wells six times as in step 3.
5. Prepare a dilution series of reference gp160 and samples in PBS, 1% Triton X-100 and leave for 30 min at 4 °C.
6. Add 50 μl of the dilution series to the wells and after 2 h incubation at room temperature, wash wells as in step 3.
7. Dilute the indicator antibody to 1 μg/ml with saturation buffer and add 50 μl to the wells.
8. Incubate for 1 h at 37 °C and wash the wells as in step 3.
9. Develop ELISA with biotinylated anti-mouse antibody and the streptavidin–peroxidase system (see Chapter 12, Volume 1).

The W61D isolate gp160 is detected according to a similar ELISA but sheep anti-recombinant CHO gp120 BH10 (MRC, ADP 401) polyclonal antiserum is used as capture antibody. As indicator system a pool of eight murine monoclonal anti-gp120 and anti-gp41 antibodies is used (33).

5.1.1 SDS–PAGE and Western blotting

SDS–slab gel electrophoresis in the presence and in the absence of a reducing agent is carried out in 10% polyacrylamide gels according to the method of Laemmli (37). Migrated proteins are visualized by silver staining after periodic acid oxidation (PAS stain) (38). Protein bands are further identified by Western blotting on nitrocellulose according to Towbin (39) and probing is with antibodies directed against either gp160, BHK-21 host cell proteins, or vaccinia proteins.

5.1.2 Leakage of antibody

As antibodies may leak from the immunoaffinity columns and contaminate the final product, trace amounts of murine antibody in purified gp160 are determined by ELISA. Leaked antibodies are captured with goat anti-mouse antibodies (Boehringer Mannheim 1097–105) and further detection is with a biotinylated anti-mouse antibody and the streptavidin–peroxidase system. Contamination of purified gp160 by leaked antibody ranges from 0.004% to 0.02% of the total amount of protein present in the samples.

5.1.3 Glycan analysis

Limited glycan analysis of gp160 is by Western blotting using the glycan differentiation kit (Boehringer Mannheim 1210 238) in combination with enzymatic deglycosylation (Endo H, glycopeptidase F, sialidase). The recombinant protein contains *N*-linked glycan antennae of high mannose and/or hybrid-type and of complex-type. Fucose and α(2–6) linked sialic acid are identified.

5.2 Oligomeric forms

Cysteines in the primary sequence of the gp160 protein allow formation of (homo-)oligomers linked by disulfide bridges. Under reducing conditions a single protein band migrating with an apparent molecular mass of 160 kDa is observed by SDS–PAGE. This shifts to a higher apparent molecular mass in the absence of a reducing agent.

The formation of putative oligomeric structures is further analysed by HPLC size exclusion chromatography of gp160 on a calibrated TSK 4000 SW (TosoHaas 05790) column (7.5 mm × 300 mm) equilibrated in 0.2 M phosphate buffer pH 7 supplemented with 1% OGP. In the presence of a detergent, but without a reducing agent, the recombinant gp160 proteins elute at a retention time corresponding to a molecular weight of 640 kDa, compatible with the formation of a tetramer.

5.3 CD4 binding assay

The interaction of purified recombinant gp160 with the CD4 receptor is measured by a modified ELISA (*Protocol 9*). Soluble CHO-derived recombinant CD4 (35) is used as a trapping agent, and HIV-positive human serum containing anti-envelope antibodies, but no anti-CD4 antibodies, is used as indicator.

Protocol 9. gp160–CD4 binding assay

Equipment and reagents

- Recombinant sCD4 (5 mg/vial—SK&F 106528) stock solution at 1 μg/ml in PBS
- Saturation buffer: PBS supplemented with 1% BSA, 4% NCS, and 0.1% Tween-20
- HIV-positive human serum (1/5000 diluted in saturation buffer)
- Purified gp160 as the reference (concentration of 1 μg/ml and serially diluted in PBS, 1% Triton X-100)
- Biotinylated anti-human IgGs (Amersham RPN 1003), 1/2000 diluted in saturation buffer
- Streptavidin–peroxidase/OPDA (Amersham RPN 1052)

Method

1. Coat microtitre plate wells overnight at 4 °C with 50 μl of recombinant sCD4 stock solution.
2. Wash wells six times with 300 μl of PBS, 0.1% Tween-20.
3. Saturate with 100 μl saturation buffer at 37 °C for 1 h.
4. Wash wells six times as in step 2.
5. Add 50 μl of gp160 to the wells.
6. Incubate for 2 h at room temperature.
7. Wash wells as in step 2.
8. Incubate with 50 μl of the HIV-positive human serum for 1.5 h at 37 °C.
9. Wash wells as in step 2.
10. Add 50 μl of the biotinylated anti-human IgGs.
11. Wash wells as in step 2.
12. Develop the plate with the streptavidin–peroxidase system according to manufacturer's instructions (see Chapter 12, Volume 1).

(a) The specificity of the assay is verified by a competition assay using either OKT 4A (or Leu 3A) or OKT 4 antibodies (40).

(b) The obtained results indicate that the interaction between purified recombinant gp160 and the CD4 receptor is entirely specific.

Acknowledgements

We thank S. de Henau, M. Delchambre, A. Delers, A. Devroede, B. Dodemont, M. C. François, St Godart, R. Meykens, and Ch. Molitor for their excellent technical assistance, F. Van Wijnendaele, J. P. Prieels, M. Slaoui, and M. De Wilde for their contribution and support, R. C. Gallo for providing the HIV-BH10 molecular clone, and T. Tersmette for providing the HIV-W61D molecular clone. This work was supported by funding from MRC AIDS Directed Programme and from Belgian Walloon Region.

References

1. Barré-Sinoussi, F., Chermann, J. C., Rey, F., Nugeyre, M. T., Chamaret, S., Gruest, J., *et al.* (1983). *Science*, **220**, 868.
2. Dalgleish, A., Beverley, P. C. L., Clapham, P. R., Crawford, D. H., Greaves, M. F., and Weiss, R. (1984). *Nature*, **312**, 763.
3. Klatzmann, D., Champagne, E., Chamaret, S., Gruest, J., Guetard, D., Hercend, T., *et al.* (1984). *Nature*, **312**, 767.
4. McDougal, J. S., Kennedy, M. S., Sligh, J. M., Cort, S. P., Mawle, A., and Nicholson, J. (1986). *Science*, **231**, 382.
5. Sodroski, J., Goh, W. C., Rosen, C., Campbell, K., and Haseltine, W. A. (1986). *Nature*, **322**, 470.
6. Lifson, J. D., Reyes, G. R., McGrath, M. S., Stein, B. S., and Engleman, E. G. (1986). *Science*, **232**, 1123.
7. Allan, J. S., Coligan, J. E., Barin, F., McLane, M. F., Sodroski, J. G., Rosen, C. A., *et al.* (1985). *Science*, **228**, 1091.
8. di Marzo Veronese, F., De Vico, A. L., Copeland, T. D., Oroszlan, S., Gallo, R. C., and Sarngadharan, M. G. (1985). *Science*, **229**, 1402.
9. Robey, W. G., Safai, B., Oroszlan, S., Arthur, L. O., Gonda, M. A., Gallo, R. C., *et al.* (1985). *Science*, **228**, 593.
10. Lee, W. R., Syu, W. J., Du, B., Matsuda, M., Tan, S., Wolf, A., *et al.* (1992). *Proc. Natl Acad. Sci. USA*, **89**, 2213.
11. Gruters, R. A., Neefjes, J. J., Tersmette, M., de Goede, R. E. Y., Tulp, A., Huisman, H. G., *et al.* (1987). *Nature*, **330**, 74.
12. Walker, B. D., Kowalski, M., Goh, W. C., Kozarsky, K., Krieger, M., Rosen, C., *et al.* (1987). *Proc. Natl Acad. Sci. USA*, **84**, 8120.
13. Thali, M., Furman, C., Ho, D. D., Robinson, J., Tilley, S., Pinter, A., *et al.* (1992). *Proc. Natl Acad. Sci. USA*, **66**, 5635.
14. Ivey-Hoyle, M., Culp, J. S., Chaikin, M. A., Hellmig, B. D., Matthews, T. J., Sweet, R. W., *et al.* (1991). *Proc. Natl Acad. Sci. USA*, **88**, 512.
15. Gelderblom, H. R., Hausmann, E. H. S., Ozel, M., Pauli, G., and Koch, M. A. (1987). *Virology*, **156**, 171.
16. Einfeld, D. and Hunter, E. (1988). *Proc. Natl Acad. Sci. USA*, **85**, 8688.
17. Rey, M. A., Krust, B., Laurent, A. G., Montagnier, L., and Hovanessian, A. G. (1989). *J. Virol.*, **63**, 647.
18. Schawaller, M., Smith, C. E., Skehel, J. J., and Wiley, D. C. (1989). *J. Virol.*, **172**, 367.

19. Gelderblom, H. R., Reupke, H., and Pauli, G. (1985). *Lancet*, **2**, 1016.
20. Hu, S.-L., Kosowski, S. G., and Dalrymple, J. (1986). *Nature*, **320**, 537.
21. Chakrabarti, S. M., Robert-Guroff, M., Wong-Staal, F., Gallo, R. C., and Moss, B. (1986). *Nature*, **320**, 535.
22. Kieny, M. P., Rautmann, G., Schmitt, D., Dott, K., Wain-Hobson, S., Alizon, M. *et al.* (1986). *Biotechnology*, **4**, 790.
23. Hahn, B. H., Shaw, G. M., Suresh, K. A., Popovic, M., Gallo, R. C., and Wong-Staal, F. (1984). *Nature*, **312**, 166.
24. Maniatis, T., Fritsch, E. F., and Sambrook, J. (ed.) (1982). *Molecular cloning, a laboratory manual.* Cold Spring Harbor Press, Cold Spring Harbor, NY.
25. Mackett, M., Smith, G. L., and Moss, B. (1984). *J. Virol.*, **49**, 857.
26. Chakrabarti, S., Brechling, K., and Moss, B. (1985). *Mol. Cell. Biol.*, **5**, 3403.
27. Davison, A. J. and Moss, B. (1990). *Nucleic Acids Res.*, **18**, 4285.
28. Kanki, P. J., McLane, M. F., King, N. W., Letvin, N. L., Hunt, R. D., Sehgal, P., *et al.* (1985). *Science*, **228**, 1199.
29. Kühlbrandt, W. (1988). *Q. Rev. Biophys.*, **21**, 429.
30. Hart, T. K., Klinkner, A. M., Ventre, J., and Bugelski, P. J. (1993). *J. Histochem. Cytochem.*, **41**, 265.
31. Beeley, J. G. (1985). In *Laboratory techniques in biochemistry and molecular biology. Vol. 16: Glycoprotein and proteoglycan techniques* (ed. R. H. Burdon and P. H. van Knippenberg), pp. 301–64. Elsevier, Amsterdam.
32. Gilljam, G. (1993). *AIDS Res. Hum. Retroviruses*, **9**, 431.
33. Thiriart, C., Francotte, M., Cohen, J., Collignon, C., Delers, A., Kummert, S., *et al.* (1989). *J. Immunol.*, **143**, 1832.
34. Bethell, G. S., Ayers, J. S., Hancock, W. S., and Hearn, M. T. W. (1979). *J. Biol. Chem.*, **254**, 2572.
35. Deen, K. C., McDougal, J. S., Inacker, R., Folena-Wasserman, G., Arthos, J., Rosenberg, J., *et al.* (1988). *Nature,* **331**, 82.
36. Bradford, M. (1976). *Anal. Biochem.*, **72**, 248.
37. Laemmli, U. K. (1970). *Nature*, **227**, 680.
38. Dubray, G. and Bezard, G. (1982). *Anal. Biochem.*, **119**, 325.
39. Towbin, H., Staehelin, T., and Gordon, J. (1979). *Proc. Natl Acad. Sci. USA*, **76**, 4350.
40. McDougal, J. S., Nicholson, J. K. A., Cross, G. D., Cort, S. P., Kennedy, M. S., and Mawle, A. C. (1986). *J. Immunol.*, **137**, 2937.

8

HIV envelope glycoprotein/CD4 interactions: studies using recombinant vaccinia virus vectors

E. A. BERGER, O. NUSSBAUM, and C. C. BRODER

1. Introduction

Since the discovery of human immunodeficiency virus (HIV) as the causative agent of acquired immune deficiency syndrome (AIDS), the mechanism of HIV entry into target cells has been a major focus of scientific investigation (see ref. 1 for review). It is difficult to overstate the importance of the fortuitous fact that HIV 'chose' as its receptor the cell surface marker CD4. This glycoprotein, present on a subset of T cells as well as on monocyte/macrophages, was already under intense scrutiny for its role in T cell function. The isolation and sequencing of the cDNA encoding human CD4 (2), the availability of a battery of anti-CD4 monoclonal antibodies (3), and the molecular cloning of numerous diverse HIV isolates (4) allowed rapid progress to be made in elucidating structural features of the binding interaction between CD4 and gp120, the external envelope glycoprotein (Env) subunit.

Insight has also been gained into the post-binding events leading to HIV entry into target cells. It is now widely accepted that the main route of CD4-dependent HIV entry is via direct pH-independent fusion between virus and cell membranes (1). An analogous (possibly identical) mechanism allows HIV-infected cells expressing surface Env to fuse with CD4-expressing cells, leading to the formation of multinucleated giant cells (syncytia). The molecular mechanisms of the fusion process are currently under intense investigation. An attractive working model is that CD4 binding to gp120 induces conformation changes in Env which activate its fusogenic property, possibly by exposing the hydrophobic 'fusion peptide' at the N-terminus of gp41, the transmembrane Env subunit. The 'fusion peptide' is then thought to interact with the membrane of the target cell to induce membrane fusion. Moreover, additional components of the CD4-positive cell critically influence the activity and specificity of Env-mediated fusion.

Many current notions about the entry process have been derived from studies using infectious HIV in cell culture. The availability of cloned DNA sequences for both Env and CD4 have enabled the development of HIV-free model systems to isolate the binding and fusion events, thereby circumventing complications associated with other aspects of the HIV infection cycle. In this chapter we discuss the use of recombinant vaccinia virus expression systems to study Env/CD4 interactions. To complement the experimental protocols presented herein, the reader is referred to recent detailed methodological descriptions of the generation of recombinant vaccinia viruses, the preparation and maintenance of high titre stocks, infection of cell cultures, and analysis of vaccinia encoded foreign gene expression (5).

2. Use of the vaccinia expression system to study Env/CD4 interactions

2.1 General features of the vaccinia expression system

Amongst the numerous systems for recombinant gene expression, vaccinia virus vectors have proven particularly versatile. Effective use of this expression system requires understanding of some basic features of vaccinia molecular biology and its application to the design of recombinant vectors (see refs 6 and 7 for comprehensive reviews):

(a) Poxviruses are unique amongst animal DNA viruses in that the replication cycle transpires entirely within the cytoplasm of the infected cell.

(b) The ~ 200 kb vaccinia genome encodes the molecular components necessary for transcription of its genes, including its own RNA polymerase, transcription factors, and capping enzyme.

(c) Upon infection, host cell protein synthesis is shut off; concomitantly vaccinia promoters direct the cytoplasmic synthesis of messenger RNAs which are translated by host cell ribosomes.

In the most efficient use of the vaccinia system, the gene of interest is expressed by infecting cells with a recombinant vaccinia vector:

(a) The DNA sequence encoding the protein of interest linked to a vaccinia promoter is introduced by homologous recombination into a non-essential site in the vaccinia genome.

(b) A vaccinia recombinant is selected, plaque purified, and amplified in tissue culture to produce a high titre stock preparation.

(c) The vector stock is used to infect the cell type of choice for individual experiments, and expression of the foreign protein ensues.

As an alternative method, cells are infected with wild-type vaccinia virus and transfected with a plasmid containing the desired coding region linked to a vaccinia promoter. This approach eliminates the need to isolate a vaccinia recombinant for each foreign gene, though the expression level may be somewhat compromised.

By either method, transcription of the foreign gene occurs in the cytoplasm and is presumably enhanced by amplification of the DNA template (through viral DNA replication in the case of a vaccinia recombinant, or through vaccinia-mediated non-specific replication of cytoplasmic plasmid DNA in the case of transfection). Another highly efficient and versatile adaptation is the vaccinia/bacteriophage T7 hybrid expression system (reviewed in ref. 8):

(a) Cells are infected with a vaccinia recombinant encoding the T7 RNA polymerase.
(b) The foreign gene linked to a T7 promoter is introduced either on a second vaccinia recombinant or on a plasmid.
(c) Foreign gene expression results from T7 polymerase-mediated transcription of the gene in the cytoplasm; enormous quantities of the foreign mRNA are produced.
(d) The translation efficiency can be enhanced by incorporating between the T7 promoter and the protein coding sequence the long 5′ untranslated region of encephalomyocarditis virus, which confers cap-independent translatability of T7 transcripts in mammalian cells.
(e) A variation of this approach employs the bacteriophage SP6 RNA polymerase and promoter (9).

There are several major advantages of the vaccinia system, some of which are of special value in the study of Env/CD4 interactions:

(a) The extremely broad host range of vaccinia virus permits foreign gene expression in a wide-range of cell types of diverse species (mammalian and non-mammalian) and lineages; vaccinia vectors can be used in continuous cell lines and in primary cells.
(b) The foreign protein undergoes the expected post-translational events characteristic of the mammalian host cell (e.g. glycosylation, fatty acylation, phosphorylation, proteolytic processing, intracellular trafficking and transport, etc.).
(c) The foreign gene is expressed uniformly throughout the cell population by using a high multiplicity of infection (m.o.i.) (⩾3).
(d) A variety of natural and synthetic vaccinia promoters (early, intermediate, late, and hybrid classes) enable high level production of the foreign protein; the expression level can be controlled by choice of promoter, time of expression, and the use of inhibitors of the vaccinia replication cycle.
(e) The transient nature of the vaccinia system circumvents problems associated with cytotoxicity of some foreign proteins (e.g. HIV Env) (10) which

often compromise the alternative strategy of selection of continuous transformed cell lines.

(f) The cytoplasmic localization of the vaccinia transcription apparatus obviates the need for accessory HIV proteins to facilitate transport of unspliced mRNAs out of the nucleus (e.g. the Rev requirement for Env expression) (11).

(g) The biohazard concerns associated with experimental use of infectious HIV are eliminated.

Having emphasized these advantages, we point out that use of the vaccinia expression system is not without limitations:

(a) Vaccinia infection shuts down host cell protein synthesis and causes cell death within a few days. Experiments must be conducted within a temporal window that allows for sufficient synthesis and analysis of the protein of interest before complications caused by host cell death. For preparative scale protein production, expression is generally performed for 24–48 hours. Analytical experiments are usually conducted over shorter periods dictated by the particular protocol. In many cases it is convenient to use cells after overnight incubation (10–16 hours post-infection).

(b) The vaccinia replicative cycle may be aborted in some cells. Effective expression therefore requires use of suitable vaccinia promoters active in the cell type of interest (e.g. early or intermediate promoters in primary human peripheral blood macrophages) (12).

(c) The level of synthesis of the vaccinia encoded foreign protein may exceed the post-translational processing capacity of the cell, particularly at late times after infection and when strong vaccinia promoters are used.

(d) Certain proteins with inherent high toxicity to eukaryotic cells may be incompatible despite the transient nature of the vaccinia system (e.g. CD4–*Pseudomonas* exotoxin chimeric proteins which inhibit eukaryotic protein synthesis) (13).

(e) Vaccinia virus is capable of forming lesions in healthy humans and can potentially cause serious pathogenesis in immunocompromised individuals. Accidental infection with a vaccinia recombinant encoding an HIV protein could lead to HIV-positive seroconversion. To minimize hazards associated with exposure, Standard Safety Level 2 (BL-2) practices and class I or II biological safety cabinets are recommended. Furthermore, vaccination using vaccine supplied by the Centers for Disease Control should be considered for individuals who work directly with infectious or contaminated material.

2.2 Application to the Env/CD4 problem

Vaccinia recombinants containing the *env* gene of HIV type-1 (HIV-1) were initially reported in 1986 (14, 15), and enabled the first demonstration that

Env is the sole HIV-1 encoded protein required for CD4-mediated fusion (16). Soon thereafter, vaccinia recombinants encoding Envs from HIV type-2 (HIV-2) (17, 18) and the related simian immunodeficiency virus (SIV) (19, 20) were reported. In each of these cases, Env was shown to be the only retroviral encoded protein required for CD4-dependent fusion. Vaccinia-mediated expression of full-length and truncated forms of human CD4 were also described during this time period (21–23). The use of vaccinia vectors by several research groups has yielded important insight into various aspects of the interaction of CD4 with HIV and SIV Envs, including the regions of CD4 (21, 22, 24–26) and Env (19, 27–36) involved in binding and fusion activities, mapping of CD4 epitopes (22, 25), biosynthesis and processing of Env (37–41), structural feature of the Env oligomer (37, 42–45), CD4-induced changes in Env structure (46, 47), kinetic features of the binding and fusion reactions (48–50), additional factors associated with fusion activity and specificity (17, 20, 23, 51–55), and pharmacological agents that act at the level of Env/CD4 interactions (13, 56–58). Beyond the fusion problem, HIV Env encoding vaccinia recombinants have been used to study Env/Gag interactions in the assembly of virus-like particles (59, 60) and to analyse Env-specific cytotoxic T cells (61); they have also received considerable attention for subunit vaccine production and for potential use as live vaccine vectors for AIDS prophylaxis and therapy (62).

The ability of vaccinia vectors to support foreign gene expression in a variety of cell types has facilitated study of cellular factors in addition to CD4 which are involved in the fusion process. An important conclusion from such analyses is that CD4 supports Env-mediated fusion when expressed on a wide array of human cell types, but not when expressed on most non-human cell types (23, 53). By contrast Env supports efficient cell fusion when expressed on either a human or non-human cell. This restriction is not due to defects in surface display or gp120 binding activity of vaccinia encoded CD4 expressed in a non-human cell. Transient cell hybrid studies have revealed the essential involvement of a human-specific accessory fusion component(s) rather than an inhibitor in the non-human cells (53). These conclusions extend concepts derived from studies of HIV infection of murine transfectant cell lines expressing human CD4 (63), and are in agreement with cell hybrid studies using non-vaccinia-based systems (64, 65). Clearly, the requirement that CD4 be expressed on a human cell type has important conceptual implications for fusion mechanisms, as well as practical consequences for application of the vaccinia vector system to the fusion problem.

A legitimate concern is whether cell fusion in the vaccinia system is mechanistically equivalent to fusion in HIV systems, particularly in view of the high protein expression levels. Undoubtedly the surface levels of each molecular component, the precise experimental conditions, and the nature of the fusion assay may determine which steps in the fusion process are rate-limiting, and therefore may influence the effects of certain experimental variables (e.g.

temperature, ionic conditions, inhibitory agents, mutations, etc.). However, results from systems using infectious HIV are likewise subject to variations arising from differences in technical protocols. The crucial point to emphasize is that several laboratories (17–20, 23, 25, 36, 46, 47, 53, 66–69) have found that the structure/function properties of cell fusion observed in the recombinant vaccinia system faithfully parallel those reported for HIV-based systems by a diversity of criteria including:

- the absolute requirement for expression of both Env and CD4
- the inhibitory effects of specific mutations in either Env or CD4
- the fusion blocking activities of monoclonal antibodies (against Env or CD4), soluble CD4, Env- or CD4-based peptides, and various pharmacological agents
- the ability of CD4 to induce structural changes in Env
- the specificity requirements for Env-mediated fusion with a suitable CD4-positive cell type

In this chapter we discuss two complementary modes in which the vaccinia expression system can be used to analyse HIV Env/CD4 interactions:

(a) Studies of the properties of cell-free preparations of the proteins, which can be either soluble (sometimes truncated) forms secreted into the medium, or cell-associated forms liberated by detergent solubilization.

(b) Analyses of these molecules on the surface of living cells.

3. Analysis of gp120/CD4 binding using the vaccinia expression system

Specific binding of gp120 to CD4 was a major focus of early biochemical investigations into the HIV Env/CD4 interaction, which used experimental systems based on infectious HIV (1). Our initial application of vaccinia vectors to study the binding process involved production of secreted soluble forms of CD4 (sCD4) containing specific truncations (21). These studies demonstrated that the gp120 binding site resides within the first two immunoglobulin-like domains of the four domain extracellular region of the CD4 molecule. Extensive site-directed mutagenesis (22) led us to conclude that the so-called CDR2 region within the first domain plays a particularly important role in gp120 binding. These findings with vaccinia vectors paralleled those from several laboratories using alternative recombinant gene expression systems to produce CD4 molecules containing specific mutations (1). A binding assay for gp160 and sCD4 is described in Chapter 7.

4. Analysis of Env/CD4-mediated cell fusion using the vaccinia expression system

The initial use of vaccinia to study membrane fusion mediated by the Env/CD4 interaction involved demonstration of syncytia formation upon infection of human CD4-positive T cell lines with a vaccinia recombinant encoding Env (16). A more flexible procedure is to prepare separate cell populations expressing Env and CD4, with one or both molecules encoded by vaccinia vectors. Upon mixing of the two populations, cell fusion begins within minutes and can be monitored by any of several complementary assays (see *Protocols 1–4*). This strategy permits better control over the initiation of the fusion reaction, and avoids complications associated with down-modulation of CD4 or Env when both molecules are expressed in the same cell.

Several assay techniques have been employed to measure Env/CD4-mediated cell fusion in the vaccinia system:

- syncytia formation assay (16–20, 23, 25, 36, 46, 47, 53, 54, 66–69)
- fluorescent dye transfer and/or redistribution (48, 53, 70)
- activation of a reporter gene selectively in fused cells (58, 71, 72)

With each of these methods it is important to verify the specificity of the fusion reaction by performing appropriate controls. Not surprisingly, we have found that the efficiency of cell fusion is greater with higher surface expression of Env and/or CD4. However, the sensitivity of the assay for detection of the inhibitory effects of various mutations or blocking agents may be compromised under high expression conditions (58). Thus, depending on the particular application, the levels of vaccinia encoded proteins should be adjusted appropriately by choice of promoters, use of inhibitors of vaccinia expression, and control of the duration of expression of the vaccinia encoded proteins. Alternatively, in some situations it is useful to use as one fusion partner a cell line producing endogenous CD4 or Env, typically at much lower levels than attained with vaccinia vectors (25, 58).

4.1 Syncytia formation assay

The earliest and simplest use of vaccinia vectors to study HIV-1 Env/CD4-mediated fusion involved analysis of syncytia formation using light microscopy (16, 23) (see Volume 1, Chapters 3 and 8 for assays with HIV-infected cells). Syncytia can be scored on a semi-quantitative +/− scale according to size, or the number of syncytia can be counted (e.g. per microscopic field or per well of a tissue culture plate).

Figure 1 illustrates the use of the vaccinia system to express CD4 and to assay its ability to mediate Env-dependent cell fusion (53). When CD4 encoding recombinant vaccinia vectors were used to infect simian BS-C-1 cells,

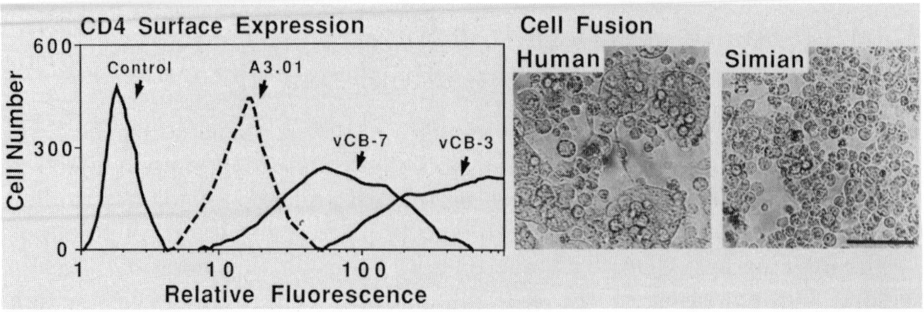

Figure 1. Syncytia formation assay for Env/CD4-mediated cell fusion using the vaccinia expression system (adapted from ref. 53). Graph—surface expression of CD4 was monitored by flow cytometry using fluorescein isothiocyanate labelled monoclonal antibody OKT4 on the A3.01 human T cell line and on HeLa cells infected with vaccinia recombinants vCB-7 or vCB-3. Photomicrographs—BS-C-1 cells expressing high levels of HIV-1 Env encoded by vSC60 were mixed with human (HeLa) or simian (BS-C-1) cells expressing high levels of CD4 encoded by vCB-3. Syncytia were monitored 8 h after cell mixing. The bar in the right panel represents 100 μm. See section 4.1 and *Protocol 1* for details.

strong surface CD4 expression was observed by flow cytometric analysis (*Figure 1*, graph). Compared to the endogenous level in the A3.01 human cell line, much higher CD4 expression was achieved with either vaccinia recombinant vCB-7 or vCB-3. The former vector employs the natural P7.5 early/late vaccinia promoter whereas the latter vector uses a recently developed synthetic strong early/strong late vaccinia promoter (Chakrabarti and Moss, personal communication). Quantitative analyses of BS-C-1 cells infected with vCB-3 indicated the presence of 10^6 surface CD4 molecules/cell (53).

The vCB-3 vector was used to express high levels of CD4 on either human (HeLa) or simian (BS-C-1) cells (53). A separate population of cells (BS-C-1) was infected with vSC60 encoding wild-type HIV-1 Env under control of the synthetic strong early/strong late vaccinia promoter (Chakrabarti and Moss, personal communication). The cells were incubated in suspension overnight to allow accumulation of the foreign proteins, then washed and suspended in fresh medium. In the syncytia formation assay (*Protocol 1*) pairwise mixtures (1:1) of CD4- and Env-expressing cells were added to individual wells of a 96-well flat-bottom tissue culture plate and the plate was incubated at 37°C. At 8 h after cell mixing large syncytia were observed, but only when the CD4 was expressed on the human cell type; the CD4-expressing simian cell type did not support syncytia formation (*Figure 1*, photomicrographs). Control experiments indicated that no syncytia were observed when CD4-expressing HeLa cells were mixed with cells infected with either wild-type vaccinia strain WR, or with vaccinia recombinant vCB-16 (Broder and Berger, unpublished) encoding a mutant uncleaved form of Env rendered non-fusogenic by deletion of the normal gp120/gp41 cleavage site (also under control of the synthetic strong early/strong late vaccinia promoter). These results and other studies

(23, 53) indicate that CD4 can mediate Env-dependent syncytia formation when expressed on a wide variety of human cell types, but not on most non-human cell types.

Protocol 1. Syncytia formation assay for HIV Env/CD4-mediated cell fusion using the vaccinia system

Equipment and reagents

- Recombinant vaccinia virus purified stocks of vCB-3 encoding human CD4, vSC60 encoding wild-type HIV-1 Env (Lai, HTLV-IIIB, BH8 isolate), and vCB-16 encoding mutant uncleaved Env (titres of purified vaccinia stocks are typically 1–10 × 10^{10} p.f.u./ml)
- Three 75 cm^2 tissue culture flasks (Costar, 3275, or equivalent) containing near confluent monolayers of BS-C-1 cells, and another containing a near confluent monolayer of HeLa cells (1 × 10^7 cells/flask)
- Medium (MEM-2.5): complete minimal essential medium (Quality Biological, Inc.) containing 2.5% fetal bovine serum
- 0.25% trypsin/0.02% ethylenediaminetetraacetic acid (EDTA) (Quality Biological), pre-equilibrated to ambient temperature
- Humidified 37°C CO_2 tissue culture incubator
- Haemocytometer and upright light microscope
- Beckman GS-6R table-top centrifuge (or equivalent)
- Polypropylene 50 ml conical centrifuge tubes (Falcon, 2070, Becton Dickinson Labware)
- 96-well flat-bottom tissue culture plates (Costar, 3596)
- Inverted tissue culture microscope (bright-field or phase-contrast) fitted with × 10 eyepiece and × 10 objective; optional photomicrographic system

Method

1. Place tubes containing each vaccinia virus into a cup sonicator containing a mixture of ice–water, and sonicate for 30 sec at maximum power.
2. Prepare inocula of each virus by dilution to 3.3 × 10^7 p.f.u./ml in MEM-2.5.
3. Aspirate media from each flask of cells. Overlay each flask with 3 ml of the indicated vaccinia virus inoculum (corresponding to m.o.i. of 10 p.f.u./cell):
   - 1 BS-C-1 flask: vSC60
   - 1 BS-C-1 flask: vCB-16
   - 1 BS-C-1 flask: vCB-3
   - 1 HeLa flask: vCB-3
4. Incubate flasks in tissue culture incubator for 1.5 h, with gentle rocking every ~ 15 min.
5. Remove inoculum by aspiration. Rinse with 3 ml trypsin/EDTA, then overlay with 2 ml trypsin/EDTA. Gently rock the flask and dislodge the monolayer by shaking, and then add 10 ml MEM-2.5.
6. Transfer cell suspensions to 50 ml polypropylene centrifuge tubes. Centrifuge at 400 *g* for 8 min at ambient temperature. Aspirate media and suspend cell pellets in 10 ml MEM-2.5.

Protocol 1. *Continued*

7. Determine cell number by counting with a haemocytometer. Centrifuge as in step 6.
8. Suspend pellets in MEM-2.5 to a density of 2.5×10^5 cells/ml and dispense each cell suspension into two 50 ml centrifuge tubes; cap the tubes loosely.
9. Transfer cell suspensions to 31 °C tissue culture incubator, placing tubes at a near horizontal angle to allow adequate gas exchange.
10. Incubate overnight (10–16 h) at 31 °C for production of the vaccinia encoded foreign proteins.
11. Centrifuge as in step 6. Wash each pellet with 10 ml MEM-2.5, then suspend in 10 ml MEM-2.5. Count cells in haemocytometer. Centrifuge cells, and suspend pellets in MEM-2.5 to a final density of 1×10^6 cells/ml.
12. Add 0.1 ml aliquots to duplicate wells of a 96-well flat-bottom tissue culture plate of each of two vaccinia-infected cell suspensions:
    - vCB-3/HeLa + vSC60/BS-C-1
    - vCB-3/BS-C-1 + vSC60/BS-C-1
    - vCB-3/HeLa + vCB-16/BS-C-1 negative control
13. Mix samples gently using micropipettor.
14. Centrifuge plate at 400 *g* for 2 min.
15. Transfer plate to 37 °C tissue culture incubator.
16. Monitor syncytia formation by light microscopic examination at periodic intervals. Syncytia are observable in the positive wells within 1 h of cell mixing, and increase in size and number over the next 8–16 h.

When performing these assays note:

(a) In our studies, vaccinia recombinants are typically prepared from the WR parental wild-type strain, with the foreign gene recombined at the thymidine kinase locus which is non-essential for vaccinia replication in cell culture. As an alternative to purified vaccinia virus stocks, crude virus preparations can be employed using established protocols (5).

(b) When different flask sizes are used, the amounts of reagents at each step are adjusted accordingly.

(c) Alternative media besides MEM-2.5 can be used, such as Dulbecco's MEM or RPMI 1640. The serum concentration should not exceed 2.5% during the virus inoculation step, but can be higher during initial growth of cell cultures or during subsequent incubation of the infected cells.

(d) Polypropylene tubes are preferred over polystyrene during the overnight incubation stage to minimize cell adherence to the plastic.

(e) Since low pH induces fusion of vaccinia-infected cells (73), it is critical to avoid acidification during the overnight incubation. For this reason the cells are suspended at a density below 3 × 10^5 cells/ml (step 8); Hepes buffer (pH 7.5) can be added to a concentration of 10 mM.

(f) Overnight incubation for production of vaccinia encoded proteins (steps 9 and 10) has recently been found to give superior cell fusion activity when conducted at 31 °C compared to 37 °C; alternatively the incubation can be performed at 37 °C for shorter periods (4–6 h) (58). The basis for the enhanced fusion activity under these conditions versus overnight incubation at 37 °C has yet to be determined; possibilities include reduced rates of decline of an essential cellular fusion component(s) upon vaccinia-mediated shut-off of host protein synthesis, as well as reduced cytopathic effect associated with vaccinia infection.

4.2 Reporter gene activation assay

The syncytia formation assay requires microscopic examination of individual samples and therefore can be subjective and tedious. We have recently devised a vaccinia-based cell fusion assay which provides a quantitative measure of reporter gene activation in fused cells (58, 71). The assay (*Protocols 2–4*) is an adaptation of the vaccinia/bacteriophage T7 hybrid expression system in which a foreign gene linked to a T7 promoter is transcribed in the cytoplasm by vaccinia encoded T7 RNA polymerase (8).

(a) Separate cell populations are infected with vaccinia vectors encoding HIV Env and CD4.

(b) One population is also infected with a vaccinia recombinant encoding T7 RNA polymerase, and the other is transfected with a plasmid containing a reporter gene linked to a T7 promoter.

(c) The two cell populations are combined; cell fusion results in mixing of the cytoplasmic contents and consequent activation of the reporter gene selectively in the cytoplasm of the fused cells.

(d) *Escherichia coli* (*E. coli*) *lacZ* is the reporter gene contained in plasmid pG1NT7β-gal (donated by R. A. Morgan, National Center for Human Genome Research, NIH, Bethesda, MD; personal communication). This plasmid contains the T7 promoter linked to the *lacZ* gene, with the untranslated region of encephalomyocarditis virus at the 5′ end of the protein coding region.

(e) β-galactosidase activity is measured either by colorimetric enzyme assay of detergent cell lysates (*Protocol 3*) or by *in situ* staining (*Protocol 4*)

The assay has proven to be highly specific and versatile, and much more sensitive and quantitative than the syncytia formation assay. The rapid throughput of the 96-well format makes this assay well suited for analysis of mechanistic features of the fusion reaction as well as for screening fusion-blocking antibodies and drugs.

Protocol 2. Reporter gene activation assay for HIV Env/CD4-mediated cell fusion using the vaccinia system

Equipment and reagents

- Purified stocks of wild-type vaccinia virus WR and vaccinia recombinants vSC60 (encoding wild-type Env), vCB-16 (encoding mutant uncleaved Env), vCB3 (encoding CD4), and vTF7–3 (encoding T7 RNA polymerase)
- Plasmid pG1NT7β-gal containing the *lacZ* gene linked to the T7 promoter, also with the 5′ untranslated region of encephalomyocarditis virus
- DOTAP transfection reagent, 1 mg/ml (Boehringer Mannheim)
- Transfection buffer: 20 mM Hepes, 150 mM NaCl pH 7.4
- Two polystyrene tubes, 12 × 75 mm (Falcon, 2058)
- Three 75 cm^2 tissue culture flasks (Costar, 3275) containing near confluent monolayers of HeLa cells
- As described for *Protocol 1*: MEM-2.5 medium, trypsin/EDTA, tissue culture incubator, haemocytometer, upright light microscope, centrifuge, polypropylene centrifuge tubes, 96-well flat-bottom tissue culture plates, inverted tissue culture microscope

Method

1. Prepare DNA for transfection: Add 5 μg plasmid pG1NT7β-gal DNA in 0.1 ml transfection buffer to one polystyrene tube. Add 30 μl DOTAP reagent in 0.1 ml transfection buffer to a second polystyrene tube. Add contents of one tube to the other and incubate for 10 min at ambient temperature. Add MEM-2.5 to a final volume of 5 ml.
2. Transfect first partner cell population: Aspirate medium from one 75 cm^2 flask of HeLa cells and overlay with DNA/DOTAP mixture. Transfer flask to 37°C tissue culture incubator. Incubate for 3–6 h. Aspirate transfection mix. Wash monolayer twice with 10 ml MEM-2.5; add 10 ml fresh MEM-2.5, and incubate 37°C for ~ 1 h. Detach the cells from the flask using trypsin/EDTA, wash them with MEM-2.5, count, and centrifuge as described in *Protocol 1*, steps 5–7. Suspend the pellet in MEM-2.5 to a density of 1×10^7 cells/ml.
3. Infect transfected cell suspensions with vaccinia viruses: Add 0.4 ml of the cell suspension to each of two 50 ml conical polypropylene centrifuge tubes. Add 4×10^7 p.f.u. vaccinia recombinant vCB3 to one tube; add 4×10^7 p.f.u. wild-type vaccinia virus strain WR to the second tube. This corresponds to m.o.i. = 10 for each virus. Cap the tubes loosely and transfer them to a tissue culture incubator for 1.5 h, shaking tubes gently every 30 min. (Note: step 4 can be initiated at this point and conducted concomitantly.) Dilute cell suspensions with MEM-2.5 to 16 ml final volume (corresponding to 2.5×10^5 cells/ml). Cap the tubes loosely. Transfer tubes to a 31°C tissue culture incubator for overnight incubation (10–16 h) at a near horizontal angle.

4. Infect second partner cell populations with vaccinia viruses: Prepare 4 ml of two mixed vaccinia virus inocula in MEM-2.5 (3.3×10^7 p.f.u./ml for each virus):
   - vTF7–3 + vSC60
   - vTF7–3 + WR

   Aspirate media from each of two 75 cm^2 flasks of HeLa cells. Overlay one flask with 3 ml of the vTF7–3 + vSC60 mixture; overlay the other flask with the vTF7–3 + WR mixture (m.o.i. = 10 for each virus). Incubate 1.5 h, remove inocula, trypsinize, wash, count cells in haemocytometer, suspend pellets in MEM-2.5 to a density of 2.5×10^5 cells/ml, and incubate at 31 °C overnight (10–16 h), as in *Protocol 1*, steps 4–10.
5. Wash and count each cell suspension, centrifuge, and suspend each pellet in MEM-2.5 to a final density of 1×10^6 cells/ml, all as in *Protocol 1*, step 11.
6. In replicate (typically duplicate or triplicate) individual wells of 96-well flat-bottom plate, prepare the following mixtures:
   (a) 0.15 ml aliquots of the vTF7–3-infected cells co-infected with the indicated virus:
   (b) 0.05 ml aliquots of the pG1NT7β-gal transfected cells infected with the indicated virus:

• vTF7–3 + vSC60	pG1NT7β-gal + vCB3	positive sample
• vTF7–3 + WR	pG1NT7β-gal + vCB3	negative control
• vTF7–3 + vSC60	pG1NT7β-gal + WR	negative control
• vTF7–3 + WR	pG1NT7β-gal + WR	negative control

7. Centrifuge plate at 400 *g* for 2 min, and transfer to 37 °C tissue culture incubator.
8. Analyse samples by either the colorimetric enzymatic assay of detergent cell lysates (*Protocol 3*) or the *in situ* assay (*Protocol 4*).

When performing these assays note:

(a) Alternative plasmid constructs besides pG1NT7β-gal containing the *lacZ* gene linked to the T7 promoter can be used (critical features are discussed in section 4.3.2).

(b) Alternative transfection reagents besides DOTAP can be employed along with the corresponding protocols (Lipofectin was used in the experiment shown in *Figure 2*).

(c) Alternatively to detaching the transfected cells by trypsinization and infecting them in suspension (steps 2 and 3), the transfected cells can be infected as monolayers, trypsinized, washed, counted, suspended, and incubated overnight as in *Protocol 1*, steps 4–10.

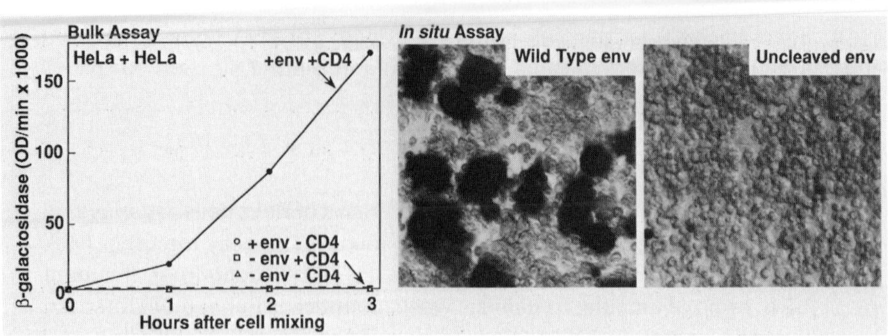

Figure 2. Reporter gene activation assay for Env/CD4-mediated cell fusion using the vaccinia expression system. Graph—time course of cell fusion analysed by the β-galactosidase colorimetric assay of detergent cell lysates. A population of HeLa cells was co-infected with vaccinia recombinants vTF7-3 encoding T7 RNA polymerase and vSC60 encoding HIV-1 Env. A second population of HeLa cells was transfected with plasmid pG1NT7β-gal containing the *lacZ* gene linked to the T7 promoter, then infected with vaccinia recombinant vCB-3 encoding CD4. As controls, cells were infected with wild-type vaccinia strain WR in place of vSC60, vCB-3, or both. T7 polymerase-containing and *lacZ*-containing cells were mixed at a ratio of 3 : 1. At the indicated times, detergent lysates were prepared and analysed by the colorimetric assay. See section 4.2 and *Protocols 2* and *3* for details. Photomicrographs—cell fusion analysed by the *in situ* β-galactosidase staining assay. HeLa cells transfected with plasmid pG1NT7β-gal and expressing high levels of CD4 encoded by vCB-3 were mixed with cells expressing T7 RNA polymerase encoded by vTF7-3 and either wild-type Env encoded by vSC60 (*left* panel) or a non-fusogenic uncleaved mutant Env encoded by vCB-16 (*right* panel). Cells were stained 3 h later with X-gal. See section 4.2 and *Protocols 2* and *4* for details.

(d) When performing co-infections (step 4), it is important to expose the cells to both viruses simultaneously; for this reason a pre-mix containing both viruses is first prepared.

(e) As an alternative negative control for Env (step 6), vCB-16 encoding mutant uncleaved Env can be used in place of WR; it is not essential to perform all three negative controls.

4.2.1 Analysis of β-galactosidase activity in detergent cell lysates

As a simple accurate method to quantitate fusion-dependent *lacZ* gene activation, we perform colorimetric assays of β-galactosidase activity in detergent cell lysates (see *Protocol 3*). We typically use as substrate chlorophenol red-β-D-galactopyranoside (CPRG), which is up to tenfold more sensitive than the commonly used 2-nitrophenyl-β-D-galactopyranoside (ONPG) (74). *Figure 2* (graph) shows results obtained with pairwise mixtures of HeLa cells expressing vaccinia encoded proteins. When Env was present on one cell partner and CD4 on the other, enzyme activity was readily detected at one hour and accumulated through three hours. Only extremely low background activities were observed when Env and/or CD4 were omitted, thereby demonstrating that enzyme accumulation resulted from Env/CD4-mediated cell fusion.

Protocol 3. Colorimetric enzymatic assay of β-galactosidase in cell lysates

Equipment and reagents

- NP-40 stock solution (Calbiochem, 492015)—20% (v/v) in H_2O
- 10 × buffered substrate stock solution: 80 mM CPRG (Boehringer Mannheim, 884308), in 0.6 M $Na_2HPO_4{\cdot}7H_2O$, 0.4 M $NaH_2PO_4{\cdot}H_2O$, 0.1 M KCl, 0.01 M $MgSO_4{\cdot}7H_2O$, 0.5 M β-mercaptoethanol (store this solution frozen at −20°C)
- Multipipettor: 50–200 μl (ICN Titertek Plus, or equivalent)
- β-galactosidase (purified from *E. coli*): standard stock solution, 1000 U/ml (600 U/mg protein, Boehringer Mannheim, 567779) (store frozen at −20°C in 50 μl aliquots to minimize repeated freeze–thawing)
- 96-well flat-bottom plate
- Stop-watch
- Microplate absorbance reader (Molecular Devices Vmax, or equivalent); equipped with a 590 nm filter

Method

1. Add 5 μl NP-40 stock solution to replicate wells (final detergent concentration 0.5%) at desired time points (typically 1–3 h after cell mixing). Mix by gentle pipetting five times. If additional time points are required in the cell fusion assay, remove the contents of the detergent-treated wells to a fresh 96-well plate, and return the original plate to the incubator.
2. When detergent has been added to all the desired wells, freeze the plate containing the samples on a bed of dry ice, then thaw (this ensures complete cell lysis).
3. Add 50 μl of each lysate to individual wells of a fresh 96-well assay plate.
4. Prepare 10^{-4} and 10^{-5} dilutions of the β-galactosidase standard stock solution, using as the diluent MEM-2.5 plus 0.5% NP-40. Add 50 μl to individual wells of the assay plate.
5. Add 50 μl of MEM-2.5 plus 0.5% NP-40 to one well of the assay plate, to serve as a blank.
6. Prepare a 2 × substrate solution by diluting one part of the 10 × buffered substrate stock solution with four parts water.
7. Equilibrate the assay plate and the 2 × substrate solution to ambient temperature.
8. Initiate the enzyme reaction by adding 50 μl of 2 × substrate solution to a row of sample wells using a multipipettor; start the stop-watch. Mix by gently pipetting three times. If additional rows of samples are to be assayed, note the time of substrate addition to each row.

Protocol 3. *Continued*

9. Centrifuge the plate for 30 sec at 800 *g* to eliminate bubbles due to detergent.
10. Determine the rate of substrate hydrolysis by measuring at various time points the optical density at 590 nm using the microplate absorbance reader; the instrument is blanked using the sample described in step 5.

When performing these assays note:

(a) Alternative non-ionic detergents besides NP-40 are suitable (step 1). Because there is negligible interference by components in the medium (including serum and phenol red) or by cellular debris, detergent can be added directly to the sample wells without time-consuming cell harvesting and washing steps.

(b) Frozen samples (step 2) can be stored at −20°C and thawed at a later time for enzyme assay.

(c) The amount of cell lysate assayed can be varied (step 3), using the appropriate diluent to maintain the same concentrations of substrate, buffer, and detergent, in a total volume of 0.1 ml/well.

(d) Some enhancement of sensitivity can be achieved by performing the assay at higher temperature (31 °C or 37 °C) (step 7).

(e) The rates of substrate hydrolysis (step 10) have been found to be linear with time and enzyme content over at least a three-log range of β-galactosidase concentration.

(f) A 590 nm filter commonly supplied with microplate absorbance readers is used (step 10). Increased sensitivity can be achieved by using a 570 nm filter which corresponds to the absorbance peak of the reaction product.

4.2.2 Analysis of β-galactosidase expression by *in situ* staining

An alternative method to detect fusion-dependent *lacZ* gene activation is to stain cultures *in situ* with the β-galactosidase substrate 5-bromo-4-chloro-3-indolyl-β-D-galactopyranoside (X-gal) (see *Protocol 4*). In the experiment shown in *Figure 2* (photomicrographs), one HeLa partner cell population expressed either wild-type HIV-1 Env or the non-fusogenic uncleaved mutant Env; the other partner HeLa cell population expressed CD4. Pairwise mixtures were prepared and after three hours the cells were fixed and stained *in situ*. When wild-type Env was expressed on one cell partner, large syncytia containing the blue β-galactosidase reaction product were observed; when the mutant uncleaved Env was expressed, only background staining of occasional cells was evident (note that these are single cells, not syncytia). These results and other controls indicate that the *in situ* staining protocol is suitable for detecting cell fusion mediated by the Env/CD4 interaction.

Protocol 4. *In situ* assay of β-galactosidase

Reagents

- 10 × fixative solution: 20% formaldehyde, 2% glutaraldehyde, diluted in phosphate-buffered saline (store at 4°C)
- X-gal (Boehringer Mannheim, 651737), 40 mg/ml stock solution in dimethyl formamide (store at 4°C)
- Staining buffer: 5 mM potassium ferricyanide, 5 mM potassium ferrocyanide, 2 mM magnesium chloride

Method

1. Prepare staining solution: Equilibrate staining buffer to 37°C. Prepare a 1:40 dilution of X-gal stock solution in staining buffer.
2. Add 20 μl 10 × fixative solution to individual wells at desired time points (typically 1–3 h after cell mixing).
3. Incubate the plate at 4°C for 5 min.
4. Remove 0.15 ml of medium with pipettor without disturbing settled cells.
5. Add 0.15 ml staining solution and return the plate to tissue culture incubator.
6. Examine periodically with an inverted tissue culture microscope for blue-stained cells. Usually staining can be detected within 1 h, but is more complete after overnight incubation. After staining, the plate can be transferred to 4°C and stored for at least one week.

4.3 Helpful hints for cell fusion assays

4.3.1 Recombinant vaccinia viruses

Certain points are critical in choosing suitable recombinant vaccinia vectors for a given experiment.

(a) Some cell types are non-permissive for certain classes of vaccinia promoters. For example primary human macrophages support expression from early and intermediate, but not from late vaccinia promoters (12). Thus the promoter must contain early and/or intermediate components; when hybrid early/late promoters are used in these cells, expression results only from the early component.

(b) While most Env and CD4 encoding vaccinia recombinants are suitable for the syncytia formation assay, certain ones can not be employed in the *lacZ* reporter gene activation assay, since they also contain the *lacZ* gene under control of a vaccinia promoter (a commonly used technique to facilitate the isolation of vaccinia recombinants, see refs 5–7).

4.3.2 Background and sensitivity

A variety of issues influence the signal/background ratio and the nature of the quantitative information obtained in the reporter gene activation assay.

1. Vaccinia recombinants containing reporter genes have proven extremely effective in the standard hybrid vaccinia/T7 expression system, in which both the T7 polymerase and the foreign gene are introduced into the same cell (8). However we have observed that with such vaccinia vectors there is a low level of β-galactosidase production in the absence of T7 polymerase produced during the initial vaccinia protein production stage, presumably due to 'read through' from upstream promoters in the vaccinia genome. This creates an undesired background level of fusion-independent *lacZ* expression. Therefore instead of using a recombinant virus, introduce the *lacZ* gene by transfection with a plasmid vector (pG1NT7β-gal). The critical features of this plasmid are the absence of flanking vaccinia sequences which might contain known or cryptic vaccinia promoters, and the presence of the 5′ untranslated region of encephalomyocarditis virus which presumably enhances the cell fusion signal by increasing the translation efficiency of T7 transcripts in the vaccinia system. Other plasmids with these features have been reported (8, 75) and presumably would be suitable in the fusion assay. It should be noted that vaccinia-based systems have special advantages for the use of plasmid transfection. Expression of the reporter gene does not require transport of the plasmid to the nucleus, since the relevant polymerases and transcription factors (in this case only T7 RNA polymerase) are present in the cytoplasm. Furthermore, the vaccinia DNA replication machinery non-specifically synthesizes plasmid DNA in the cytoplasm of infected cells; it is resonable to presume that replication of the *lacZ* containing plasmid results in signal amplification in the cell fusion assay.

2. The cell fusion reaction is generally terminated within three hours after mixing the fusion partners. A strong signal is observed over this time period with minimal background in controls wherein one or both partners are fusion incompetent (due to absence of functional Env or CD4). If the fusion reaction is allowed to proceed longer, the signal increases dramatically; however there is also an unacceptable increase in fusion-independent reporter gene activation, due primarily to superinfection of the *lacZ* gene-containing cells by the T7 RNA polymerase encoding vaccinia virus used to infect the partner population. This background can be greatly reduced (58) by using either of two recently described vaccinia recombinants: vP11gene 1 or vT7lac OI (76). In these vectors the T7 RNA polymerase gene is linked to the vaccinia P11 late promoter. Since expression from late promoters can be blocked by cytosine arabinoside (40 μg/ml) (6), addition of this inhibitor at the time of cell mixing prevents T7 polymerase expression upon superinfection, resulting in significantly reduced background β-galactosidase expression. The inhibitor provides

little benefit when vTF7–3 is used, since T7 RNA polymerase production upon superinfection still occurs from the early component of the 7.5K early/late promoter in this vector. An additional level of control is provided with vT7lacOI, in which polymerase expression driven by the P11 promoter is also regulated by the *E. coli* lac operator/lac repressor system. Addition of the inducer isopropyl-β-D-thiogalactopyranoside during the initial expression stage permits polymerase accumulation in one cell population; omission of inducer after cell mixing prevents polymerase production upon superinfection of the partner population. Using vT7lacOI with cytosine arabinoside present and inducer absent during the cell mixing stage, the signal/background ratio is greatly enhanced and the cell fusion reaction can be extended for much longer times (> 10 h) than with vTF7–3.

4.3.3 Quantitation

The colorimetric enzymatic assay of detergent cell lysates is superior to the *in situ* assay for quantitation, reproducibility, objectivity, and simplicity, particularly when large numbers of samples are involved. In considering the meaning of the quantitative information obtained, the β-galactosidase activity measured in cell lysates reflects a composite of enzyme accumulated in cells which have fused at various times after mixing of the two fusion partner cell populations. It is difficult to interpret such results in terms of the precise numbers of cell fusion events. By contrast, the *in situ* assay reveals the number of fused cells, though only those cells which have accumulated sufficient enzyme levels will be detected. By either method, uncertainty is introduced by the fact that a given cell can fuse with multiple partners, and it is not intuitively clear how such multiple fusion events will affect quantitation. This ambiguity can be minimized to some extent by performing the fusion assay with an excess (three- to ninefold) of one cell type over the other. Under such conditions, we have observed a close correlation between the total enzyme activity measured in the colorimetric lysate assay and the number of fused cells detected *in situ* (58). We have also assessed the sensitivity of the assay by varying the fraction of cells which are fusion competent. A fusion signal can be readily detected within two hours after cell mixing when as few as 0.1% of the *lacZ*-containing cells express vaccinia encoded CD4 (i.e. with only 100 CD4-positive per well); under these and less stringent conditions the syncytia formation assay does not reveal fusion events (58).

4.3.4 Variations

Several considerations must be noted when exploiting the versatility of the reporter gene activation assay.

(a) When choosing the Env-expressing cell partner:
  - it can be any of a wide variety of cell types (human or non-human)
  - it is preferable to use a cell type which does not express endogenous CD4

(b) Different classes of Envs can be employed, including those from:
- T cell line-tropic HIV isolates
- macrophage-tropic HIV isolates

(c) When choosing the CD4-expressing cell partner:
- it must be of human origin
- it can be either a continuous cell line grown as a monolayer or in suspension (HeLa cells, T lymphocyte cell lines, etc.), or a primary cell type (activated peripheral blood mononuclear cells, primary macrophages, etc.)
- it must be permissive for fusion mediated by the particular Env (58, 71)

CD4-expressing continuous cell lines support fusion by Envs from T cell line-tropic isolates but not by Envs from macrophage-tropic isolates; primary macrophages show the opposite specificity. Peripheral blood mononuclear cells support fusion by Envs of both classes (58, 71, 78).

(d) Generally there is flexibility in choosing which cell population (Env- or CD4-expressing) contains T7 RNA polymerase and which contains the reporter gene (58). However restrictions apply when cell types are used which are not fully permissive for vaccinia virus. For example, when performing fusion assays with primary macrophages which fail to mediate expression from late vaccinia promoters and also do not support vaccinia DNA replication (12), it is preferable to infect the macrophages with the vaccinia recombinant encoding T7 RNA polymerase (linked to an early/late vaccinia promoter) and to use as the *lacZ* gene-containing fusion partner a cell type which is permissive for vaccinia.

(e) Most experiments are performed in complete medium, though alternative buffered salts solutions are suitable. In such cases sufficient calcium ion levels (2 mM) must be included (54).

(f) As an alternative to the colorimetric lysate assay and the *in situ* assay for β-galactosidase, it is possible to stain cells with a fluorogenic substrate and analyse the population by flow cytometry (77). This approach has the advantage of the *in situ* assay in monitoring individual fused cells, coupled with the potential to obtain measurements of large cell numbers.

(g) Alternative reporters besides β-galactosidase presumably can be employed (firefly luciferase, β-glucuronidase, cell surface antigens, etc.).

Acknowledgements

The outstanding technical assistance of P. E. Kennedy is gratefully acknowledged. This work was supported in part by an award to E. A. B. from the NIH Intramural AIDS Targeted Antiviral Program. C. C. B. was a recipient of a National Research Council Research Associateship and an NIH Intramural Research Training Award.

References

1. Moore, J. P., Jameson, B. A., Weiss, R. A., and Sattentau, Q. J. (1993). In *Viral fusion mechanisms* (ed. J. Bentz), pp. 233–89. CRC Press, Boca Raton.
2. Maddon, P. J., Littman, D. R., Godfrey, M., Maddon, D. E., Chess, L., and Axel, R. (1985). *Cell*, **42**, 93.
3. McMichael, A. J. (ed.) (1987). *Leukocyte typing III: White cell differentiation antigens*. Oxford University Press, Oxford.
4. Myers, G., Berzofsky, J. A., Korber, B., Pavlakis, G. N., and Smith, R. F. (ed.) (1992). *Human retroviruses and AIDS 1992*. Los Alamos National Laboratory, Los Alamos.
5. Earl, P. L., Cooper, N., and Moss, B. (1991). In *Current protocols in molecular biology*. Vol. 2 (ed. F. M. Ausubel, R. Brent, R. E. Kingston, D. D. Moore, J. G. Seidman, J. A. Smith, and K. Struhl), pp. 16.15.1–16.18.10. John Wiley & Sons, New York.
6. Moss, B. (1992). In *Recombinant poxviruses* (ed. M. M. Binns and G. L. Smith), pp. 45–80. CRC Press, Boca Raton.
7. Smith, G. L. and Mackett, M. (1992). In *Recombinant poxviruses* (ed. M. M. Binns and G. L. Smith), pp. 81–122. CRC Press, Boca Raton.
8. Moss, B., Elroy-Stein, O., Mizukami, T., Alexander, W. A., and Fuerst, T. R. (1990). *Nature*, **348**, 91.
9. Usdin, T. B., Brownstein, M. J., Moss, B., and Isaacs, S. N. (1993). *Biotechniques*, **14**, 222.
10. Gallaher, W. R., Henderson, L. A., Fermin, C. D., Montelaro, R. C., Martin, A., Qureshi, M. N., *et al.* (1992). In *Membrane interactions of HIV: Implications for pathogenesis and therapy in AIDS; Advances in membrane fluidity*, Vol. 6 (ed. R. C. Aloia and C. C. Curtain), pp. 113–42. Wiley-Liss, New York.
11. Cullen, B. R. and Garrett, E. D. (1992). *AIDS Res. Hum. Retroviruses*, **8**, 387.
12. Broder, C. C., Kennedy, P. E., Michaels, F., and Berger, E. A. (1994). *Gene*, **142**, 167.
13. Chaudhary, V. K., Mizukami, T., Fuerst, T. R., FitzGerald, D. J., Moss, B., Pastan, I., and Berger, E. A. (1988). *Nature*, **335**, 369.
14. Chakrabarti, S., Robert-Guroff, M., Wong-Staal, F., Gallo, R. C., and Moss, B. (1986). *Nature*, **320**, 535.
15. Hu, S.-L., Kosowski, S. G., and Dalrymple, J. M. (1986). *Nature*, **320**, 537.
16. Lifson, J. D., Feinberg, M. B., Reyes, G. R., Rabin, L., Banapour, B., Chakrabarti, S., *et al.* (1986). *Nature*, **323**, 725.
17. Chakrabarti, S., Mizukami, T., Franchini, G., and Moss, B. (1990). *Virology*, **178**, 134.
18. Mulligan, M. J., Kumar, P., Hui, H. X., Owens, R. J., Ritter, G. D., Jr., Hahn, B. H., and Compans, R. W. (1990). *AIDS Res. Hum. Retroviruses*, **6**, 707.
19. Bosch, M. L., Earl, P. L., Fargnoli, K., Picciafuoco, S., Giombini, F., Wong-Staal, F., and Franchini, G. (1989). *Science*, **244**, 694.
20. Koenig, S., Hirsch, V. M., Olmsted, R. A., Powell, D., Maury, W., Rabson, A., *et al.* (1989). *Proc. Natl Acad. Sci. USA*, **86**, 2443.

21. Berger, E. A., Fuerst, T. R., and Moss, B. (1988). *Proc. Natl Acad. Sci. USA*, **85**, 2357.
22. Mizukami, T., Fuerst, T. R., Berger, E. A., and Moss, B. (1988). *Proc. Natl Acad. Sci. USA*, **85**, 9273.
23. Ashorn, P. A., Berger, E. A., and Moss, B. (1990). *J. Virol.*, **64**, 2149.
24. Schubert, M., Joshi, B., Blondel, D., and Harmison, G. G. (1992). *J. Virol.*, **66**, 1579.
25. Broder, C. C. and Berger, E. A. (1993). *J. Virol.*, **67**, 91.
26. Golding, H., Blumenthal, R., Manischewitz, J., Littman, D. R., and Dimitrov, D. S. (1993). *J. Virol.*, **67**, 6469.
27. Willey, R. L., Smith, D. H., Lasky, L. A., Theodore, T. S., Earl, P. L., Moss, B., *et al.* (1988). *J. Virol.*, **62**, 139.
28. Dedera, D., Gu, R., and Ratner, L. (1992). *J. Virol.*, **66**, 1207.
29. Dedera, D. A., Gu, R. L., and Ratner, L. (1992). *Virology*, **187**, 377.
30. Perez, L. G., O'Donnell, M. A., and Stephens, E. B. (1992). *J. Virol.*, **66**, 4134.
31. Mulligan, M. J., Yamshchikov, G. V., Ritter, G. D., Gao, F., Jin, M. J., Nail, C. D., *et al.* (1992). *J. Virol.*, **66**, 3971.
32. Mulligan, M. J., Ritter, G. D., Chaikin, M. A., Yamshchikov, G. V., Kumar, P., Hahn, B. *et al.* (1992). *Virology*, **187**, 233.
33. Travis, B. M., Dykers, T. I., Hewgill, D., Ledbetter, J., Tsu, T. T., Hu, S.-L., and Lewis, J. B. (1992). *Virology*, **186**, 313.
34. Otteken, A., Voss, G., and Hunsmann, G. (1993). *Virology*, **194**, 37.
35. Ritter, G. D., Mulligan, M. J., Lydy, S. L., and Compans, R. W. (1993). *Virology*, **197**, 255.
36. Nehete, P. N., Arlinghaus, R. B., and Sastry, K. J. (1993). *J. Virol.*, **67**, 6841.
37. Owens, R. J. and Compans, R. W. (1990). *Virology*, **179**, 827.
38. Earl, P. L., Moss, B., and Doms, R. W. (1991). *J. Virol.*, **65**, 2047.
39. Hallenberger, S., Bosch, V., Angliker, H., Shaw, E., Klenk, H. D., and Garten, W. (1992). *Nature*, **360**, 358.
40. Bernstein, H. B. and Compans, R. W. (1992). *J. Virol.*, **66**, 6953.
41. Spies, C. P. and Compans, R. W. (1993). *J. Virol.*, **67**, 6535.
42. Earl, P. L., Doms, R. W., and Moss, B. (1990). *Proc. Natl Acad. Sci. USA*, **87**, 648.
43. Doms, R. W., Earl, P. L., Chakrabarti, S., and Moss, B. (1990). *J. Virol.*, **64**, 3537.
44. Earl, P. L., Doms, R. W., and Moss, B. (1992). *J. Virol.*, **66**, 5610.
45. Cornet, B., Decroly, E., Thines-Sempoux, D., Ruysschaert, J. M., and Vandenbranden, M. (1992). *AIDS Res. Hum. Retroviruses*, **8**, 1823.
46. Berger, E. A., Lifson, J. D., and Eiden, L. E. (1991). *Proc. Natl Acad. Sci. USA*, **88**, 8082.
47. Berger, E. A., Sisler, J. R., and Earl, P. L. (1992). *J. Virol.*, **66**, 6208.
48. Dimitrov, D. S., Golding, H., and Blumenthal, R. (1991). *AIDS Res. Hum. Retroviruses*, **7**, 799.
49. Dimitrov, D. S., Hillman, K., Manischewitz, J., Blumenthal, R., and Golding, H. (1992). *J. Virol.*, **66**, 132.
50. Dimitrov, D. S., Hillman, K., Manischewitz, J., Blumenthal, R., and Golding, H. (1992). *AIDS*, **6**, 249.
51. Golding, H., Dimitrov, D. S., and Blumenthal, R. (1992). *AIDS Res. Hum. Retroviruses*, **8**, 1593.

52. Andeweg, A. C., Groenink, M., Leeflang, P., DeGoede, R. E. Y., Osterhaus, A. D. M. E., Tersmette, M., and Bosch, M. L. (1992). *AIDS Res. Hum. Retroviruses*, **8**, 1803.
53. Broder, C. C., Dimitrov, D. S., Blumenthal, R., and Berger, E. A. (1993). *Virology*, **193**, 483.
54. Dimitrov, D. S., Broder, C. C., Berger, E. A., and Blumenthal, R. (1993). *J. Virol.*, **167**, 1647.
55. Johnston, P. B., Dubay, J. W., and Hunter, E. (1993). *J. Virol.*, **167**, 3077.
56. Berger, E. A., Clouse, K. A., Chaudhary, V. K., Chakrabarti, S., FitzGerald, D. J., Pastan, I., and Moss, B. (1989). *Proc. Natl Acad. Sci. USA*, **86**, 9539.
57. Srinivas, R. V., Mulligan, M. J., Owens, R. J., Srinivas, S. K., Anantharamaiah, G. M., Segrest, J. P., and Compans, R. W. (1992). In *Membrane interactions of HIV: Implications for pathogenesis and therapy in AIDS; Advances in membrane fluidity*, Vol. 6 (ed. R. C. Aloia and C. C. Curtain), pp. 187–202. Wiley-Liss, New York.
58. Nussbaum, O., Broder, C. C., and Berger, E. A. (1994). *J. Virol.*, **68**, 5411.
59. Haffar, O., Garrigues, J., Travis, B., Moran, P., Zarling, J., and Hu, S.-L. (1990). *J. Virol.*, **64**, 2653.
60. Owens, R. J., Dubay, J. W., Hunter, E., and Compans, R. W. (1991). *Proc. Natl Acad. Sci. USA*, **88**, 3987.
61. Walker, B. D. and Plata, F. (1990). *AIDS*, **4**, 177.
62. Barrett, P. N. and Dorner, F. (1993). *Int. Arch. Allergy Immunol.*, **100**, 93.
63. Maddon, P. J., Dalgleish, A. G., McDougal, J. S., Clapham, P. R., Weiss, R. A., and Axel, R. (1986). *Cell*, **47**, 333.
64. Weiner, D. B., Huebner, K., Williams, W. V., and Greene, M. I. (1991). *Pathobiology*, **59**, 361.
65. Dragic, T., Charneau, P., Clavel, F., and Alizon, M. (1992). *J. Virol.*, **66**, 4794.
66. Freed, E. O., Myers, D. J., and Risser, R. (1989). *J. Virol.*, **63**, 4670.
67. Freed, E. O., Myers, D. J., and Risser, R. (1990). *Proc. Natl Acad. Sci. USA*, **87**, 4650.
68. Freed, E. O., Myers, D. J., and Risser, R. (1991). *J. Virol.*, **65**, 190.
69. Freed, E. O. and Myers, D. J. (1992). *J. Virol.*, **66**, 5472.
70. Puri, A., Dimitrov, D. S., Golding, H., and Blumenthal, R. (1992). *J. Acq. Immune Defic. Syndr.*, **5**, 915.
71. Berger, E. A., Broder, C. C., and Nussbaum, O. (1993). *5th Int. Conf. NCDDG-HIV*, Abstr. p. 34.
72. Broder, C. C., Nussbaum, O., Gutheil, W. G., Bachovchin, W. W., and Berger, E. A. (1994). *Science*, **264**, 1156.
73. Doms, R. W., Blumenthal, R., and Moss, B. (1990). *J. Virol.*, **64**, 4884.
74. Eustice, D. C., Feldman, P. A., Colberg-Poley, A. M., Buckery, R. M., and Neubauer, R. H. (1991). *Biotechniques*, **11**, 739.
75. Deng, H., Wang, C., Acsadi, G., and Wolff, J. A. (1991). *Gene*, **109**, 193.
76. Alexander, W. A., Moss, B., and Fuerst, T. R. (1992). *J. Virol.*, **66**, 2934.
77. Fiering, S. N., Roederer, M., Nolan, G. P., Micklem, D. R., Parks, D. R., and Herzenberg, L. A. (1991). *Cytometry*, **12**, 291.
78. Broder, C. C. and Berger, E. A. (1995). *Proc. Natl Acad. Sci.* USA, in press.

9

RNA binding assays for the regulatory proteins Tat and Rev

J. KARN, M. J. CHURCHER, K. RITTNER, A. KELLEY, P. J. G. BUTLER, D. A. MANN, and M. J. GAIT

1. Introduction

The replication of the human immunodeficiency virus (HIV) is controlled by two viral proteins, Tat, the *trans*-activator protein and Rev, the regulator of virion expression. The two proteins play complementary roles in the HIV life cycle. In permissive cells, transcription of the integrated proviral genome is initiated by host cell transcription factors (1–3). Initially, only small, multiply spliced mRNAs including the mRNAs for Tat and Rev are produced (4). Tat synthesized during this primary round of transcription then establishes a 'positive feedback loop' which boosts mRNA production 200- to 1000-fold (5–7). Once a critical Rev threshold is surpassed, HIV gene expression switches away from the production of the mRNAs for the regulatory proteins towards the production of the unspliced or partially spliced mRNAs encoding the virion proteins (8–10).

The Tat and Rev proteins provide rare examples of regulation by RNA binding proteins. More typically, viral regulatory proteins exert their effects through DNA elements. Tat activity requires the *trans*-activation-responsive region (TAR), an RNA regulatory element of 59 nucleotides (nt) located immediately downstream of the initiation site for transcription (11–14). Because of its position in the HIV genome, each viral mRNA carries a copy of TAR at its 5′ end. Unlike enhancer elements, the TAR element is only functional when it is placed downstream of the start of HIV transcription and in the correct orientation.

Only the short fully spliced viral mRNAs appear in the cytoplasm of Rev-minus cells (9, 15, 16). It is now known that both positive and negative regulatory sequences required for Rev function are present in the regions of the *gag* and *env* mRNAs which are removed by splicing (see Chapter 1). A *cis*-acting sequence, called the Rev-responsive element (RRE), acts as a positive control element that is absolutely required for Rev activity. The RRE is located within the *env* reading frame (17–19). The RRE RNA is specifically bound by Rev *in vitro* (20–23), but the binding reaction is complex and involves an initial interaction with a high affinity site followed by the co-operative addition of Rev monomers to lower affinity sites (21). The Rev-dependent mRNAs also carry negative regulatory elements, now called instability (INS) sequences (17, 24, 25). In the absence of either Rev, or a functional RRE sequence, Rev-dependent mRNAs are highly unstable in the cytoplasm. Their cytoplasmic levels rise dramatically if Rev is added in *trans* (24, 26) or when the INS sequences are inactivated by a series of mutations (25).

2. RNA binding by the Tat and Rev proteins

2.1 Recognition of TAR RNA by Tat

As shown in *Figure 1*, TAR RNA forms a highly stable, nuclease-resistant, stem–loop structure. Point mutations which disrupt base pairing in the upper TAR RNA stem invariably abolish Tat-stimulated transcription (1, 14, 27).

In 1989 we provided the first demonstration that recombinant Tat expressed in *Escherichia coli* could bind specifically to bases in the stem of the TAR RNA (28). The binding site for Tat includes a U-rich trinucleotide bulge found in the upper stem of TAR RNA (*Figure 1*). Residues essential for Tat recognition include the first residue in the bulge, U_{23}, and the two base pairs immediately above the bulge, G_{26}:C_{39} and A_{27}:U_{38} (29–32). The two base pairs below the bulge make a comparatively small contribution to Tat binding specificity and are probably not points of direct contact with the protein (29, 31). The other residues in the bulge, C_{24} and U_{25}, appear to act predominantly as spacers and may be replaced by other nucleotides, or even by non-nucleotide linkers (33).

2.2 Rev binding to the RRE

The genetically defined RRE element is 351 nucleotides, and is thus much larger than TAR. As shown in *Figure 2*, the apex of the RRE forms a series of complicated stem–loop structures (15). Rev interacts with the RRE both through high affinity binding to a specific site and by an RNA-mediated oligomerization reaction. The simplest way to visualize this process is by gel mobility shift assays. As shown in the top panel of *Figure 2*, the Rev binding reaction produces a series of complexes. There is a progressive increase in the formation of the highest molecular complexes as the molar ratio of Rev to RRE RNA increases (21, 34–36).

We were able to map the high affinity Rev binding site by constructing artificial stem–loop structures carrying fragments of the RRE (34). The binding site contains an unusual purine-rich ‘bubble’ containing bulged GG and GUA residues that was not predicted in the original models for the RRE structure (15). The identification of the ‘bubble’ sequence as a high affinity

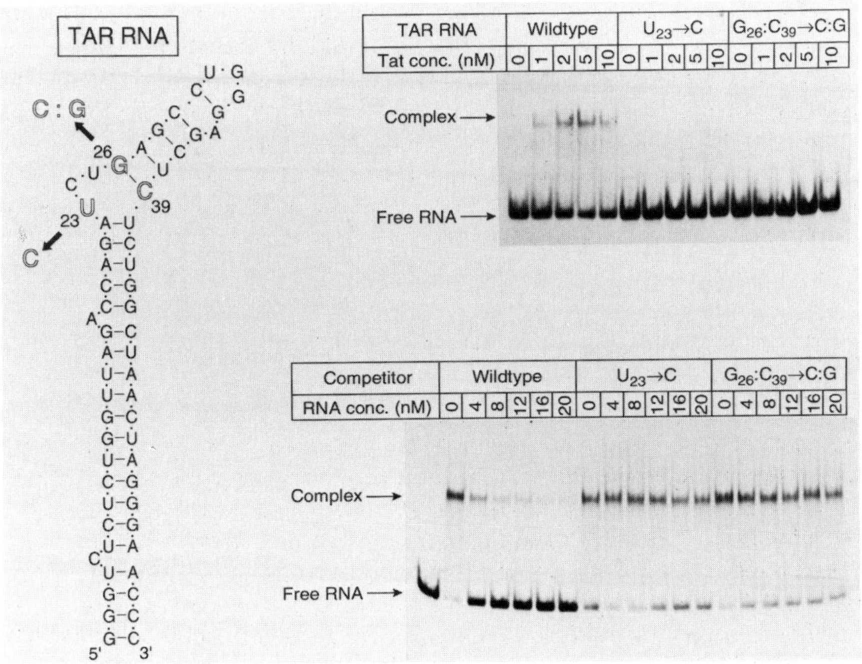

Figure 1. Structure of TAR and gel retardation assays. *Left*: Structure of TAR RNA and sequence requirements for Tat binding. The top of the TAR RNA stem–loop structure is drawn to indicate bending of the stem induced by the bulged nucleotides U_{23} to U_{25} and the structure of the apical loop. Two mutations in the stem known to inhibit Tat binding are highlighted. *Right*: Gel mobility shift assays for Tat binding to TAR RNA in which free RNA was fractionated from the Tat/TAR RNA complex by electrophoresis on non-denaturing polyacrylamide gels. *Top* panel: Saturation binding experiment. Complexes were formed between 2 nM ^{35}S-labelled TAR RNA (wild-type or mutant) and 1, 2, 5, or 10 nM *Tat* protein. *Bottom* panel: Competition binding experiment. Complexes were formed between 2 nM ^{35}S-labelled TAR RNA and 5 nM *Tat* protein 0–20 nM unlabelled competitor TAR RNA. TAR RNAs carrying the $U_{23} \rightarrow C$ and G_{26}: $C_{39} \rightarrow$ C: G mutations competed with more than tenfold lower affinity than the wild-type sequence.

binding site has now been confirmed by chemical footprinting data (37, 38), as well as by the selection of high affinity binding sites for Rev from pools of randomly generated mutants (39).

Although the 'bubble' is located in the RRE towards the base of a stem-loop near a three-way junction (*Figure 2*), Rev does not appear to recognize features of the junction and this sequence can be replaced by any short stretch of duplex RNA. An example of one of these artificial binding sites, RBC5L, is shown in *Figure 2* (bottom). RBC5L is able to bind a monomer of Rev with a binding constant of approximately 2 nM but is unable to form oligomeric complexes with Rev.

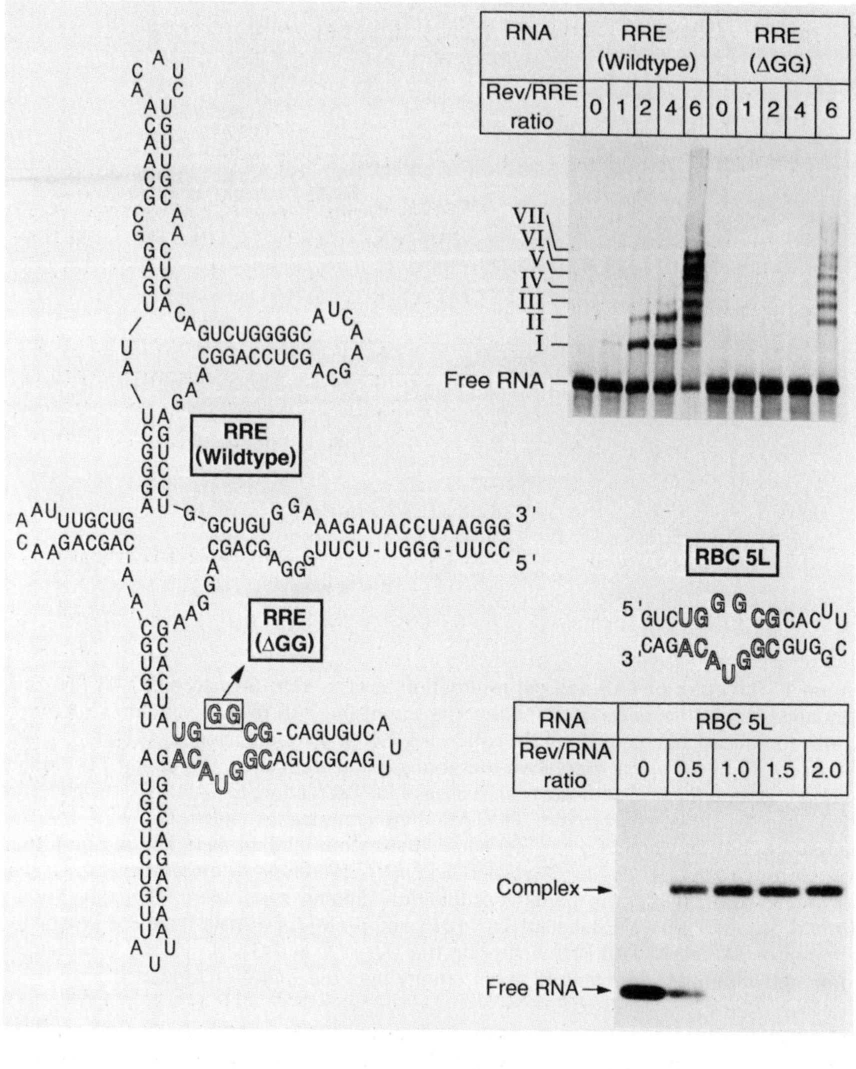

Figure 2. Binding of Rev to synthetic RNA binding sites and RRE RNA transcripts. The predicted secondary structures for a 238 nt long fragment of the RRE and a short synthetic sequence, RBC5L, are shown at *left*. Residues in the high affinity site shown by mutagenesis experiments to be essential for Rev recognition are highlighted. *Top* panel: 50 nM ^{35}S-labelled RRE RNA (residues 70 to 279) or 50 nM ^{35}S-labelled RRE RNA carrying the ΔGG mutation in the high affinity site was bound to Rev at protein to RNA ratios of between one and six. Note that no RNA–protein complexes are obtained until there is a sixfold molar excess of Rev in the reaction with the RRE carrying the ΔGG mutation. *Bottom* panel: Binding of Rev to a minimal synthetic structure. Binding reactions contained 10 nM ^{35}P-labelled RNA and between 0–44 nM of recombinant Rev protein. Note that even in the presence of excess Rev only a single complex is formed.

3. Expression of the regulatory proteins in *E. coli*

3.1 Purification of Tat protein

The Tat protein is unusually difficult to purify and to store:

(a) The central region of Tat contains a metal binding domain with seven closely spaced cysteine residues. Recombinant Tat protein prepared from *E. coli* readily forms intermolecular disulfide bonds due to oxidation of these cysteine residues in the absence of added zinc.

(b) Tat carries an arginine-rich domain which normally participates in nucleic acid binding. This region of the molecule binds a variety of low molecular weight acidic components in *E. coli*, and causes most Tat preparations to be heterogeneous in charge and elute from ion-exchange columns as a broad peaks.

It is not surprising that initial attempts to demonstrate specific binding of Tat to TAR RNA were unsuccessful due to difficulties in preparing the protein (40, 41)!

Initially, we prepared HIV-1 Tat as a fusion protein using a variety of partners including β-galactosidase, human growth hormone, and glutathione-S-transferase (28–30, 42). In each of these methods the fusion protein was cleaved with cyanogen bromide in order to release the Tat moiety. Although time-consuming, this approach does produce protein of high quality since the acidic environment required for the cyanogen bromide cleavage step both reduces the cysteines and removes charged contaminants from the Tat protein. The cleaved Tat protein is then purified by ion-exchange chromatography in the presence of 6 M urea and refolded by stepwise dialysis against buffers containing 100 μM $ZnSO_4$ (29, 30).

Recently, we have also successfully prepared Tat following expression of the unfused protein under the control of a bacteriophage T7 promoter. The Tat expression plasmid tat/ex1/kan/22,23 carries the first exon of the HIV-1_{SF2} *tat* gene (kind gift of Dr R. Gaynor, Southwestern Medical School, Dallas, Texas). To improve translational efficiency a two cistron expression system, supplying an optimal ribosome binding site for the first cistron was inserted upstream of the *tat* gene (43). A kanamycin resistance gene was also inserted into the plasmid.

Special care must be taken during the growth of Tat-expressing cells:

(a) The *tat* gene is reasonably well-tolerated during logarithmic growth, but is highly toxic to *E. coli* when cells reach saturation. Almost all the plasmid-containing cells are killed when a saturated culture is incubated for one or two hours. Cultures are maintained in logarithmic phase by careful titration of the inocula.

(b) Avoid the use of ampicillin as a selective marker. Ampicillin is rapidly removed from liquid medium due to the enzymatic activity of the β-lactamase secreted by resistant bacteria. Cultures that are selected in ampicillin alone, rapidly become overgrown with cells that lack plasmid. To circumvent this problem, select bacteria with the non-hydrolysable antibiotic kanamycin.

Protocol 1 gives a simple purification method for the Tat protein, which can be performed without HPLC. The following points should be borne in mind:

(a) A denaturation and renaturation cycle is required to remove contaminants from *E. coli* which inhibit RNA binding.
(b) Label the cultures with [^{35}S]cysteine (100 μCi/ml) during the induction period. Since Tat is the major cysteine-containing protein found in the inclusion bodies, virtually all the cysteine label is associated with Tat.
(c) Don't use dithiothreitol (DTT) as a reducing agent. DTT chelates zinc with high affinity. Preferred reducing agents are β-mercaptoethanol or dithioerythritol (DTE).

Approximately 2 mg of purified Tat protein was obtained from six litres of induced cells. The concentrations of Tat protein stocks were determined by amino acid analysis. To avoid oxidation, it is important to store Tat in small aliquots in liquid nitrogen and to avoid repeated cycles of freezing and thawing of the stocks.

Protocol 1. Purification of Tat protein

Equipment and reagents

- HPLC chromatography equipment and columns (Pharmacia)
- Superdex S-200 (Pharmacia)
- Sepharose-S FF (Pharmacia)
- Sorvall centrifuge
- 15% SDS–polyacrylamide gels and electrophoresis apparatus
- 3 × D Amp/Kan media: 10.5 g Na_2HPO_4, 4.5 g KH_2PO_4, 0.6 g NH_4Cl, 15 g Casamino acids, 15 g yeast extract, 19 ml glycerol per litre supplemented after autoclaving with 100 μg ml/ampicillin, 25 μg/ml kanamycin, 10 μg/ml thiamine, and 10 mM $MgSO_4$
- [^{35}S]cysteine (Amersham)
- Buffer A: 20 mM Tris–HCl pH 8.0, 100 mM NaCl, 1 mM EDTA, 1 mM DTE
- Buffer B: 25% (w/v) sucrose, 20 mM Tris–HCl pH 8.0, 100 mM NaCl, 0.1 mM EDTA, 1 mM DTE
- Buffer C: 2% Triton, 20 mM Tris–HCl pH 8.0, 100 mM NaCl, 0.1 mM EDTA, 1 mM DTE
- Buffer D: 6 M guanidine–HCl, 50 mM Tris–HCl pH 8.0, 100 mM DTT
- Buffer E: 150 mM guanidine–HCl, 50 mM Tris–HCl pH 8.0, 0.1 mM EDTA, 1 mM DTE
- Buffer F: 150 mM NaCl, 20 mM Tris–HCl pH 8.0, 1.0 mM DTE
- Buffer G: 1 M NaCl, 20 mM Tris–HCl pH 8.0, 0.1 mM EDTA, 1 mM DTE
- Refolding buffer: 20 mM Tris–acetate pH 5.0, 150 mM NaCl, 1 mM β-mercaptoethanol, 20 μM $ZnSO_4$, 10% (v/v) glycerol
- Tat storage buffer: 20% glycerol (v/v), 50 mM Tris–HCl pH 8.0, 1.0 mM DTT, 100 μM $ZnSO_4$

Method

1. Grow BL21/*lys*S cells carrying the tat/ex1/kan/22/23 plasmid. Streak out bacteria from frozen stocks on plates containing 100 μg/ml ampicillin and 25 μg/ml kanamycin. Incubate plates at room temperature overnight. Pick a colony and disperse into 30 ml 3 × D Kan/Amp medium. Dilute bacteria 1:100, 1:1000, 1:10 000 in 10 ml 3 × D Kan/Amp medium and grow overnight at 30 °C. Pick cultures that are still in log phase and inoculate 6 litres of 3 × D Kan/Amp medium with 1 ml.
2. Induce and label cells for 1 h at log phase ($A_{590} = 0.6$). Add 10 ml 100 mM IPTG (Sigma) per litre. Add 1 mCi [^{35}S]cysteine per litre.
3. Collect cells by centrifuging for 10 min at 5000 r.p.m. in a Sorvall GSA rotor. Wash cells by resuspending in 500 ml buffer A and repeat centrifugation.
4. Freeze pellets and store overnight at −20 °C. Note that cell pellets containing Tat should not be stored frozen for more than two days as the protein present in the inclusion bodies tends to form oxidation products, including covalently linked dimers.
5. Lyse cells. Resuspend cells in 40 ml buffer B in ice-bath. Add 4 ml lysozyme (10 mg/ml) and incubate for 15 min. Add 40 ml of buffer C to lyse cells. Add 4 ml of 100 mM MgOAc and 4 ml of DNase I (1 mg/ml). Incubate 30 min at 0 °C, or until the solution is no longer viscous.
6. Recover inclusion bodies by centrifuging at 20 000 r.p.m. for 10 min in a Sorvall centrifuge. Wash pellet three times by resuspending in 40 ml buffer C. Repeat centrifugation, 20 000 r.p.m. for 10 min.
7. Dissolve inclusion body pellets in 10 ml buffer D. Centrifuge at 20 000 r.p.m., 10 min to remove insolube material. Heat supernatant for 5 min at 70 °C.
8. Apply to 350 ml Superdex S-200 column in buffer D. Tat protein elutes as a monomer well separated from nucleic acid in the breakthrough peak and low molecular weight contaminants. Determine location of Tat protein by electrophoresis on 15% SDS–PAGE or by scintillation counting.
9. Dialyse Tat protein overnight against 2 litres buffer E. Remove precipitate by centrifugation at 12 000 r.p.m. in Sorvall.
10. Apply Tat protein to 20 ml Sepharose-S FF column equilibrated with buffer E. Wash with 25 ml 0.2 M NaCl in buffer F. Elute with a 40 ml linear gradient from 0.2 M–1.0 M NaCl.
11. Refold Tat protein. Add DTE to 100 mM and heat for 10 min at 70 °C. Dialyse 2 × 90 min against 1 litre refolding buffer. Note that this buffer is at pH 5.0 to facilitate introduction of the zinc. Dialyse 2 × 90 min against Tat storage buffer.
12. Store Tat frozen in 400 μl aliquots in liquid nitrogen.

3.2 Purification of Rev protein

The Rev protein is generally easier to handle than the Tat protein and several groups have developed satisfactory expression systems (44–46). We have found that a synthetic *rev* gene (21) cloned between the *Nde*I and *Hin*dIII sites of the pT7-SC expression vector (United States Biochemical Corp.) gives a high level of Rev expression. The pT7-SC vector includes strong termination signals to suppress readthrough transcription.

However, the purification of Rev presents some special difficulties:

- Rev aggregates and precipitates in buffers containing less than 200 mM NaCl
- Rev binds nucleic acids present in *E. coli*

Protocol 2 presents a simple purification scheme for an unfused Rev protein. The scheme takes advantage of the fact that, Rev–RNA complexes obtained from *E. coli* can be purified by anion-exchange chromatography (45) and eluted with 0.8 M NaCl. Because of the very high affinity of Rev for heparin, the protein can then be separated from the nucleic acid by binding to heparin–Sepharose columns in 0.8 M NaCl and eluted in 2 M NaCl.

Protocol 2. Purification of Rev protein

Equipment and reagents

- FPLC chromatography equipment and columns (Pharmacia)
- Sepharose-Q FF (Pharmacia)
- Heparin–Sepharose (Pharmacia)
- Sonicator
- Sorvall centrifuge
- 18% SDS–polyacrylamide gels and electrophoresis apparatus
- 3 × D Amp media: 10.5 g Na_2HPO_4, 4.5 g KH_2PO_4, 0.6 g NH_4Cl, 15 g Casamino acids, 15 g yeast extract, 19 ml glycerol per litre supplemented after autoclaving with 100 μg/ml ampicillin, 10 μg/ml thiamine, and 10 mM $MgSO_4$
- Buffer A: 400 mM NaCl, 50 mM Tris–HCl pH 8.0, 1 mM EDTA, 1 mM DTT, 0.5 mM phenylmethylsulfonyl fluoride (PMSF)
- Buffer B: 800 mM NaCl, 50 mM Tris–HCl pH 8.0, 1 mM EDTA, 1 mM DTT
- Buffer C: 1 M NaCl, 50 mM Tris–HCl pH 8.0, 1 mM EDTA, 1 mM DTT
- Buffer D: 2 M NaCl, 50 mM Tris–HCl pH 8.0, 1 mM EDTA, 1 mM DTT
- Rev storage buffer: 10% glycerol (v/v), 2 M NaCl, 50 mM Tris–HCl pH 8.0, 1 mM EDTA, 1 mM DTT

Method

1. Grow cells in 3 litres 3 × D Amp medium.
2. Induce cells for 1 h at log phase (A_{590} = 0.6) by addition of 10 ml/litre 100 mM IPTG.
3. Centrifuge 10 min at 5000 r.p.m. in a Sorvall centrifuge.
4. Freeze pellets and store overnight at −20°C.

5. Lyse cells. Resuspend in 45 ml buffer A, and sonicate 3 × 30 sec. Freeze in dry ice. Thaw, and sonicate as above.
6. Centrifuge 10 min at 10 000 r.p.m. in a Sorvall centrifuge and save the supernatant.
7. Resuspend pellet in 40 ml buffer A, repeat steps 5 and 6, and combine the supernatants.
8. Chromatograph on 230 ml column of Q-Sepharose FF equilibrated in buffer A. Wash with 800 ml buffer A, or until the A_{260} of the eluant is below 0.05. Elute Rev protein with buffer B. Check peak fractions by SDS–PAGE using 18% gels.
9. Chromatograph on 10 ml column of heparin–Sepharose equilibrated in buffer B. Wash with 50 ml buffer B, or until the A_{260} of the eluant is below 0.05. Wash with 50 ml buffer C. Elute Rev protein with buffer D. Check peak fractions by SDS–PAGE using 18% gels.
10. Dialyse peak fractions against Rev storage buffer and freeze 400 μl aliquots in liquid nitrogen.

4. Preparation of labelled RNAs by transcription with bacterial polymerases

Labelled RNA templates for use in RNA binding assays are conveniently prepared by transcription of cloned genes using either T3 or T7 RNA polymerase. Synthetic oligonucleotides corresponding to the sequences of interest can be simply inserted in pBS+ plasmid series (Stratagene). However, when preparing these types of templates particular attention needs to be paid to the removal of 3′ or 5′ tails which can result from inclusion of polylinker sequences in the transcripts.

(a) RNA transcripts carrying 'tails' tend to aggregate and produce a low level of dimeric RNAs under the electrophoresis conditions used in the gel mobility shift assays.

(b) Transcripts carrying 'tails' give rise to a faster migrating species which appears to be alternatively folded forms.

To circumvent these problems, transcripts should be inserted immediately downstream of the T3 or T7 transcription initiation sites.

(a) It is often convenient to include a *Sma*I (CCCGGG) site at the 3′ end of the sequence. After cleavage with this flush-cutting enzyme, the 3′ CCC sequence is able to hybridize to the GGG sequences present in the optimal start sites for both T3 and T7 RNA polymerases.

(b) In more specialized applications, involving short RNA transcripts, it is sometimes useful to prepare transcripts from synthetic single-stranded DNA templates (47), or by chemical synthesis (48).

Protocol 3 gives a method for the large scale synthesis of labelled transcripts:

(a) This procedure yields between 1–2 nmol of labelled RNA with specific activities of 6000 c.p.m./pmol for the ^{35}S-labelled RNA and 4000 c.p.m./pmol for the ^{3}H-labelled RNA.
(b) It is convenient to prepare large quantities of RNA when performing a series of binding experiments. The reactions can be scaled down easily, if less material is required.
(c) It is convenient to radiolabel the RNAs in order to accurately calculate their concentrations in binding reactions. The calculated values for the RNA concentration, must be corrected for the number of U residues in the sequence. For example, the 59-mer TAR RNA sequence includes 15 U residues and its concentration is therefore the concentration of incorporated uridine divided by 15.

Protocol 3. Preparation of uniformly labelled RNA by T3 or T7 transcription

Equipment and reagents

- 1 mg/ml linearized plasmid DNA
- ACG stock: 10 mM ATP, 10 mM CTP, 10 mM GTP
- 1 mM UTP
- 1 mg/ml T3 (or T7) RNA polymerase
- RNasin (Promega)
- Uridine-5′-[α-^{35}S]thiophosphate (37 TBq/mmol, Amersham)
- [5, 6-^{3}H]UTP (1.5 to 2.2 TBq/mmol, Amersham)
- 10 × transcription buffer: 400 mM Tris–HCl pH 7.4, 500 mM NaCl, 80 mM $MgCl_2$
- TKE buffer: 10 mM Tris–HCl pH 7.4, 50 mM KCl, 0.1 mM EDTA
- TBE buffer: 45 mM Tris base, 45 mM boric acid, 10 mM EDTA, pH 8.3
- Gel elution buffer: 0.3 M sodium acetate, 10 mM Tris–HCl pH 7.4, 1 mM EDTA, 0.5% (w/v) SDS

Method

1. Set-up 1 ml transcription reactions (1 ml) containing:

• 10 × transcription buffer	100 μl
• 100 mM DTT	100 μl
• ACG stock	100 μl
• 1 mM UTP	100 μl
• RNasin	20 μl
• 40 μCi uridine-5′-[α-^{35}S]thiophosphate or 100 μCi [5, 6-^{3}H]UTP	10 μl
• linearized plasmid DNA (1 mg/ml)	100 μl

- T3 RNA polymerase (1 mg/ml) 40 μl
- H_2O *q.s.* 1000 μl

2. Count three aliquots of 1 μl each to calculate specific activity of label. Each reaction contains approx. 2 nmol UTP in 50 μl.
3. Incubate 2 h 37°C.
4. Extract with an equal volume of phenol. Precipitate by addition of 1/10 vol. 5 M Na acetate pH 5.0 and 3 vol. of EtOH. Incubate on dry ice for 15 min. Spin in microcentrifuge for 5 min at 15000 r.p.m. Remove supernatant. Wash pellet with 70% (v/v) EtOH. Remove supernatant and re-spin nearly empty tube to remove traces of remaining EtOH. Dissolve pellet in 500 μl (large reactions) H_2O.
4. Fractionate RNA on 10% (w/v) polyacrylamide gels containing 8 M urea in TBE buffer.
5. Elute RNA. Visualize band by UV shadowing. Cut band from gel and immerse in gel elution buffer. Incubate at 37°C for 12–18 h. Extract with phenol and precipitate with ethanol (as in step 4). Resuspend in 500 μl TKE buffer.
6. Calculate yield and specific activity of RNA.
7. Refold RNAs by heating to 80°C for 2 min and rapid cooling in an ice-bath immediately before use in binding reactions.

5. RNA binding assays

The most convenient and frequently used assays for RNA binding are:

- gel retardation
- filter binding

Both assays should be performed in parallel since the filter binding assay lends itself to the quantitative determination of binding constants, while the gel retardation assay provides useful information about the stoichiometry of binding and the numbers of different complexes that can be formed.

5.1 Gel retardation assays

The gel retardation assay is based on the use of non-denaturing gel electrophoresis to separate free RNA from protein–RNA complexes. *Figure 1* shows some examples of gel retardation assays performed with purified Tat protein and ^{35}S-labelled TAR RNA. The Tat protein forms specific one to one complexes with TAR RNA with a K_d of approximately 3 nM. Saturation was reached at approximately 20 nM, with 82% of the wild-type TAR RNA bound. Under these conditions there was negligible binding to TAR RNAs carrying mutations in the Tat binding site.

A more accurate estimate of the relative binding of Tat to mutant TAR RNAs can be obtained from competition experiments. *Figure 1* (bottom panel) shows an example of a competition binding experiment similar to the experiments we used to define the Tat binding site on TAR RNA (29). Unlabelled TAR RNA acts as an effective competitor with a $D_{1/2}$ (the concentration of competitor required to reduce binding by 50%) of approximately 3 nM. TAR RNAs carrying mutations in the binding site, such as the $U_{23} \rightarrow C$ and $G_{26}:C_{39} \rightarrow C:G$ mutations were poor competitors for protein binding and there was virtually no loss of radioactivity from the Tat–TAR complex in the presence of 20 nM competitor RNA ($D_{1/2} > 20$ nM). Following electrophoresis, the relative amounts of free and bound RNA can be determined by densitometry of autoradiographs of the gels (49, 50).

Gel retardation analyses of the Rev–RRE interaction is similar to that of Tat–TAR interaction, but because of the ability of Rev to multimerize on the RRE multiple complexes are formed using long templates. As shown in the top panel of *Figure 2*, there is a progressive increase in the formation of the highest molecular complexes as the molar ratio of Rev to RRE RNA increases (21, 34–36). By contrast, the artificial binding site, RBC5L, is able to bind a monomer of Rev with a binding constant of approximately 2 nM, but even in the presence of a tenfold molar excess of Rev, only a single complex is formed.

Specific Rev binding to longer RRE sequences requires the presence of an intact high affinity binding site. When the high affinity site is disrupted by the ΔGG mutation, no specific complexes are seen (*Figure 2*, top). However, when an excess of Rev (< sixfold) is present in the reaction, some low affinity, non-specific binding occurs and gives rise to the additional complexes seen in *Figure 2* (top).

Protocol 4 gives binding and gel electrophoresis conditions that work well for both Tat and Rev binding. In general, gel retardation experiments are easy to perform and reproducible. However, a few simple technical 'tricks' will markedly improve the results:

(a) Include a non-ionic detergent such as 0.1% Triton X-100 to minimize protein aggregation in the binding reactions and the gels. This results in a lower apparent K_d because the 'free' protein concentration is increased in the reaction.

(b) Remove 'tails' from the RNA transcripts to minimize aggregation and low affinity binding reactions.

(c) Refold the RNA by heating and rapid cooling immediately prior to binding.

(d) Perform binding reactions with both the protein and RNA concentrations near the K_d for binding (i.e. in the nanomolar range).

(e) Add protein as the last component of the binding reaction.

(f) Optimize and control the temperature of the binding reactions and the gel running conditions.

(g) Do not include EDTA in the gel buffer.

Protocol 4. Gel retardation assays

Equipment and reagents

- Binding buffer: 250 mM Tris–HCl pH 7.4, 100 mM KCl, 10 mM DTT, 0.5% Triton X-100
- RNasin (Promega)
- Loading buffer: 50% glycerol, 0.1% bromophenol blue, 0.1% xylene cyanol FF
- TB gel buffer: 44.5 mM Tris base, 44.5 mM boric acid, 0.1% Triton X-100 adjusted to pH 8.3 with HCl
- Tat or Rev protein, approx. 300 μg/ml (*Protocol 1* or *2*)
- ^{35}S-labelled RNA, approx. 70 000 c.p.m. per reaction (*Protocol 3*)

Method

1. Set up 25 μl reactions containing:
   - binding buffer 5 μl
   - RNasin (Promega) 1 μl
   - ^{35}S-labelled RNA 2 nM (≈ 70 000 c.p.m.)
   - Tat or Rev protein (*Protocols 1* and *2*) 0–400 nM

   For competition binding assays use ^{35}S-labelled wild-type RNA at 2 nM and up to 100 nM ^{3}H-labelled or unlabelled competitor RNA (*Protocol 3*), and 5 nM Tat or Rev protein.
2. Incubate for 30 min on ice.
3. Add 5 μl gel loading buffer.
4. Apply to non-denaturing polyacrylamide gels (20 cm × 20 cm). For short transcripts use 7.5% gels, for long transcripts use 4% polyacrylamide gels. Pre-electrophorese for 1 h. Run at 20 mA at 4°C or until the xylene cyanol is approximately half-way down the gel. Do not recirculate buffer when using TB gel buffer.

5.2 Filter binding assays

It is difficult to measure binding constants accurately using gel retardation methods. This is especially true in the case of Tat binding to TAR RNA and its mutants where the affinities usually differ by less than ten fold. A better method is the dual-label filter binding assay outlined in *Protocol 5* (29).

The theory behind the dual-label assay is as follows. The equilibrium mixture of ^{35}S-labelled TAR RNA (R_1) and ^{3}H-labelled competitor RNA (usually mutant TAR RNA) (R_2) binding to Tat protein (P), can be described by the equations:

$$R_1 + P \rightleftharpoons R_1P;\ R_2 + P \rightleftharpoons R_2P. \tag{1}$$

Therefore,

$$K_1 = \frac{[R_{1,\text{free}}]\,[P_{\text{free}}]}{[R_1P]} \text{ and } K_2 = \frac{[R_{2,\text{free}}]\,[P_{\text{free}}]}{[R_2P]} \tag{2}$$

and,

$$\frac{K_2}{K_1} = \frac{[R_{2,\text{free}}]\,[R_1P]}{[R_{1,\text{free}}]\,[R_2P]} = \frac{([R_2] - [R_2P])\,[R_1P)}{([R_1] - [R_1P])\,[R_2P)}, \tag{3}$$

where K_1, and K_2 are the two dissociation constants, $[R_1]$, $[R_2]$ are the total RNA concentrations, and $[R_1P]$, $[R_2P]$ are the bound RNA concentrations. Eqn 3 can be used to calculate the dissociation constants of mutant TAR RNAs from the dual-label filter binding experiments. The analysis was based on the usual assumptions that the counts retained on the filter correspond to the RNA–protein complex present at equilibrium in solution and that the free RNA can be estimated as the difference between the total and bound RNA concentrations.

Figure 3 gives an example of a dual-label experiment, in which ^{35}S-labelled TAR RNA is competed against mutant TAR RNAs labelled with ^{3}H. In the self-competition experiment (*Figure 3*, top), the K_2/K_1 ratio remains constant over a broad range of competitor RNA concentrations and gives the expected value of 1.02 (± 0.08). When mutant competitor RNAs are used more ^{35}S-labelled TAR RNA is retained on the filter in the presence of the ^{3}H-labelled competitor RNA and the total RNA bound in the experiment is reduced compared to the self-competition experiments. The K_2/K_1 ratios remain reasonably constant giving values of 2.99 (± 0.45) for the $A_{22} \cdot U_{40}$ to U · A mutation (*Figure 3*, middle), and 15.7 (± 1.5) for the $G_{26} \cdot C_{39}$ to C · G mutation (*Figure 3*, bottom).

When using the dual-label assay the following points should be borne in mind:

(a) The high level of precision of the dual-label binding assay is only possible because the concentrations of free and bound RNA are determined directly from radioactive counts, and the ratio of dissociation constants is therefore independent of any errors in the measurements of RNA-specific activities.

(b) The most serious potential errors in the method lie in possible loss of binding of the RNA–protein complex to the filter membrane or in artifactual binding of a fraction of the free RNA to the membrane.

(c) The high precision of the dual-label binding assay makes it sensitive to contaminants that are occasionally present in TAR RNA preparations

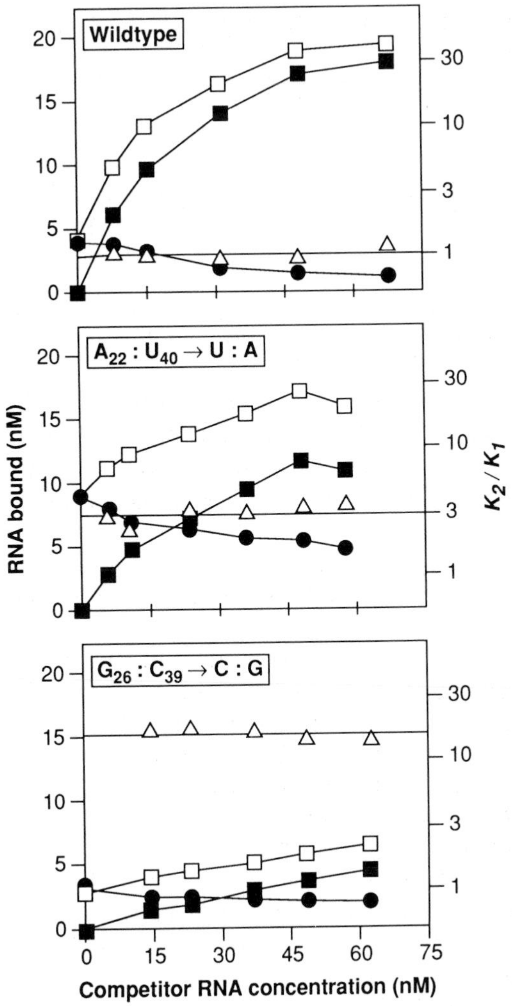

Figure 3. Dual-label filter binding assay. Competition experiments using mutant TAR RNAs. Filter binding assays were set-up as described in *Protocol 5* containing 10 nM Tat protein, ^{35}S-labelled TAR RNA at the concentrations indicated below, and up to 70 nM ^{3}H-labelled competitor TAR RNA. *Top*: Self-competition using 4.9 nM ^{35}S-labelled TAR RNA and ^{3}H-labelled wild-type TAR RNA competitor. *Middle*: 10.8 nM ^{35}S-labelled TAR RNA and ^{3}H-labelled A_{22}: U_{40} → U: A competitor. *Bottom*: 3.7 nM ^{35}S-labelled TAR RNA and ^{3}H-labelled G_{26}C: C_{39} → C: G competitor. (■) Concentration (nM) of ^{3}H-labelled TAR competitor RNA bound by Tat. (●) Concentration (nM) of ^{35}S-labelled TAR RNA bound by Tat. (□) Total RNA bound by Tat. (Δ) Ratio of dissociation constants (K_2/K_1) calculated from the filter binding data using Eqn 3. Data from Churcher *et al.* (29).

purified by polyacrylamide gel electrophoresis. This leads to anomalously high values for the retention of competitor RNA on the filters and a consequent distortion of the K_2/K_1 ratios.

(d) Run self-competition experiments as calibration curves to select batches of ^{35}S-labelled TAR RNA with the expected binding behaviour.

(e) Analyse the binding curves obtained with mutant TAR RNAs to make certain that the total RNA bound gives values that are consistent with the total concentration of Tat protein present in the binding reactions.

Protocol 5. Dual-label filter binding assays

Equipment and reagents

- TK buffer: 50 mM Tris–HCl pH 7.4, 20 mM KCl
- Nitrocellulose filters, Millipore GS washed in TK buffer and kept wet
- Filtration manifold (Millipore)
- Tat or Rev protein, approx. 300 μg/ml (*Protocols 1* and *2*)
- ^{35}S-labelled RNA, 20 nM/reaction (*Protocol 3*)
- ^{3}H-labelled RNA, 0–100 nM/reaction (*Protocol 3*)

Method

1. Set-up 500 μl binding reactions on ice containing:
   - TK buffer
   - 20 nM ^{35}S-labelled TAR RNA (*Protocol 3*)
   - 100 nM ^{3}H-labelled competitor RNAs (*Protocol 3*)
   - 10 nM Tat protein (*Protocol 1*)
   - 1 mM DTT
   - 0.5 μg calf thymus DNA
   - 0.2 μg yeast tRNA
   - 20 U RNasin (Promega)
2. Incubate for 30 min on ice.
3. Count 50 μl of each reaction mix.
4. Filter the remaining 450 μl under reduced pressure through 25 mm prewashed individual nitrocellulose filters (0.2 μm pore size, Millipore GS filters).
5. Wash the filters once with 500 μl of TK buffer.
6. Dry the filters and count by liquid scintillation.

6. Principles of RNA recognition

Studies of the interactions of Tat and Rev with RNA have provided new insights into the chemistry of nucleic acid recognition. There are many examples of RNA binding proteins that recognize bases displayed in single-stranded loop and bulge structures (51). By contrast, both Tat and Rev share an arginine-rich basic domain. There is little base-specific recognition due to sequences from this region and the basic domains of the two proteins can be interchanged, or even replaced by random sequences containing one or more arginine residues (52). It seems likely that the basic domain inserts into the distorted RNA helices found at the recognition sites for Tat and Rev. However, this interaction appears to represent only one of several types of contacts made between RNA and the regulatory proteins. In addition to the base-specific contacts, both proteins make phosphate contacts on both strands of the flanking RNA duplexes. There is also a subtle interplay between the RNA structures and both the Tat and Rev proteins. Both proteins induce conformational changes in the RNA structure allowing functional groups on Watson–Crick base pairs to be exposed within the major groove of a distorted double-stranded RNA. This new principle of RNA recognition is likely to extend to many other protein–RNA interactions.

Acknowledgements

We thank our colleagues at LMB, Drs J. Butler, A. Klug, and G. Varani for helpful discussions, T. Smith, J. Fogg, and R. Grenfell for oligonucleotide synthesis, and the MRC AIDS-Directed Programme for support.

References

1. Garcia, J. A., Harrich, D., Soultanakis, E., Wu, F., Mitsuyasu, R., and Gaynor, R. B. (1989). *EMBO J.*, **8**, 765.
2. Jones, K., Kadonaga, J., Luciw, P., and Tjian, R. (1986). *Science*, **232**, 755.
3. Nabel, G. and Baltimore, D. A. (1987). *Nature*, **326**, 711.
4. Kim, S., Byrn, R., Groopman, J., and Baltimore, D. (1989). *J. Virol.*, **63**, 3708.
5. Arya, S. K., Guo, C., Josephs, S. F., and Wong-Staal, F. (1985). *Science*, **229**, 69.
6. Sodroski, J., Patarca, R., Rosen, C., Wong-Staal, F., and Haseltine, W. A. (1985). *Science*, **229**, 74.
7. Sodroski, J. G., Rosen, C. A., Wong-Staal, F., Salahuddin, S. Z., Popovic, M., Arya, S., *et al.* (1985). *Science*, **227**, 171.
8. Sodroski, J., Goh, W. C., Rosen, C. A., Dayton, A., Terwilliger, E., and Haseltine, W. A. (1986). *Nature*, **321**, 412.
9. Feinberg, M. B., Jarrett, R. F., Aldovini, A., Gallo, R. C., and Wong-Staal, F. (1986). *Cell*, **46**, 807.
10. Pomerantz, R. J., Trono, D., Feinberg, M. B., and Baltimore, D. (1990). *Cell*, **61**, 1271.
11. Cullen, B. R. (1986). *Cell*, **46**, 973.
12. Rosen, C. A., Sodroski, J. G., and Haseltine, W. A. (1985). *Cell*, **41**, 813.
13. Muesing, M. A., Smith, D. H., and Capon, D. J. (1987). *Cell*, **48**, 691.

14. Feng, S. and Holland, E. C. (1988). *Nature*, **334**, 165.
15. Malim, M. H., Hauber, J., Le, S.-Y., Maizel, J. V., and Cullen, B. R. (1989). *Nature*, **338**, 254.
16. Arrigo, S. J. and Chen, I. S. Y. (1991). *Genes Dev.*, **5**, 808.
17. Rosen, C. A., Terwilliger, E., Dayton, A. I., Sodrowski, J. G., and Haseltine, W. A. (1988). *Proc. Natl Acad. Sci. USA*, **85**, 2071.
18. Malim, M. H., Bohnlein, S., Hauber, J., and Cullen, B. R. (1989). *Cell*, **58**, 205.
19. Dayton, E. T., Powell, D. M., and Dayton, A. I. (1989). *Science*, **246**, 1625.
20. Daly, T. J., Cook, K. S., Gary, G. S., Maione, T. E., and Rusche, J. R. (1989). *Nature*, **342**, 816.
21. Heaphy, S., Dingwall, C., Ernberg, I., Gait, M. J., Green, S. M., Karn, J., *et al.* (1990). *Cell*, **60**, 685.
22. Malim, M. H., Tiley, L. S., McCarn, D. F., Rusche, J. R., Hauber, J., and Cullen, B. R. (1990). *Cell*, **60**, 675.
23. Zapp, M. L. and Green, M. R. (1989). *Nature*, **342**, 714.
24. Hadzopoulou-Cladaras, M., Felber, B. K., Cladaras, C., Athanassopoulos, A., Tse, A., and Pavlakis, G. N. (1989). *J. Virol.*, **63**, 1265.
25. Schwartz, S., Campbell, M., Nasioulas, G., Harrison, J., Felber, B. K., and Pavlakis, G. N. (1992). *J. Virol.*, **66**, 7176.
26. Schwartz, S., Felber, B. K., and Pavlakis, G. N. (1992). *J. Virol.*, **66**, 150.
27. Selby, M. J., Bain, E. S., Luciw, P., and Peterlin, B. M. (1989). *Genes Dev.*, **3**, 547.
28. Dingwall, C., Ernberg, I., Gait, M. J., Green, S. M., Heaphy, S., Karn, J., *et al.* (1989). *Proc. Natl Acad. Sci. USA*, **86**, 6925.
29. Churcher, M., Lamont, C., Dingwall, C., Green, S. M., Lowe, A. D., Butler, P. J. G., *et al.* (1993). *J. Mol. Biol.*, **230**, 90.
30. Dingwall, C., Ernberg, I., Gait, M. J., Green, S. M., Heaphy, S., Karn, J., *et al.* (1990). *EMBO J.*, **9**, 4145.
31. Delling, U., Reid, L. S., Barnett, R. W., Ma, M. Y.-X., Climie, S., Sumner-Smith, M., *et al.* (1992). *J. Virol.*, **66**, 3018.
32. Roy, S., Delling, U., Chen, C.-H., Rosen, C. A., and Sonenberg, N. (1990). *Genes Dev.*, **4**, 1365.
33. Sumner-Smith, M., Roy, S., Barnett, R., Reid, L. S., Kuperman, R., Delling, U., *et al.* (1991). *J. Virol.*, **65**, 5196.
34. Heaphy, S., Finch, J. T., Gait, M. J., Karn, J., and Singh, M. (1991). *Proc. Natl Acad. Sci. USA*, **88**, 7366.
35. Kjems, J., Brown, M., Chang, D. D., and Sharp, P. A. (1991). *Proc. Natl Acad. Sci. USA*, **88**, 683.
36. Malim, M. H. and Cullen, B. R. (1991). *Cell*, **65**, 241.
37. Kjems, J., Calnan, B. J., Frankel, A. D., and Sharp, P. A. (1992). *EMBO J.*, **11**, 1119.
38. Tiley, L. S., Malim, M. H., Tewary, H. K., Stockley, P. G., and Cullen, B. R. (1992). *Proc. Natl Acad. Sci. USA*, **89**, 758.
39. Bartel, D. P., Zapp, M. L., Green, M. R., and Szostak, J. W. (1991). *Cell*, **67**, 529.
40. Frankel, A. D., Bredt, D. S., and Pabo, C. O. (1988). *Science*, **240**, 70.
41. Marciniak, R. A., Garcia-Blanco, M. A., and Sharp, P. A. (1990). *Proc. Natl Acad. Sci. USA*, **87**, 3624.

42. Dingwall, C., Ernberg, I., Gait, M. J., Heaphy, S., Karn, J., and Skinner, M. A. (1991). In *Advances in molecular biology and targeted treatment for AIDS* (ed. A. Kumar), pp. 133–43. Plenum, New York.
43. Schoner, B. E., Belagaje, R. M., and Schoner, R. G. (1990). In *Methods in enzymology* (ed. D. V. Goeddel), Vol. 185, pp. 94–103. Academic Press, San Diego, Calif.
44. Nalin, C. M., Purcell, R. D., Antelman, D., Mueller, D., Tomchak, L., Wegrzynski, B., *et al.* (1990). *Proc. Natl Acad. Sci. USA*, **87**, 7593.
45. Wingfield, P. T., Stahl, S. J., Payton, M. A., Venkatesan, S., Misra, M., and Steven, A. J. (1991). *Biochemistry*, **30**, 7527.
46. Iwai, S., Pritchard, C., Mann, D. A., Karn, J., and Gait, M. J. (1992). *Nucleic Acids Res.*, **20**, 6465.
47. Milligan, J. F., Groebe, D. R., Witherell, G. W., and Uhlenbeck, O. C. (1987). *Nucleic Acids Res.*, **15**, 8783.
48. Gait, M. J., Pritchard, C., and Slim, G. (1991). *Oligonucleotides and analogues: a practical approach* (ed. F. Eckstein), pp. 25–48. IRL Press, Oxford.
49. Sulston, J. E., Mallet, F., Staden, R., Horsnell, T., and Coulson, A. (1988). *CABIOS*, **4**, 125.
50. Smith, J. M. and Thomas, D. J. (1990). *CABIOS*, **6**, 93.
51. Nagai, K. (1992). *Curr. Opin. Struct. Biol.*, **2**, 131.
52. Calnan, B. J., Biancalana, S., Hudson, D., and Frankel, A. D. (1991). *Genes Dev.*, **5**, 201.

10

Cellular and cell-free assays for Tat

M. F. LASPIA

1. Introduction

The human immunodeficiency virus 1 (HIV-1) encodes a novel regulatory protein known as Tat. Tat greatly increases the expression of genes linked to the long terminal repeat (LTR) of the virus. It does so by binding to a structured RNA element, called TAR RNA (see Chapter 9), located in the 5′ untranslated region of all HIV mRNAs (1–3). The mechanism of *trans*-activation by Tat appears complex, but the principal level at which Tat acts is transcriptional. Based on results obtained in a model system consisting of COS cells transiently expressing replicating plasmids, Peterlin and colleagues (4) were the first to obtain evidence that Tat stimulates transcriptional elongation. By contrast, Tat was found to increase both transcriptional initiation and the efficiency of transcriptional elongation in a recombinant adenovirus HeLa cell model system (5, 6). In this system, in the absence of Tat the level of HIV-1 transcription is low and the density of transcribing RNA polymerases declines as a function of the distance downstream of the promoter. Tat increases the number of RNA polymerases in the immediate vicinity of the promoter and also suppresses transcriptional polarity in downstream sequences leading to a large increase in transcription rates overall (*Figure 1*). A general *trans*-activator protein, the adenovirus E1A 13S protein, elevates HIV-1 promoter proximal transcription, but does not significantly stimulate elongation. Thus, Tat appears to stimulate both transcriptional initiation and elongation. Recently, a cell-free transcription system has been developed (7) that faithfully reproduces *in vitro* important aspects of *trans*-activation by Tat observed *in vivo* (8).

The HIV-1 Tat protein is 86 amino acids in length (HBX-2 isolate) and is encoded by two exons. The first exon of Tat (amino acids 1–72) is sufficient for *trans*-activation and a truncation of Tat after residue 58 is capable of partially stimulating LTR-directed gene expression in plasmid transfection assays. Mutational analysis of the first 58 amino acids of Tat suggests that it consists of four essential regions: the amino-terminus, a cysteine-rich region, a conserved region, and a basic region. The amino-terminus, the

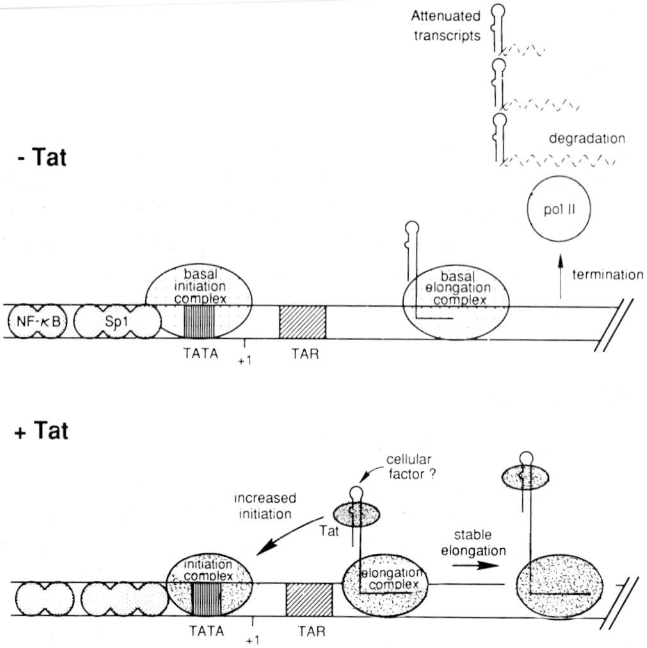

Figure 1. Model of *trans*-activation of HIV-1-directed transcription by Tat. In the absence of Tat, the level of transcriptional initiation at the HIV-1 promoter is low and transcriptional elongation is unstable such that only a small fraction of the initiating RNA polymerases transcribe to the end of the transcription unit. The binding of Tat to TAR RNA, in co-operation with a putative cellular factor(s), results in increased initiation of transcription by RNA polymerase II and also improves the efficiency of transcriptional elongation. Reprinted with permission of the publisher (16).

cysteine-rich region, and the conserved region constitute an essential activation domain, while the basic region is responsible for binding to TAR RNA.

A minimal TAR stem–loop is formed by nucleotides +14 to +44, relative to the site of transcriptional initiation. Mutations that disrupt base paring in the stem or change the sequence of the trinucleotide bulge or the hexanucleotide loop greatly reduce *trans*-activation by Tat *in vivo*. As discussed in Chapter 9, Tat is capable of binding to TAR RNA via contact with residues within and immediately above the bulge. It appears that the principal role of TAR RNA is to provide a promoter proximal binding site for Tat. A chimeric protein consisting of Tat fused to other RNA, or even DNA, binding domains is capable of partially *trans*-activating a modified promoter containing the appropriate RNA or DNA target sequence in place of TAR. On the other hand, while the apical loop of TAR RNA is dispensable for Tat binding *in vitro*, mutations in the loop greatly reduce *trans*-activation *in vivo* in plasmid transfection assays. This implies that a cellular protein(s) that interacts with the apical loop of TAR RNA may be important for *trans*-activation by Tat.

This chapter describes:

- cellular assays for Tat
- cell-free assays for Tat

Cellular assays usually measure the effect of Tat on HIV-1 gene expression in cultured cells. Typically, an LTR-directed reporter gene is introduced, by infection or transfection into cells that either do or do not express Tat. Cell-free assays for Tat either measure the ability of purified Tat to bind to TAR RNA (Chapter 9) or, as described here, measure the ability of purified Tat to *trans*-activate HIV transcription *in vitro*.

2. Cellular systems

2.1 Plasmid transfection assays

Trans-activation by Tat is usually measured in a transient expression assay with an HIV-1 LTR-directed reporter gene, such as the gene for chloramphenicol acetyltransferase (CAT). Cells are transfected with the reporter plasmid following treatment with calcium phosphate or DEAE–dextran (9). To study the effects of Tat, cells are co-transfected with a plasmid in which Tat expression is directed by a strong promoter such as the cytomegalovirus immediate early promoter or the SV40 early promoter (see Chapter 11 for methods of transfecting cells with reporter plasmids). Alternatively, a HeLa/*tat* cell line, which contains an integrated Tat expression plasmid and expresses Tat constitutively, may be used (10).

The drawbacks of transfection are:

- only a low percentage of cells that become competent to take up DNA (5–20%)
- the number of DNA copies per cell is not constant since the transfected DNA forms large concatemers
- the efficiency of this process is dependent on the cell type that is used

While calcium phosphate or DEAE–dextran-mediated transfection remain the most widely used methods to introduce DNA other methods exist such as electroporation, liposomal fusion, and microinjection. Expression levels may be increased by the use of plasmids that contain an SV40 origin of DNA replication and that undergo replication when transfected into COS cells, which express T antigen.

2.2 Viral vectors

The use of viral vector systems to study HIV gene expression offers several important advantages over plasmid transfection assays:

- cells do not have to be made competent

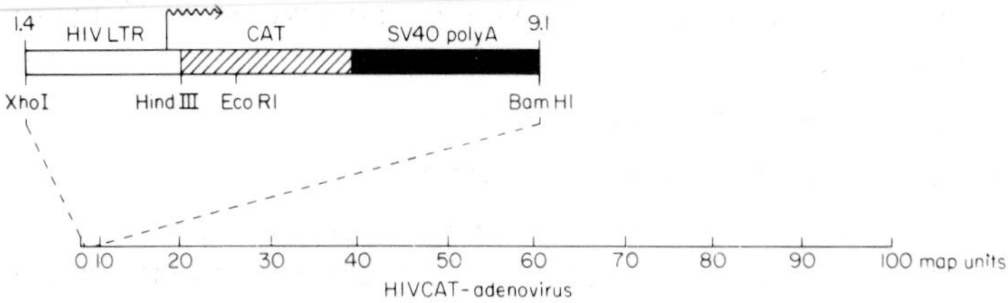

Figure 2. Schematic of the recombinant adenovirus. See text for a description.

- most of the cells become infected
- the number of copies introduced per cell is discrete compared to transfections

A recombinant adenovirus system has been particularly useful in analysing the mechanism of *trans*-activation by Tat. The recombinant adenovirus contains the HIV-1 LTR fused to the reporter gene CAT cloned into adenovirus 5 between 1.4 and 9.1 map unit in place of the E1 region (*Figure 2*). This results in the formation of a defective viral vector which can be used to introduce an HIV LTR-directed reporter gene fusion into human cells in a controlled and efficient manner. An important advantage of this system is the ability to infect cells at high multiplicity of infection (m.o.i.) which permits quantitation of low basal levels LTR-directed gene expression. The construction of the recombinant adenovirus has been described (11). *Protocols 1* and *2* describe the preparation of recombinant adenovirus stocks and the infection of HeLa or HeLa/*tat* cells.

Protocol 1. Preparation of recombinant adenovirus stocks in 293 cells[a]

Equipment and reagents

- 293 cells
- Recombinant adenovirus (HIV-1CATad)
- Dulbecco's modified medium (DMEM)
- Calf serum (CS)
- Gamma globulin-free calf serum (γG-F-CS)

Method

1. Grow 293 cells at 37 °C in monolayer cultures to approx. 75% confluency ($1–1.5 \times 10^7$ cells/10 cm dish) in DMEM supplemented with 10% CS.
2. Remove the medium from the cells and replace it with 4 ml of DMEM supplemented with 5% γG-F-CS containing the recombinant adenovirus at an m.o.i. of 10.
3. Incubate 45 min at 37 °C.
4. Add an additional 5 ml of DMEM/5% γG-F-CS to the plates and in-

cubate 24–48 h at 37°C. Cells should exhibit cytopathic effects (rounding and piling).

5. Freeze plates at −20°C, and thaw.
6. Remove cells and medium by pipetting.
7. Freeze–thaw the lysate three times to release the virus. Virus titres should be between 4×10^8 and 1.5×10^9 plaque-forming units/ml.
8. Store virus stocks at −20°C.

[a] Recombinant adenovirus are propagated in 293 cells which express the adenovirus E1A and E1B proteins.

Protocol 2. Infection with recombinant adenovirus

Equipment and reagents

- HeLa cells
- HeLa/*tat* cells (10)
- DMEM
- γG-F-CS
- Fetal bovine serum (FBS)
- Geneticin (Gibco)
- Phosphate-buffered saline (PBS)

Method

1. Grow HeLa/*tat* cells and parental HeLa cells to approx. 7.5×10^6 cells/10 cm dish in DMEM plus 10% fetal calf serum.
2. Remove the medium and replace with 5 ml of DMEM supplemented with 2% γG-F-CS containing recombinant adenovirus at an m.o.i. of 50 or 100.
3. Replace the medium 2 h post-infection with fresh medium containing 10 mM hydroxyurea and return the plates to the incubator.
4. Harvest the cells 24 h post-infection by scraping with a rubber policeman in 1 ml of ice-cold PBS.

If comparison to another *trans*-activator of HIV-1 gene expression is desired, such as the adenovirus E1A protein or phorbol-12-myristate-13-acetate (PMA), treat the cells with the inducer in parallel:

(a) E1A is supplied by co-infection at time zero with a phenotypically wild-type adenovirus (*dl*309) at an m.o.i. of 50. Hydroxyurea is added to medium two hours post-infection at a concentration of 10 mM to prevent replication of the recombinant adenovirus.

(b) For PMA induction PMA is added is added to the cells two hours post-infection at a concentration of 200 ng/ml.

2.3 Analysis of HIV-directed RNA levels

The effects of Tat on the accumulation of LTR-directed RNA levels may be accurately measured by RNase protection assay (*Protocol 3* and *Figure 3*):

(a) Cytoplasmic RNA is purified from recombinant adenovirus-infected cells and hybridized to a ^{32}P-labelled, antisense, RNA probe homologous to the HIV-1 leader region.

(b) Samples are treated with single-stranded RNA endonucleases, RNase A, and RNase T1.

(c) The protected species are resolved by electrophoresis under denaturing conditions.

In the absence of Tat there are two classes of correctly initiated RNA (*Figure 3*):

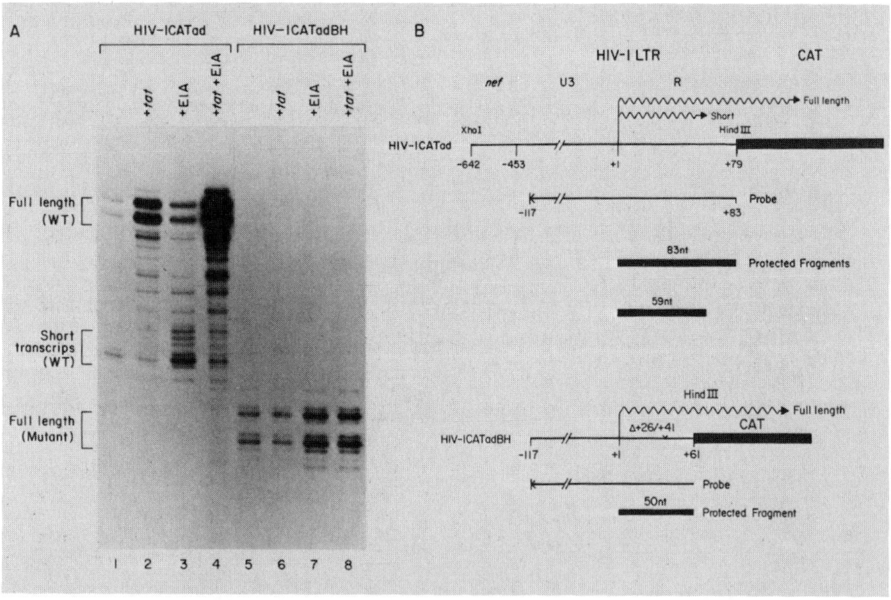

Figure 3. Analysis of HIV-1-directed cytoplasmic RNA levels by RNase protection assay. A. HeLa cells, HeLa cells expressing Tat (+*tat*), HeLa cells expressing E1A (+E1A), or HeLa cells expressing both Tat and E1A (+*tat* +E1A) were infected at an m.o.i. of 100 with a recombinant adenovirus bearing a wild-type copy of the HIV-1 LTR CAT gene (HIV-1CATad; *Figure 1*) or a recombinant adenovirus with a mutated TAR element (HIV-1CATadBH) and cytoplasmic RNA isolated 24 h post-infection. RNA was hybridized to homologous, anti-sense, radiolabelled RNA probes depicted in B. Protected probe fragments were resolved by electrophoresis in a urea–polyacrylamide gel. Protected fragments corresponding to full-length RNA and the prematurely terminated short transcripts are marked. Reprinted with permission of the publisher (16).

(a) Full-length, polyadenylated mRNA that protects an 83 nucleotide probe fragment.

(b) Prematurely terminated, non-polyadenylated, possibly processed mRNA that protects probe fragments of 55–59 nucleotides in length.

While Tat increases the accumulation of full-length RNA 20-fold, it does not increase the accumulation of the prematurely terminated RNA. By contrast, the adenovirus E1A 13S gene product, increases the accumulation of full-length RNA transcripts and the prematurely terminated transcripts to the same extent. Together, Tat and E1A increase LTR-directed RNA levels synergistically. *Trans*-activation by Tat but not E1A, and synergy between Tat and E1A, is dependent on TAR.

The fact that Tat increases the level of full-length transcripts without concomitantly increasing the level of the prematurely terminated transcripts (while E1A increases both) implies that Tat affects transcriptional elongation. That Tat increases the total amount of HIV-directed RNA suggests that it does not act solely to overcome premature termination, but that it also stimulates transcriptional initiation.

Protocol 3. RNase protection assay of LTR-directed cytoplasmic RNA levels

Equipment and reagents[a]

- TSM: 10 mM Tris pH 8.0, 150 mM NaCl, 2 mM $MgCl_2$
- 10% Nonidet P-40 (NP-40)
- TSES: 10 mM Tris pH 8.0, 150 mM NaCl, 5 mM EDTA, 1.0% SDS
- 100% ethanol, −20°C
- 70% ethanol, −20°C
- Redistilled phenol equilibrated with 0.25 M Tris pH 7.6
- Chloroform/isoamyl alcohol (24:1 v/v)
- Equilibrated phenol/chloroform–isoamyl alcohol (1:1 v/v)
- 1 mg/ml DNase I (RNase, protease-free; Worthington Biochemicals): resuspend in water
- DNase I buffer: 20 mM Hepes pH 7.5, 5 mM $MgCl_2$, 1 mM $CaCl_2$
- 10 mCi/ml [α-^{32}P]UTP (3000 Ci/mmol)
- SP6 RNA polymerase (Promega)
- pGEM 23 (plasmid containing HIV-1 sequence −117 (*Xba*I) to +83 (*Hin*dIII) (oriented to produce antisense RNA directed by the SP6 promoter)
- *Xba*I
- 3 M sodium acetate
- 10 mg/ml calf liver or *E. coli* tRNA (Boehringer Mannheim): dissolve in H_2O; phenol extract several times, precipitate, and resuspend in H_2O
- 10 × TBE: 0.89 M Tris, 0.89 M boric acid, 20 mM EDTA
- Loading dye: 80% formamide, 0.5 × TBE, 0.1% bromophenol blue, 0.1% xylene cyanol FF
- Elution buffer: 10 mM Tris pH 8.0, 1 mM EDTA, 0.5% SDS, 20 μg calf liver tRNA
- ^{32}P-labelled low molecular weight DNA markers
- Hybridization buffer: 40 mM Pipes pH 6.7, 400 mM NaCl, 1 mM EDTA, 80% formamide
- Digestion buffer: 10 mM Tris pH 7.5, 300 mM NaCl, 5 mM EDTA
- 10 mg/ml RNase A (Pharmacia): dissolve 10 mM Tris pH 7.5, 15 mM NaCl, and heat to 100°C for 15 min
- TE: 10 mM Tris pH 7.6, 1 mM EDTA
- 10 mg/ml RNase T1 in TE (Pharmacia)
- 10% SDS
- 10 mg/ml proteinase K (Boehringer Mannheim): dissolve in 0.5% SDS and heat to 37°C for 30 min

Protocol 3. *Continued*

Method

1. Resuspend the PBS washed cell pellet from one 10 cm dish of recombinant adenovirus-infected cells in 100 μl of ice-cold TSM.
2. Add 10 μl of 10% NP-40.
3. Mix and place on ice 10 min.
4. Centrifuge 30 sec in a microcentrifuge, and transfer the supernatant to a fresh tube containing an equal volume of TSES at room temperature. Extract with an equal volume of equilibrated phenol, then with an equal volume of phenol/chloroform, and finally extract with an equal volume of chloroform. Precipitate the RNA with 2.5 vol. of ice-cold 100% ethanol. Place in a dry ice–ethanol bath 15 min. Spin 15 min in a microcentrifuge at 4°C. Carefully remove the supernatant and add 1 ml of 70% ethanol. Re-spin. Remove the supernatant and air dry the pellet. The expected yield is approx. 50–100 μg of cytoplasmic RNA/10 cm plate.
5. Prepare an antisense radioactive RNA probe homologous to HIV mRNA as follows. Linearize pGEM23 with *Xba*I. Purify the DNA by phenol/chloroform extraction and ethanol precipitation. In a 40 μl volume transcribe 1 μg of pGEM 23 with 1.5 μl SP6 RNA polymerase in the buffer recommended by the manufacturer in the presence of 150 μCi of [α-^{32}P]UTP (3000 Ci/mmol) for 1 h at 40°C. Add 100 μl DNase I buffer. Add 10 μl DNase I and incubate for 15 min at 37°C. Extract with 1 vol. phenol/chloroform, then, with 1 vol. chloroform. Add sodium acetate to a final concentration of 0.3 M. Add 2.5 vol. of ice-cold ethanol and precipitate the RNA as described in step 4.
6. Resuspend the RNA in loading dye. Electrophorese probe in an 8% polyacrylamide, 7 M urea, 0.5 × TBE sequencing gel at 1600 V, constant voltage, until the bromophenol blue has migrated to the bottom of the gel. Separate the glass plates and expose the gel covered with Saran Wrap to X-ray film for 20 sec to locate the RNA probe using molecular weight markers as a guide. Cut out the gel slice containing the RNA probe and place in a microcentrifuge tube. Add 400 μl elution buffer. Place tube on a rotator and elute the probe overnight. Purify the probe by phenol/chloroform extraction and ethanol precipitation as in step 4.
7. Dry 5 μg of cytoplasmic RNA and 5 × 10^5 c.p.m. of the RNA probe in a lyophilizer.
8. Resuspend the RNA in 30 μl of hybridization buffer. Heat the samples to 85°C for 5 min. Incubate the samples at 42°C overnight.
9. Add 300 μl digestion buffer containing 10 μg/ml RNase A and 10 μg/ml

RNase T1 and incubate 1 h at room temperature. Add 10 μl of 10% SDS and 5 μl of 10 mg/ml proteinase K to each sample and incubate for 15 min at 37°C. Extract with phenol/chloroform, add 20 μg of calf liver tRNA carrier. Precipitate the nucleic acid as described in step 4.

10. Resuspend the samples in loading buffer, denature at 95°C for 3 min, and analyse by electrophoresis in an 8% polyacrylamide, 7 M urea, 0.5 × TBE sequencing gel. Protected species are visualized by autoradiography with an intensifying screen.

[a] Solutions should be prepared from autoclaved stocks. Do not autoclave solutions containing nucleic acid, DNase I, RNase A, RNase T1, proteinase K, nucleotides, DTT, or formamide.

2.4 Analysis of LTR-directed transcription rates

The effects of Tat on HIV-1 transcription may be examined directly by measuring the rates of HIV-directed transcription using nuclear run-on assays (*Protocol 4, Figure 4*).

(a) Nuclei are isolated from recombinant adenovirus-infected HeLa cells.
(b) The nuclei are incubated in a reaction buffer with unlabelled nucleotides and [α-^{32}P]UTP for short intervals, which permits elongation of previously initiated RNA polymerase II complexes and pulse labels nascent transcripts.
(c) Transcripts are purified, fragmented with sodium hydroxide, and hybridized to short, single-stranded, DNA probes corresponding to various portions of the LTR–reporter gene fusion.
(d) Filters are treated with RNase A and subjected to autoradiography, or the amount of radioactivity hybridizing to each probe quantified with a phosphorimager.

In nuclear run-on assays:

(a) The amount of radioactivity hybridizing to each DNA probe, normalized for the uridine content of the fragment and standardized to a constitutively expressed gene, is a measure of the relative amount of transcription occurring in that corresponding portion of the gene.
(b) With short pulse labelling times, the relative transcription rates in different portions of the gene provides an estimate of distribution of elongating RNA polymerases.

In the absence of Tat, transcription in the promoter proximal region is low and decreases sharply in the reporter gene (*Figure 5*). This indicates that there is a polarity of HIV-1 transcription such that only a fraction of the initiating RNA polymerases transcribe into promoter distal portions of the gene. Tat produces a large increase in transcription rates over the first 83

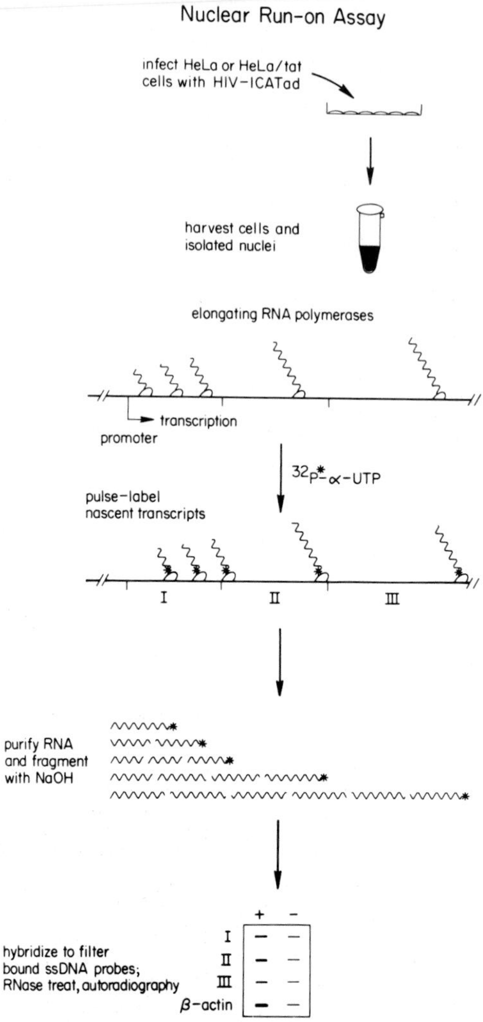

Figure 4. Schematic of nuclear run-on transcription assay.

nucleotides of the HIV leader region and increases transcription over the first 24 nucleotides of the leader as well suggesting that it elevates transcriptional initiation. Tat also suppresses transcriptional polarity indicating that it stabilizes elongation. E1A, like Tat, increases promoter proximal transcription, but it does not significantly suppress transcriptional polarity. Thus, Tat appears to have two effects on HIV-1 transcription. One effect is to increase promoter proximal transcription by increasing transcriptional initiation by RNA polymerase II. In addition, Tat suppresses polarity by improving the efficiency of transcriptional elongation.

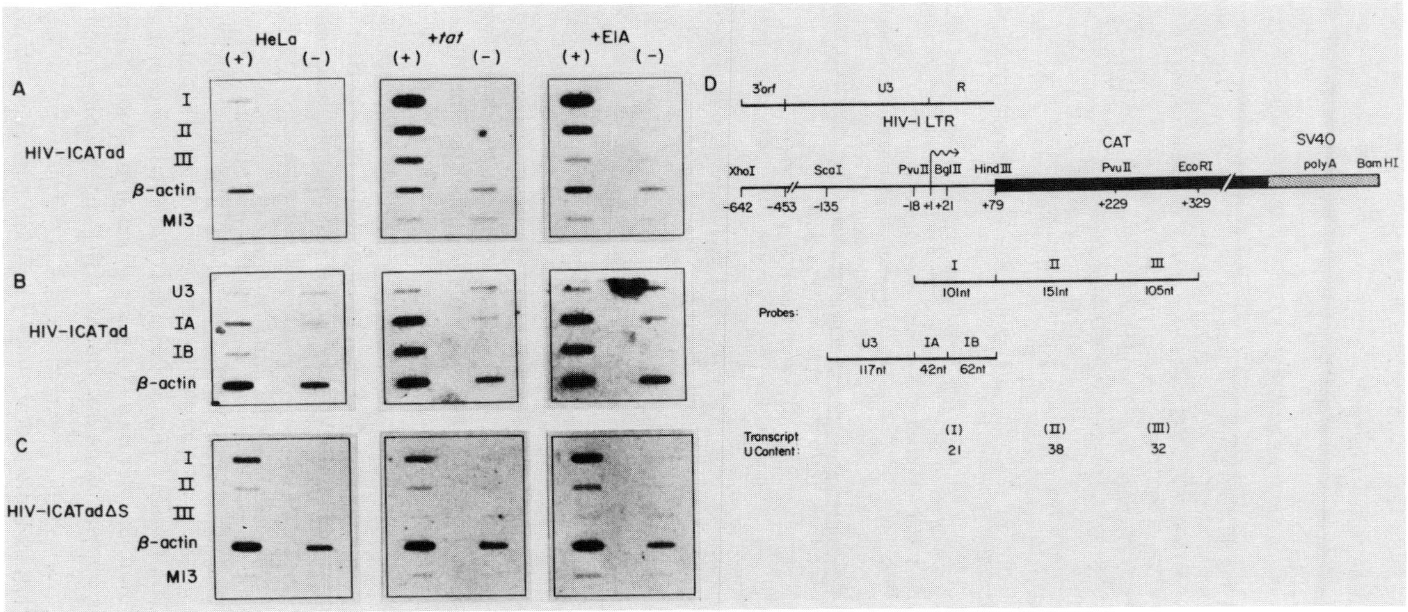

Figure 5. Analysis of HIV-1 transcription by nuclear run-on assay. HeLa cells, HeLa cells expressing Tat (+*tat*), or HeLa cells expressing E1A (+E1A) were infected at an m.o.i of 100 with a recombinant adenovirus bearing a wild-type copy of the HIV-1 LTR CAT gene (HIV-1CATad; A, B) or a recombinant adenovirus containing a mutated TAR element (HIV-1CATΔS; C). HIV-1CATΔS contains a four nucleotide deletion at the *Sac*I site within TAR. D. Schematic of the DNA probes used in the nuclear run-on assay. Transcription of β-actin is measured as a control for RNA recovery. Adapted with permission of the publisher (5).

Protocol 4. Nuclear run-on assay[a]

Equipment and reagents[b]

- PBS
- Lysis buffer: 10 mM Tris pH 7.4, 10 mM NaCl, 3 mM $MgCl_2$, 0.5% NP-40
- Glycerol storage buffer: 50 mM Tris pH 8.3, 5 mM $MgCl_2$, 0.1 mM EDTA, 40% glycerol
- 2 ml screw-cap tubes
- 15 ml polypropylene tube
- Liquid nitrogen storage
- 2 × reaction buffer: 10 mM Tris pH 8.0, 5 mM $MgCl_2$, 0.3 M KCl
- 100 mM DTT
- 10 mM stocks of ATP, CTP, GTP (Pharmacia): dissolve in water, adjust to pH 7.0
- 10 mCi/ml [α-^{32}P]UTP (3000 Ci/mmol)
- HSB: 10 mM Tris pH 7.4, 500 mM NaCl, 50 mM $MgCl_2$, 2 mM $CaCl_2$
- 1 mg/ml DNase 1[c]
- α-amanitin (Sigma Chemical Co.)
- SDS–Tris: 500 mM Tris pH 7.4, 125 mM EDTA, 5% SDS
- 10 mg/ml proteinase K[c]
- GF/A filters (Whatman)
- Filtration apparatus (Millipore)
- 10% trichloroacetic acid, 60 mM sodium pyrophosphate, 0°C
- 5% trichloroacetic acid, 30 mM sodium pyrophosphate, 0°C
- 10% SDS
- Elution buffer: 10 mM Tris pH 7.5, 5 mM EDTA, 1.0% SDS
- 25 mg/ml DNase I in DNase I buffer[c]
- 30 ml glass Corex tubes
- 1 M NaOH
- 1 M Hepes (free acid)
- 3 M sodium acetate
- 100% ethanol, −20°C
- TE[c]
- TES: 10 mM Tris pH 7.4, 10 mM EDTA, 0.2% SDS
- Single-strand antisense and sense DNA probes corresponding to the HIV-1-directed reporter gene cloned into M13
- Slot blot apparatus (Schleicher and Schuell)
- 0.45 μM nitrocellulose sheets cut to fit the slot blot apparatus
- Vacuum drying oven
- Seal-a-meal bags and sealer
- 20 × SSC: 3.0 M NaCl, 0.3 M sodium citrate, pH 7.0
- TES, 0.6 M NaCl: 10 mM Tris pH 7.4, 0.6 M NaCl, 10 mM EDTA, 0.2% SDS
- 10 mg/ml RNase A[c]

Method

1. Resuspend the PBS washed cell pellet from five 10 cm dishes in 4 ml of ice-cold lysis buffer. Vortex on low speed to loosen the pellet. Add the lysis buffer while vortexing and continue to vortex for 10 sec more. Incubate 5 min on ice, and centrifuge at 500 *g* for 5 min at 4°C.
2. Remove the supernatant (contains cytoplasmic RNA) and carefully aspirate any remaining supernatant. Resuspend the nuclei pellet in 200 μl glycerol storage buffer. Vortex first, then completely resuspend the nuclei by gently pipetting with a P-1000 Pipetman. Freeze the nuclei in 2 ml screw-cap tubes in liquid nitrogen.
3. Thaw frozen nuclei on ice and dispense 200 μl to a 15 ml polypropylene tube. Add 200 μl of 2 × reaction buffer containing 1 mM DTT, 10 mM ATP, 10 mM GTP, 10 mM CTP, and 200 μCi of [α-^{32}P]UTP. Incubate at 30°C for 2–10 min. Shorter pulse labelling times give a more representative measurement of the distribution of RNA polymerases along the template.
4. Stop the reaction by the addition of 600 μl of HSB containing 40 μg/

ml RNase-free DNase I and 200 μg/ml α-amanitin. Pipette with P-1000 pipette (carefully) to disperse nuclei. Incubate for 5 min at 37°C.

5. Add 200 μl of SDS–Tris and 20 μl of 10 mg/ml proteinase K. Vortex. Incubate for 30 min at 42°C. Vortex during the incubation to obtain a uniform solution.
6. Extract the solution with 1 ml phenol/chloroform.
7. Add 2 ml water, 3 ml 10% TCA, 60 mM sodium pyrophosphate, and 10 μg calf liver tRNA to the aqueous phase and incubate 30 min on ice.
8. Filter the TCA precipitation through GF/A filters, and wash three times with 5% TCA, 30 mM sodium pyrophosphate.
9. Place filters in glass scintillation vials. Add 1.5 ml 25 μg/ml DNase I in DNase I buffer. Incubate 30 min at 37°C. Add 45 μl 0.5 M EDTA and 68 μl of 20% SDS, and heat to 65°C for 10 min to elute the RNA.
10. Remove the solution containing the eluted RNA. Add 1.5 ml elution buffer to each filter. Heat to 65°C for 10 min.
11. Combine the second eluate with the first. Add 9 μl of 10 mg/ml proteinase K, and incubate 30 min at 37°C. Extract with phenol, then phenol/chloroform, and finally extract with chloroform.
12. Add 0.75 ml of 1 M NaOH to the solution of RNA in a glass Corex tube. Incubate on ice 15 min. The purpose of this step is to digest the RNA into approximately 150–200 nucleotide pieces to prevent over-representation of promoter proximal transcription rates. Then neutralize with 1.5 ml of 1 M Hepes (free acid).
13. Add 0.1 vol. 3 M sodium acetate and 2.5 vol. of ethanol. Precipitate the RNA on dry ice. Recover the RNA by spinning for 20 min at 17 000 *g*. Dry the pellet in a lyophilizer, taking care not to over dry. Resuspend the nucleic acid in 1 ml of TE. The expected yield is 1–5 × 10^6 c.p.m./ml.
14. Prepare nitrocellulose filters containing single-stranded, sense and antisense, M13 DNA probes as follows. Assemble a slot blot filtration apparatus with 0.45 μm nitrocellulose pre-soaked in 6 × SSC, fill wells with 6 × SSC, and apply vacuum. Add 5 μg single-strand M13 DNA (purified by phenol extraction and ethanol precipitation) in 250 μl of 6 × SSC to the wells and apply vacuum. Wash the bound samples by filling the wells with 6 × SSC and filtering. Air dry the filters. Bake for 2 h at 80°C in a vacuum drying oven.
15. Mix 1 ml the purified, labelled RNA in TES with 1 ml TES, 0.6 M NaCl and add to nitrocellulose strips and containing bound single-strand DNA probes in seal-a-meal bags. (It is not necessary to normalize counts before adding the RNA to the filters, rather it is better to measure the amount of radioactivity that hybridizes to a human β-actin cDNA probe as internal control for recovery.)

Protocol 4. *Continued*

16. Hybridize at 65°C for 36 h. Remove filters. Wash filters in 250 ml of 2 × SSC. Repeat wash with fresh SSC.
17. Incubate the filters in 50 ml of 2 × SSC containing 10 μg/ml RNase A for 30 min at 37°C. Wash the filters in 50 ml of 2 × SSC for 30 min at 37°C. Air dry the filters.
18. Expose the filters to X-ray film with an intensifying screen from overnight to four days at −70°C.

[a] This protocol is a modification of the nuclear run-on assay developed by Greenberg and Ziff (12).
[b] See *Protocol 3*; footnote [a].
[c] See *Protocol 3, Equipment and reagents.*

3. Cell-free transcription assays for Tat

A more detailed understanding of the mechanism of the action of Tat and the identification of cellular factor(s) that may play a role in *trans*-activation is most likely to emerge from a cell-free analysis. An *in vitro* transcription system has been developed consisting of a nuclear extract prepared from HeLa cells, purified bacterially-expressed Tat protein, and a linearized HIV-1 DNA template (7). In this system, Tat produces a large increase in transcription from the wild-type HIV-1 promoter but not from a control HIV-1 promoter with a mutated TAR element or from a heterologous promoter (*Figure 6*) (8). Transcription is stimulated by Tat at concentrations ranging from 125–1000 ng per 20 μl reaction. In addition to the full-length run-off RNA, there is a heterogeneous class of short, Tat responsive, HIV RNA, approximately 83 nucleotides long, whose level decreases approximately threefold in the presence of Tat. This short class of RNA is likely to result from premature transcription termination and processing in the *in vitro* system.

RNase protection analysis of LTR-directed RNA accumulation following hybridization to non-radioactive RNA probes corresponding to the HIV-1 leader and the CAT gene indicates that Tat does not affect the accumulation of promoter proximal RNA, but that it does produce a large stimulation in promoter distal RNA accumulation (7, 8). In addition, a direct examination of HIV-directed transcription rates confirms that transcriptional elongation is also subject to polarity *in vitro* (8). While Tat does not significantly increase the number of promoter proximal transcription complexes, it increases the number of promoter distal complexes 15-fold. Together, these results indicate that the principal effect of Tat *in vitro* is to stabilize elongation. The inability to detect an effect of Tat on initiation may be a consequence of high basal levels of HIV transcription or inefficient reinitiation.

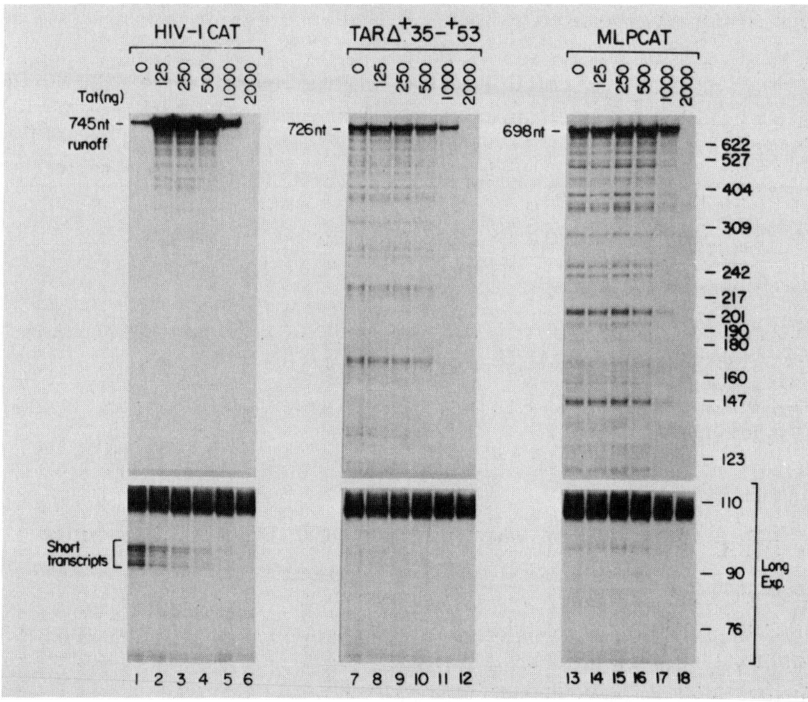

Figure 6. Stimulation of HIV-1 transcription *in vitro* by purified Tat and a HeLa cell nuclear extract. Transcription assays contained 250 ng pU3R111 CAT, linearized +745 nucleotides downstream of the transcription start site, purified Tat in the amounts indicated, and 10 μl HeLa cell nuclear extract 10 μl. Transcription was initiated at time zero by the combination the nuclear extract, DNA template, and Tat in reaction buffer. After 30 min of pre-transcription, transcription was labelled with 10 μCi [α-^{32}P]UTP for 30 min and then the reactions stopped. Transcripts were purified and resolved by electrophoresis on urea–polyacrylamide gels. The positions of the full-length 745 nucleotide run-off transcript as well as the short prematurely terminated transcripts are indicated. The broad 110 nucleotide radioactive band represents endogenous labelled RNA and is not template-dependent. Adapted with permission of the publisher (8).

(a) The magnitude of *trans*-activation is greatest at later times during the transcription reaction.

(b) Transcription is radioactively labelled during the final 30 min of a 60 min reaction.

(c) The increased stimulation by Tat accompanying 'pre-transcription' is due to a reduction in the efficiency of basal transcriptional elongation rather than an increase in transcription levels in the presence of Tat. This implies that 'pre-transcription' causes the accumulation of transcription complexes that require Tat for efficient elongation (8).

A potential explanation for these observations is that Tat acts to overcome inefficient elongation caused by a general cellular inhibitor(s) of transcriptional elongation.

Protocol 6 give the conditions for cell-free transcription. Bacterially-expressed Tat for cell-free transcription can be purified according to Chapter 9, or alternatively the following purification (*Protocol 5*) may be utilized (13).

Protocol 5. Purification of Tat

Equipment and reagents

- pET-Tat/BL21(DE3) *lys*S
- Isopropylthio-β-D-galactoside (IPTG)
- 10% Polymin P, pH 7.9 (Sigma)
- Ammonium sulfate
- 6 M guanidine–HCl, 0.5 M DTT
- C18 HPLC column (10 ml bed volume, 15 μm, 300 Å pore size)
- Acetonitrile
- Trifluoroacetic acid (TFA)

Method

1. Induce Tat expression in *E. coli* strain BL21 (DE3) *lys*S harbouring a pET-Tat expression vector with IPTG.
2. Precipitate the nucleic acids and Tat from *E. coli* crude lysates by adding Polymin P to the lysate dropwise until a concentration of 0.5% (v/v) is reached. Recover precipitate by centrifugation at 12 000 *g* for 15 min. Extract it three times with 50 ml of 10% ammonium sulfate in 50 mM Tris pH 8.0, 2 mM EDTA (use a dounce homogenizer to disperse the pellet). Combine the extracts and increase the ammonium sulfate concentration to 40% to precipitate Tat. Recover the precipitated protein by centrifugation. Resuspend in 10 ml of 6 M guanidine–HCl, 0.5 M DTT.
3. Apply the protein in 0.1% TFA to a C18 reverse phase high-pressure liquid chromatography column (10 ml bed volume, 15 μm, 300 Å pore size). Elute protein with a gradient of 0–60% acetonitrile, 0.1% TFA (approx. 200 ml). Tat will elute from the column at approx. 38% acetonitrile. Lyophilize fractions and resuspend in sterile water.

Protocol 6. Cell-free transcriptions assay

Equipment and reagents[a,b,c]

- Nuclear extract (14, 15)
- Buffer D: 20 mM Hepes pH 7.9, 100 mM KCl, 20% glycerol, 0.5 mM DTT, 0.5 mM PMSF, 0.2 mM EDTA
- Purified Tat (see *Protocol 5* or Chapter 9)
- Dilution buffer: 10 mM Tris pH 7.9, 50 mM KCl, 1 mM EDTA, 5 mM DTT
- pU3RIII
- *Sca*I
- 1 mg/ml poly(dI–dC) · poly(dI–dC) in TE, 15 mM NaCl (Pharmacia)
- 500 mM KCl
- 20 mM DTT
- 400 mM creatine phosphate

- 10 mCi/ml [α-^{32}P]UTP (3000 Ci/mmol)
- Stop buffer: 10 mM Tris pH 7.5, 300 mM NaCl, 5 mM EDTA, 0.5% SDS
- 10 mg/ml proteinase K[b]
- 10 mg/ml calf liver tRNA[b]
- Nucleotide mix: 18 mM ATP, 18 mM CTP, 18 mM GTP, 1.2 mM UTP in water pH 7.0[c]
- Loading dye[b]

Method

1. Thaw nuclear extracts prepared as described by Dignam *et al.* (14, 15) on ice. Extracts should have a protein concentration of 5–10 mg/ml and have been previously dialysed against buffer D.
2. Linearize HIV-1 DNA template at a restriction site approx. 700 nucleotides from the start of transcription (for pU3RIII, *Sca*I is 745 nucleotides from the transcription start site). Purify the DNA by phenol/chloroform extraction and ethanol precipitation. Resuspend the DNA in water at a concentration of 0.5 mg/ml.
3. Assemble the following 10 μl reaction mixture in microcentrifuge tubes:

• linearized template DNA	0.5 μl
• 1 mg/ml poly(dI–dC) · poly(dI–dC)	0.25 μl
• 500 mM KCl	1.0 μl
• 20 mM DTT	0.75 μl
• nucleotide mix	0.66 μl
• 400 mM creatine phosphate	0.2 μl
• sterile water	6.39 μl
• Tat (130–500 ng) or dilution buffer	0.25 μl

   Spin 10 sec in microcentrifuge.
4. Start transcription reactions by the addition of 10 μl of nuclear extract. Pipette up and down with a Pipetman to mix. Place tubes at 30°C.
5. Add 1 μl of 1 mCi/ml [α-^{32}P]UTP (3000 Ci/mmol) after 30 min of pre-transcription.
6. Incubate 30 additional min of incubation at 30°C.
7. Terminate reactions by the addition of 100 ml of stop buffer. Add 5 μl of 10 mg/ml proteinase K. Add 1 μl 10 mg/ml calf liver tRNA. Incubate the reactions 10 min at room temperature. Add 300 μl 10 mM Tris pH 7.5, 300 mM NaCl, 5 mM EDTA. Extract with phenol/chloroform. Add 1 ml of ice-cold 100% ethanol and precipitate the RNA. Resuspend the RNA pellet in 3 μl H_2O and add 3 μl of loading dye.
8. Denature samples at 95°C for 3 min. Resolve transcription products by electrophoresis in a 6% polyacrylamide 7 M urea sequencing gel. Visualize transcripts by autoradiography with an intensifying screen.

[a] See *Protocol 3*, footnote [a].
[b] See *Protocol 3, Equipment and reagents.*
[c] See *Protocol 4, Equipment and reagents.*

Transcription in both the presence and absence of Tat should be optimized with each preparation of nuclear extract for the following:

(a) Concentration of template and non-specific competitor poly(dI–dC) · poly(dI–dC), KCl, $MgCl_2$, and the amount of nuclear extract.

(b) In our hands optimal conditions are: 10 mM Hepes pH 7.9, 75 mM KCl, 7.5 mM $MgCl_2$, 1 mM DTT, 10% glycerol, 600 μM ATP, 600 μM GTP, 600 μM CTP, 40 mM UTP, 4 mM creatine phosphate, 250 ng linearized template DNA, 250 ng poly(dI–dC):poly(dI–dC), 75–100 mg nuclear extract, 250 ng of purified Tat.

4. Conclusions

Tat plays a pivotal role in the life cycle of HIV-1 by regulating HIV-1 mRNA expression. It is essential for a productive infection. The precise mechanism by which Tat elevates HIV transcription is not clearly understood, particularly the role that cellular cofactors might play. Tat is unique in so far as it binds to an RNA target (TAR) and it elevates both transcriptional initiation and elongation. Accordingly, approaches which interfere with the action of Tat and TAR may have therapeutic potential. Hopefully, a detailed understanding of the mechanism of *trans*-activation may help in the development of new strategies or refine ones that interfere with HIV-1 gene expression and mitigate the pathobiological consequences of infection.

References

1. Cullen, B. R. (1986). *Cell*, **46**, 973.
2. Sharp, P. A. and Marciniak, R. A. (1989). *Cell*, **59**, 229.
3. Karn, J. and Graeble, M. A. (1992). *Trends Genet.*, **8**, 365.
4. Kao, S.-Y., Calman, A. F., Luciw, P. A., and Peterlin, B. M. (1987). *Nature*, **330**, 489.
5. Laspia, M. F., Rice, A. P., and Mathews, M. B. (1989). *Cell*, **59**, 283.
6. Laspia, M. F., Rice, A. P., and Mathews, M. B. (1990). *Genes Dev.*, **4**, 2397.
7. Marciniak, R. A., Calnan, B. J., Frankel, A. D., and Sharp, P. A. (1990). *Cell*, **63**, 791.
8. Laspia, M. F., Wendel, P. A., and Mathews, M. B. (1993). *J. Mol. Biol.*, **232**, 732.
9. Cullen, B. R. (1987). In *Methods in enzymology* (ed. R. Wu), Vol. 152, pp. 684–704. Academic Press, London.
10. Valerie, K., Delers, A., Bruck, C., Thiriart, C., Rosenberg, H., Debouck, C., *et al.* (1988). *Nature*, **333**, 78.
11. Rice, A. P. and Mathews, M. B. (1988). *Proc. Natl Acad. Sci. USA*, **85**, 4200.
12. Greenberg, M. E. and Ziff, E. B. (1984). *Nature*, **311**, 433.
13. Frankel, A. D. and Pabo, C. O. (1988). *Cell*, **55**, 1189.

14. Dignam, J. D., Lebovitz, R. M., and Roeder, R. G. (1983). *Nucleic Acids Res.*, **11**, 1475.
15. Dignam, J. D., Martin, P. L., Shastry, B. S., and Roeder, R. G. (1983). In *Methods in enzymology* (ed. R. Wu), Vol. 101, pp. 582–98. Academic Press, London.
16. Laspia, M. F., Gunnery, S., Kessler, M., Rice, A. P., and Mathews, M. B. (1991). In *Advances in Molecular Biology and Target Treatment for AIDS*. (ed. A. Kumar), pp. 93–105. Plenum Press, New York.

11

Cellular assays for Rev

M. H. MALIM

1. Introduction

The human immunodeficiency virus type-1 (HIV-1) Rev *trans*-activator is a nuclear phosphoprotein absolutely essential for virus replication. It is expressed from fully spliced (~ 2 kb) viral mRNAs and is required for the expression of viral proteins, e.g. Gag, Pol, and Env, from the unspliced (~ 9 kb) and singly spliced (~ 4 kb) mRNAs. The *cis*-acting target for Rev, the Rev-response element (RRE), is an elaborate RNA stem and loop structure that is located within the *env* gene and is therefore present in all 9 kb and 4 kb transcripts. Although Rev has been the focus of considerable research effort in recent years, its precise mechanism of action still remains elusive; specifically, models invoking the inhibition of splicing, the activation of nucleocytoplasmic RNA transport, the stabilization of RNA, and the induction of translation in the cytoplasm have all been proposed (reviewed in refs 1 and 2).

In this chapter a variety of *in vivo* assays that may be used for studying this fascinating post-transcriptional modulator of viral gene expression are described. The experimental approaches are broadly classified as:

- assays that can be used for looking at Rev function in general
- assays for the identification of the important components of Rev, the RRE, and the responsive mRNA itself
- strategies for the detailed analysis of a single aspect of Rev (or RRE) function, for example following mutagenesis

None of the methods described here make use of infectious material or full-length proviral DNA clones; as a result, they do *not* require a specialized isolation facility and may be performed under standard laboratory conditions.

2. Rev and RRE function—experimental approaches

All of the assays described in this chapter rely upon the use of eukaryotic expression vectors and their introduction, by transient transfection, into tissue culture cell lines. Other than the 'expression cassettes' themselves, the 'backbones' of the vectors will not be discussed in any great detail. Suffice to say, however, that the use of high efficiency promoters, for example the Rous sarcoma virus (RSV) long terminal repeat (LTR) or the human cytomegalovirus major immediate early (CMV-IE) promoter, will be desirable in most cases. Another consideration is whether, or not, to include the simian virus 40 (SV40) origin of DNA replication in the expression vectors. Because vectors that carry this sequence have the ability to replicate to a high copy number once introduced into the SV40 T antigen-expressing African green monkey kidney cell line COS (3), their use results in particularly high levels of expression and is often advantageous.

In common with the study of many *trans*-acting proteins (or factors) and their *cis*-acting targets (e.g. DNA, RNA, or protein), experiments regarding Rev function *in vivo* are most easily performed by co-transfection. One vector expresses the Rev protein whereas the second vector contains the RRE and a suitable reporter gene. An obvious advantage of this type of approach is that the vectors can be individually altered, or mutated.

There are a number of options to be considered when choosing a reporter system for Rev function:

- viral genes as reporters
- reporter genes without introns
- reporter genes with introns

Three of the more commonly used alternatives are illustrated in *Figure 1*.

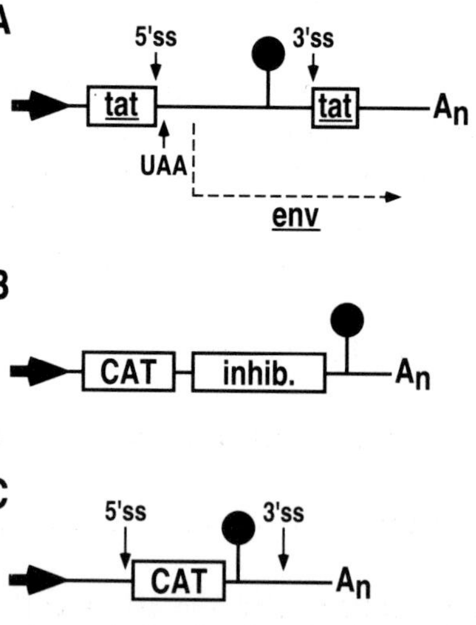

Figure 1. Responsive RNAs for analysing Rev function in transfected cells. Solid arrows, efficient promoter sequences for driving transcription; open boxes, reporter genes and inhibitory sequences (B only); solid circles, RRE; ss, splice site; A_n, polyadenylated 3′ terminus. Refer to the text for detailed explanations.

Since the precise details of the vector backbones have little impact upon their utility as reporters, only the expression cassettes themselves will be discussed here.

2.1 Transfection of cells

There are many methods, and variations thereof, for the transient introduction of expression plasmids into cultured cell lines. Basic procedures for DEAE–dextran and calcium phosphate-mediated transfection are described in *Protocols 1* and *2*. *Protocols 3* and *4* are two methods adapted for lymphoid cells, one is an optimized DEAE–dextran protocol and one involves electroporation.

Protocol 1. Transient transfection using DEAE–dextran[a]

Equipment and reagents

- Plasmid DNAs at 0.25 μg/μl and in sterile water
- 1 × phosphate-buffered saline (PBS, calcium-free, magnesium-free)
- Iscove's modified Dulbecco's medium (IMDM) supplemented with 10% fetal bovine serum, 1% fungizone, 0.1% gentamycin sulfate
- DEAE–dextran (Pharmacia Biotech Inc.) at 20 mg/ml in 1 × PBS and sterilized by autoclaving (this is a 40 × stock)
- Chloroquine (Sigma Chemical Co.) at 52 mg/ml in 1 × PBS, made fresh and filtered through a 0.45 μm acrodisc (this is a 1000 × stock)
- Complete medium supplemented with 1/100 vol. fungizone and 1/1000 vol. chloroquine

Method

1. Gelatinize the plates with a 0.1% sterile gelatin solution. Wash once with PBS. Plate out the cells; if using COS, seed 2.5 × 10^6 cells/100 mm dish or 2.5–3 × 10^5 cells/35 mm dish. Incubate overnight.
2. Prepare the transfection cocktails in the hood. The cocktail volumes required for each sample are: 200 μl/35 mm dish or 2 ml/100 mm dish.
   - 25 μl DEAE–dextran
   - 5 μl of each plasmid (= 1.25 μg)
   - 965 μl PBS
3. Aspirate the medium from the cells and wash once with pre-warmed PBS.
4. Carefully add the cocktail to the dish and distribute evenly across the cells.
5. Incubate at 37 °C for 30 min, tilt the dish every 5 min to avoid letting the cells dry out.
6. Add the chloroquine supplemented medium: 20 ml/100 mm dish or 2 ml/35 mm dish.

Protocol 1. *Continued*

7. Incubate at 37°C for 2.5 h.
8. Aspirate the medium and add medium supplemented with 10% dimethyl sulfoxide (DMSO) (10 ml/100 mm dish or 2 ml/35 mm dish), leave for 2.5 min in the hood.
9. Aspirate and add complete medium to the dish (20 ml/100 mm dish or 2 ml/35 mm dish).
10. Incubate at 37°C for between 48–72 h and then analyse for gene expression.

[a] This method works particularly well for COS cells where up to 5% of the cells within a culture may be transfected.

Protocol 2. Transient transfection of adherent cells using calcium phosphate[a]

Equipment and reagents

- Plasmid DNAs that are in water and have been purified from caesium chloride gradients
- 10 × NTE: 150 mM NaCl, 10 mM Tris–HCl pH 7.4, 1 mM EDTA
- complete medium
- 2 × transfection buffer, made fresh just prior to use: for 10 ml use 1 ml 0.5 M Hepes–NaOH pH 7.1 +/− 0.05 (the pH of this solution is critical), 8.1 ml water, 0.9 ml 2 M NaCl, 20 μl 1 M Na_2HPO_4 (pH to 7.0, using phosphoric acid)

Method

1. Seed the cells the day before the transfection, for HeLa use 4×10^5 cells/35 mm dish or 2.5×10^6 cells/100 mm dish.
2. Replace the medium with fresh complete medium approximately 4 h before the transfection.
3. Prepare the DNA cocktail in an Eppendorf tube and mix well. For a 35 mm dish this should be:
   - 4–5 μg (total) DNA
   - 10 μl 10 × NTE
   - 12.5 μl 2 M $CaCl_2$
   - sterile water to the final volume of 100 μl

   For transfection of 100 mm cultures make up 1 ml cocktails containing 25 μg DNA.
4. Add the DNA cocktail dropwise to an equal volume of 2 × transfection buffer. An Eppendorf is convenient for 100 μl volumes but a 15 ml conical tube is needed for larger volumes.

5. Gently blow a stream of bubbles through the mixture five to ten times using a Pipetman (Eppendorf sized samples) or vortex gently while adding the cocktail down the side of the tube. The precipitate should be barely visible.
6. Leave the precipitate at room temperature for 15–30 min.
7. Add the precipitate dropwise to the medium and swirl around.
8. Incubate at 37°C for 4–15 h.
9. Rinse the cells once with medium and add fresh medium.
10. Incubate at 37°C for between 48–72 h and then analyse for gene expression.

[a] This procedure is probably the most widely used method for transfection. A version that works well for such cell lines as HeLa (human), CV-1 (monkey), and L (mouse) is described here. These three lines all grow well in IMDM (refer to *Protocol 1*).

Protocol 3. DEAE–dextran-mediated transfection of lymphoid cell lines[a]

Equipment and reagents

- Plasmid DNAs that are in water and have been purified from caesium chloride gradients
- 10 × stock of TD buffer: 1 × TD is 25 mM Tris, 137 mM NaCl, 5 mM KCl, 0.37 mM Na_2HPO_4, pH 7.4
- Ca/Mg solution: 2.0 g $MgCl_2$, 2.65 g $CaCl_2.2H_2O$ in 200 ml H_2O
- TS buffer: 1 × TD buffer, 1% Ca/Mg solution
- DEAE–dextran and chloroquine as in *Protocol 1*
- RPMI 1640 medium containing L-glutamine and supplemented with 10% fetal bovine serum and 0.1% gentamycin sulfate

Method

1. Dilute DEAE–dextran 1/20 in TS buffer.
2. Prepare DNA cocktails containing 1–5 μg (total) DNA.
3. In a 50 ml conical tube, add the DNA cocktail to 250 μl of the TS/DEAE–dextran mix.
4. Collect cells by centrifugation at ~ 300 *g* and wash once with TS buffer at room temperature, ~ 10^7 cells are required for each transfection.
5. Resuspend the cell pellet in TS buffer at 4 × 10^7 cells/ml.
6. Add 250 μl cells (i.e. 10^7 total cells) to the 250 μl of TS/DEAE–dextran/DNA mix.
7. Leave at room temperature for 15–30 min.
8. Add 5 ml complete medium supplemented with 1/1000 vol. of chloroquine.

Protocol 3. *Continued*

9. Incubate at 37 °C for between 30–60 min.
10. Spin down the cells and wash once with 10 ml medium.
11. Plate out the cells in 10–50 ml medium and return cultures to 37 °C.
12. Analyse the cells for expression after 24–48 h. Note that by this time quite a lot of the cells will have died, the amount of DNA used for the transfection influences this and so it is recommended that a DNA titration be performed prior to the experiment itself.

[a] This procedure has been shown to work well for a range of human CD4-positive lymphoid (non-adherent) cell lines, for example H9, CEM, and Jurkat, that are known to be permissive for HIV-1 replication.

Protocol 4. Electroporation of lymphoid cell lines[a]

Equipment and reagents

- Gene Pulser and Capacitance Extender, these can be obtained from Bio-Rad Laboratories
- Plasmid DNAs and culture medium as in *Protocol 3*

Method

1. Harvest the cells at room temperature by centrifugation at ~ 300 *g*. Wash once with PBS. Resuspend in complete medium at 1×10^7 cells/ml.
2. Transfer 400 μl cells (i.e. 4×10^6 total cells) to an electroporation cuvette. Add 10 μg DNA. Mix by flicking the cuvette.
3. Leave at room temperature for 10 min.
4. Shock the cells once at 250 V and 960 μF. Between 20–60% of the cells should be killed for optimum electroporation efficiencies, this is determined empirically for different cell lines.
5. Leave at room temperature for 10 min.
6. Add the cells to ~ 5 ml medium and return to 37 °C.
7. Analyse the cells for expression after 24–48 h.

[a] The following conditions are optimized for the human CD4-positive lymphoblastoid line CEM-SS.

2.2 The viral *tat* gene as a reporter

Transcripts expressed from the region of the provirus that spans the viral *tat*, *rev*, and *env* genes are regulated by Rev (*Figure 1A*). The reporter gene is the viral *tat* gene, both coding exons of which are present in this vector. These two exons are separated by an intron that naturally contains the RRE as well as much of the overlapping viral *env* gene (see below). Once transfected into cultured cells, the intron-containing primary transcript is inefficiently spliced (HIV splice sites are inherently suboptimal) to yield a fully processed mRNA. As a result, significant levels of both the unspliced and spliced RNA species are present within the cell. In the spliced version, the two exons of *tat* are ligated together and an 86 aa form of the protein is expressed. Since Rev is not required for the expression of proteins from fully spliced mRNA, this full-length Tat protein is expressed constitutively. The positioning of a stop codon immediately downstream from the 5′ splice site (or splice donor) in the unspliced transcript results in this intron-containing mRNA encoding a truncated Tat protein of only 72 aa. Because the function of Rev is to activate the functional expression of unspliced viral mRNAs, this 72 aa Tat protein is expressed exclusively in the presence of Rev.

The assay proceeds as follows:

(a) Co-transfect the reporter plasmid (the original was termed pgTAT, refer to ref. 4) into COS cells with (or without) a Rev expression vector.

(b) Metabolically label the culture after 48–72 h.

(c) Monitor synthesis of the 72 aa Tat protein by immunoprecipitation using an anti-Tat-specific antibody (4).

A typical assay of this type is shown in *Figure 2A*, expression of the foreshortened Tat protein is clearly restricted to the culture containing Rev (lane 2). *Protocol 5* gives a method for labelling and immunoprecipitating the Tat protein.

Protocol 5. Radioimmunoprecipitation[a]

Equipment and reagents

- RIPA buffer: 150 mM NaCl, 10 mM Tris–HCl pH 7.5, 1 mM EDTA, 0.1% SDS, 1% Triton X-100, 1% Na deoxycholate
- 100 mM phenylmethylsulfonyl fluoride (PMSF) in isopropanol (400 × stock), store at 4°C and warm up to room temperature before use
- 2 × dissociation buffer: 125 mM Tris–HCl pH 6.8, 20% glycerol, 4% SDS, 2 mg/ml bromophenol blue, 10% β-mercapto-ethanol (add just prior to use)
- Dulbecco's modified Eagle's medium lacking either cysteine (use when assessing both Tat and Rev expression) or cysteine and methionine (use when examining Tat expression alone), supplemented with 10% dialysed fetal bovine serum
- Protein A (or G) agarose beads (Gibco–BRL Inc.)
- Amplify (Amersham Corp.)

Method

1. Starve the cells by washing twice with PBS, once with the appropriately depleted medium and incubating at 37°C for 1 h in depleted medium.

Protocol 5. *Continued*

2. Label the culture for ~ 2 h in 1 ml medium supplemented with radiolabelled amino acids; for Tat and Rev use 300 μCi [^{35}S]cysteine and for Tat alone use 100 μCi [^{35}S]methionine and cysteine (Tran ^{35}S-label, ICN Biomedicals).
3. Remove the medium and lyse in 1 ml RIPA buffer freshly supplemented with 250 μM PMSF by gently rocking the plate at room temperature for 5 min.
4. Transfer the lysate to an Eppendorf tube (it may help to scrape the dish first) and clarify by centrifugation at 17 000 r.p.m. (35 000 *g*) in an SS34 rotor (Sorvall Du Pont Co.) for 30 min at 4°C.
5. Harvest the supernatant (at this point, the lysate may be stored at −70°C) and add the antibody or antiserum; for a typical polyclonal serum raised in rabbits, use a dilution of between 1/100 and 1/500.
6. Rotate the tubes overnight at 4°C.
7. Precipitate the immune complexes by adding ~ 20 μl pre-washed (with RIPA buffer) protein A (or G) agarose beads and rotating continuously for 1 h at 4°C.
8. Collect the beads by spinning at full speed in a microcentrifuge for ~ 10 sec at room temperature, wash them three times with 500 μl RIPA by vortexing and re-pelleting.
9. Add ~ 15 μl 2 × dissociation buffer to the pellet. Freeze in a dry ice/ethanol bath for 15 min. Thaw at 37°C. Vortex vigorously. Boil for 2 min. Load on to a sodium dodecyl sulfate–polyacrylamide gel. Use ^{14}C-labelled protein standards (Gibco–BRL Inc.) as relative molecular mass markers.
10. After electrophoresis, fix the gel in 25% isopropanol, 10% acetic acid, 65% water for 30 min at room temperature.
11. Fluorograph for 30 min using Amplify supplemented with 10% glycerol.
12. Dry the gel and visualize by autoradiography.

[a] This protocol works well for 35 mm cultures of COS cells 48–72 h after transient transfection and was used for the experiment shown in *Figure 2A*.

There are a number of advantages to this experimental approach:

(a) The low background of 72 aa Tat expression in the absence of Rev makes this system very sensitive to low levels of Rev activity.

(b) The absolute level of Tat expression (i.e. 86 and 72 aa forms combined) is not affected by Rev.

(c) The ratio between the two forms of Tat reflects Rev function. Thus, the

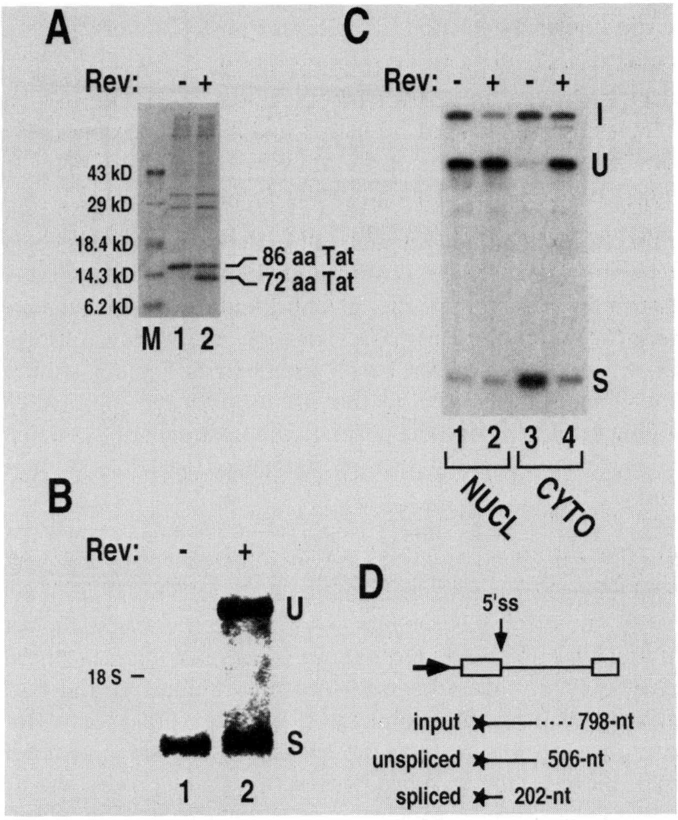

Figure 2. Immunoprecipitation, Northern, and quantitative nuclease S1 protection analyses of Rev function in COS cells transiently transfected with reporter A. A. 35 mm sub-confluent cultures of COS were transfected using DEAE–dextran and chloroquine with 0.25 μg pgTAT (4) and either 0.25 μg control plasmid (lane 1) or 0.25 μg pcREV (lane 2) (4). At 60 h, the cultures were metabolically labelled with [^{35}S]cysteine, subjected to immunoprecipitation with an anti-Tat-specific rabbit serum, and the precipitated proteins resolved on a sodium dodecyl sulfate 14% polyacrylamide gel. The bands that correspond to the 86 and 72 aa Tat proteins (relative molecular masses of ~ 15.5 and ~ 14 kDa, respectively) are indicated. B and C. 100 mm dishes of COS were transfected as above with 2.5 μg pgTAT and either 2.5 μg control plasmid (lanes 1 and 3) or 2.5 μg pcREV (lanes 2 and 4) and separated into nuclear and cytoplasmic fractions at 48 h for RNA isolation. For Northern analysis (B), cytoplasmic RNA samples were resolved on a 1% denaturing agarose gel, transferred to nitrocellulose, and hybridized with a labelled probe specific for the *tat* mRNAs. The bands corresponding to the unspliced (U) and spliced (S) *tat* mRNAs are indicated as is the position of the 18S ribosomal RNA band. For nuclease S1 protection (C), nuclear and cytoplasmic RNA samples were hybridized to a 798 nucleotide end-labelled probe (I) that spanned the 5′ splice site at the 3′ end of the first coding exon of *tat* and was rescued as fragments of 506 and 202 nucleotides by the unspliced (U) and spliced (S) *tat* mRNAs, respectively (4). A diagram illustrating this probe strategy is shown in D, the tag of unrelated sequences at the 3′ terminus of the probe is shown as a broken line. Rescued DNA fragments were resolved on a denaturing 5% polyacrylamide gel.

assays are internally controlled for both transfection efficiency and sample recovery.

(d) When [^{35}S]cysteine is used for labelling, the cell lysate can be divided and analysed separately for both Rev and Tat expression. This is useful for confirming that a series of mutant Rev proteins are being expressed at equivalent levels (for example, refer to ref. 5).

As mentioned earlier, the *env* gene is overlapping with this region of the provirus. Since Env expression is absolutely dependent upon Rev, it is also possible to use the induction of Env glycoprotein expression as a measure of Rev activity (6–8). Both immunoprecipitation and Western blotting-based assays can be used to monitor Env expression and many antibodies are currently available for this. Vectors that are used for this purpose often have the first coding exon of *tat* and the 5′ splice site upstream of *env* deleted (refer to *Figure 1A*), this presumably helps to maximize translation of the *env* open reading frame.

The mRNA species expressed by reporter constructs carrying viral genes are structurally similar to those found in infected cells, and perhaps not surprisingly, they also behave in a similar fashion. This type of reporter is therefore well suited for experiments in which the transcripts themselves are being analysed. In HIV-1-infected human T cells lacking Rev, the unspliced viral transcripts (9 kb and 4 kb mRNAs) are confined to the nucleus and therefore excluded from the cytoplasm (9). In the presence of Rev, these mRNAs are able to translocate to the cytoplasm and serve as templates for translation. Likewise, the unspliced transcript that encodes the 72 aa Tat protein is only found in the cytoplasm in cells that express Rev. This observation provides a mechanistic explanation for the dependence of 72 aa Tat synthesis upon Rev (1).

A typical experiment used to determine the effects of Rev on mRNA expression consists of:

- transfecting COS cells as above
- preparing RNA either from total cell lysates or from nuclear and cytoplasmic subcellular fractions after ~ 48 h

Protocol 6 describes the preparation of RNA from nuclear and cytoplasmic fractions.

Of the many alternatives for the analysis of RNA samples, Northern blotting and quantitative nuclease S1 protection assays have both been used successfully in such studies (4, 5). In addition to evaluating the expression levels of the RNAs being examined, these two assays can provide quite different information about the RNAs themselves.

Northern blotting involves the electrophoretic fractionation of intact RNA samples and can be used to determine the approximate length of an mRNA species.

Protocol 6. Preparation of RNA from unfractionated and fractionated cells[a,b]

Equipment and reagents

- TL-100 bench-top ultracentrifuge (Beckman Instruments Inc.)
- 4 M and 8 M guanidine isothiocyanate (GTC), made up in water: the 8 M stock is a solid at room temperature; before starting the RNA isolation, melt the stock solution (a microwave is convenient for this) and aliquot ~ 350 μl volumes into Eppendorf tubes, the solution will then re-solidify
- NP-40 lysis buffer: 10 mM NaCl, 10 mM Tris–HCl pH 7.4, 3 mM $MgCl_2$, 0.5% (w/v) Nonidet P-40
- Cushion buffer: 5.7 M CsCl, 25 mM NaOAc, pH 5.0
- 30% formamide, 70% ethanol mix (store at −20°C)
- 100 mm dish of adherent cells, for example 48 h after transient transfection

Method

1. Discard the medium and rinse the cells twice with cold PBS.
2. Scrape the cells into 10 ml PBS. Transfer to a 15 ml conical tube. Pellet the cells at 1.5 k.r.p.m. (500 *g*) in a clinical centrifuge at 4°C.
3. Resuspend the cells in ~ 1 ml PBS and transfer to an Eppendorf tube. By keeping the tubes on ice as much as possible, this procedure can be carried out at room temperature.
4. Pellet the cells by spinning for 5 sec in a microcentrifuge and discard the supernatant.
5. For preparing total RNA, add 600 μl 4 M GTC, vortex vigorously, and sonicate for ~ 10 sec to shear the chromosomal DNA. The sample changes from a very viscous liquid to a more fluid one. Proceed to step 11.
6. For nuclear/cytoplasmic fractionations, resuspend the cell pellet in 350 μl lysis buffer by gentle pipetting, vortex at medium speed for no more than 5 sec, and leave on ice for 5 min.
7. Centrifuge for ~ 5 sec at full speed and transfer the supernatant, which represents the cytoplasmic fraction, to a fresh tube.
8. Wash the nuclear pellet. Resuspend and vortex in ~ 500 μl lysis buffer. Briefly spin this sample and, at the same time, the supernatant from step 7.
9. Add the cytoplasmic sample to an aliquot of 8 M GTC and vortex until homogeneous.
10. To the nuclei add 600 μl 4 M GTC and treat as in step 5.
11. Layer all samples over 500 μl of the caesium chloride/NaOAc cushion buffer.
12. Using a TLA 100.2 rotor (Beckman Instruments Inc.), centrifuge at 100 000 r.p.m. (436 000 *g*) for 3 h at 25°C.

Protocol 6. *Continued*

13. Aspirate the supernatant. Remove debris from the side of the tube using a cotton tipped applicator. Wash once with the formamide/ethanol mix. Invert tube to dry.
14. Resuspend the RNA pellet in ~ 100 μl RNase-free water. Heat at 65°C for 5 min. Store at either −20°C or −70°C.

[a] This method works well for adherent cells, for example HeLa and COS, as well as for suspension cell lines such as CEM-SS. RNA preparations are obtained relatively quickly and are of 'good' quality.
[b] This procedure may be too harsh for use with lymphocytes and can result in nuclear lysis or 'leaking'. However, changing the lysis buffer to 50 mM KCl, 10 mM Hepes–NaOH pH 7.6, 5 mM $MgCl_2$, 0.05% (w/v) Nonidet P-40 has been shown to work well in experiments using CEM-SS cells.

Protocol 7. Northern hybridization

Equipment and reagents

- 10 × Northern running buffer: 100 mM Na_2HPO_4, 10 mM EDTA, pH to 7.0 using phosphoric acid
- 10 × Northern loading buffer: 50% glycerol, 1 mM EDTA, 4 mg/ml xylene cyanol FF, 4 mg/ml bromophenol blue
- Northern hybridization buffer: 4 × SSC, 7 mM Tris–HCl pH 7.5, 1 × Denhardt's solution (100 × stock: 2 g Ficoll, 2 g bovine serum albumin (fraction V), 2 g polyvinyl-polypyrrolidone per 100 ml water, filter sterilized and stored at −20°C), 50 μg/ml sonicated salmon sperm DNA, 10% dextran sulfate, 40% formamide
- Reagents and purified DNA fragment(s) for probe preparation. The oligolabelling kit available from Pharmacia Biotech Inc. is a particularly good one for random primed labelling and the G-50 columns from 5 Prime->3 Prime, Inc. work well for the removal of the unincorporated radiolabelled deoxynucleoside triphosphates: in each case, follow the manufacturers' instructions.
- 20 × SSC: 3 M NaCl, 0.3 M Na citrate, pH 7.0

Method

1. Prepare the gel, preferably in a chemical hood with the airflow turned off.
   - 68.3 ml water
   - 10 ml 10 × running buffer
   - 21.7 ml formaldehyde (this is usually purchased as a ~37% solution containing 10–15% methanol)
   - 1 g agarose
2. Denature RNA prior to electrophoresis by preparing a 100 μl cocktail comprising:
   - 50 μl deionized formamide
   - 17.5 μl formaldehyde
   - 10 μl 10 × running buffer

- 12.5 μl RNA (the quantity of RNA to be used is empirically determined and it should be heated to 65°C for 5 min prior to sampling to ensure that it is fully in solution).

Incubate this mixture at 56°C for 15 min. Chill on ice for 5 min. Add 10 μl loading buffer.

3. Load the gel unsubmerged, add electrophoresis buffer (1 litre = 100 ml 10 × running buffer, 225 ml formaldehyde, 675 ml water) until only the ends of the gel are in contact with the buffer, i.e. the top of the gel remains dry.
4. Run the gel at 100 mA (~ 30 min) or until the xylene cyanol FF has migrated ~ 1 cm.
5. Submerge the gel and run overnight with buffer circulation (~ 50 mA). The xylene cyanol FF migrates with RNAs that are ~ 1.4 kb in length.
6. Soak the gel at room temperature in the following solutions: water for 5 min with several changes; 50 mM NaOH, 10 mM NaCl for 45 min; 100 mM Tris–HCl pH 7.5 for 45 min; 20 × SSC for 1 h.
7. Transfer the RNA to 0.1 μm nitrocellulose (Schleicher & Schuell, Inc.) that has already been boiled in water for 5 min and soaked in 20 × SSC for 45 min by capillary action using 20 × SSC as the transfer buffer (use a standard 'blotting' set-up).
8. Leave for at least 18 h to allow the RNA to transfer.
9. Remove the nitrocellulose filter, air dry, and bake for 4 h at 80°C under vacuum.
10. Pre-hybridize the filter by wetting with water and incubating in a sealed bag at 42°C for at least 4 h in enough hybridization buffer to amply cover the filter.
11. Meanwhile, prepare and purify the DNA probe using [α-^{32}P]dCTP and monitor for the incorporation of radioactivity by trichloroacetic acid (TCA) precipitation using sonicated salmon sperm DNA as the carrier.
12. Mix:
    - 0.1–0.5 × 10^6 Cherenkov c.p.m./ml hybridization buffer
    - 100 μl 10 mg/ml sonicated salmon sperm DNA
    - water to a final volume of 1 ml
13. Denature the probe cocktail by adding, in this order and while vortexing at low speed:
    - 25 μl 10 M NaOH
    - 75 μl 2 M Tris–HCl pH 7.5
    - 250 μl 1 M HCl
14. Add the denatured probe to the hybridization bag, re-seal, and incubate at 42°C for at least 16 h.

Protocol 7. *Continued*

15. Remove the filter from the bag and wash sequentially in the following solutions for ~ 1 h each at the indicated temperatures, the loss of counts from the filter during the washing should be monitored with a Geiger counter and the washing halted as necessary: 2 × SSC, 0.1% SDS, 42°C; 0.5 × SSC, 0.1% SDS, 56°C; 0.1 × SSC, 0.1% SDS, 68°C.

16. Wrap the moist filter in Saran Wrap and visualize by autoradiography.

Figure 2B shows a Northern analysis of cytoplasmic RNA isolated from transfected COS cells, the exclusive expression of the ~ 3 kb unspliced *tat* mRNA in the cytoplasms of cells containing Rev is clear (lane 2). It should be noted, however, that Northern analyses are comparatively insensitive and often require the prolonged autoradiographic exposure of filters.

Nuclease S1 protection assays are considerably more sensitive than Northerns. Moreover, they are ideal for differentiating between the unspliced and spliced versions of a primary transcript, as is frequently the case when addressing Rev function.

Protocol 8. Quantitative nuclease S1 protection

Equipment and reagents

- S1 hybridization buffer: 80% formamide, 0.4 M NaCl, 1 mM EDTA, 40 mM Pipes–NaOH pH 6.8, store in the dark at 4°C
- Nuclease S1 buffer: 30 mM NaOAc, 250 mM NaCl, 1 mM $ZnSO_4$, pH 4.6
- Salmon sperm carrier DNA: 80% formamide, 1 mg/ml sonicated salmon sperm DNA, 10 mM Tris–HCl pH 8.0, 1 mM EDTA, store at −20°C and denature at 70°C for 5 min prior to each use
- S1 stop solution: 300 mM NaOAc pH 6.0, 10 mM EDTA, 0.2 mg/ml yeast tRNA, store at −20°C
- S1 gel loading buffer: 80% formamide, 0.2 × TBE (standard recipe), 0.5 mg/ml xylene cyanol FF, 0.5 mg/ml bromophenol blue, store at −20°C
- Gel purified DNA probe end-labelled with either [α-^{32}P]dNTPs using Klenow (this is preferred because probes with higher specific activities are generally obtained) or [γ-^{32}P]ATP using T4 polynucleotide kinase; the protocols for this are standard practice and will not be described here
- Nuclease S1 (United States Biochemical Corp.)

Method

1. For each RNA to be assayed make up a 231 μl cocktail comprising:
   - RNA sample (refer to *Protocol 7*, step 2)
   - the probe (for example, 2 × 10^4 Cherenkov c.p.m./reaction)
   - 25 μl 3 M NaOAc pH 6.0
   - 1 μl 10 mg/ml yeast tRNA
   - water to 231 μl
2. Add 750 μl ethanol. Precipitate the RNA/probe mix at −70°C. Wash the pellet once with 70% ethanol (30% water) at room temperature. Dry to completion.

3. Resuspend the pellet in 2 μl water by vigorous vortexing.
4. Add 18 μl hybridization buffer. Vortex to mix. Overlay with four drops of light mineral oil.
5. Heat at 70°C for 5 min, immediately transfer tubes to a 49°C water-bath and incubate for at least 15 h (the precise temperature of hybridization may need optimization depending upon the base compositions of the probe and the RNA).
6. To each sample add:
   - 200 μl nuclease S1 buffer
   - 2 μl salmon sperm carrier DNA
   - 0.5 μl nuclease S1

   This is accomplished easily by preparing enough mix for all samples and then aliquoting it out.
7. Vortex gently and incubate at 37°C for 1 h.
8. Add 50 μl S1 stop solution and 750 μl ethanol. Precipitate the nucleic acids at −70°C. Wash the pellet once with 100% ethanol (this floats above the residual mineral oil). Dry thoroughly.
9. Resuspend the pellet in ~8 μl S1 gel loading buffer by vortexing, denature at 90°C for 2 min, and place on ice.
10. Load samples on to a 1.5 mm thick and ~ 14 cm long denaturing polyacrylamide gel (29:1 ratio of acrylamide:bis) containing 8 M urea and 1 × TBE that has been pre-run for 15 min. The strength of the gel is determined by the length of the fragments that need to be resolved.
11. Electrophorese at ~ 300 V at room temperature.
12. Dry the gel after soaking it in water for 30 min with one change.
13. Visualize by autoradiography.

A point that is critical to the success of a particular S1 assay is the design of the probe itself:

(a) Fragments should be readily separated on standard denaturing polyacrylamide gels.

(b) A G + C content of at least 50% is desirable and long stretches of As and/or Ts should be avoided (such stretches may be susceptible to Nuclease S1 digestion because of the weaker RNA–DNA duplexes they form).

A typical S1-based analysis of nuclear and cytoplasmic RNA samples from fractionated COS cells is shown in *Figure 2C* and the accompanying probe strategy is depicted in *Figure 2D*. The input probe was uniquely end-labelled within the first coding exon of *tat* and extended across the 5′ splice site and into the intron. The further extension of the probe at its 3′ terminus with an unrelated sequence ensured that the fragment rescued by unspliced mRNA could be distinguished from the input probe. The presence of unspliced mRNA is manifested as rescue of the 506 nucleotide fragment in *Figure 2C*, it is readily detected in the cytoplasm in the presence of Rev (lane 4) but is almost entirely retained within the nucleus in the absence of Rev (lanes 1 and 3).

2.3 CAT as a reporter gene—introns absent

This system may be regarded as the least authentic of the three reporters shown in *Figure 1*. In this case, the reporter gene is the bacterial chloramphenicol acetyltransferase (*cat*) gene. Downstream from the promoter and the *cat* gene are the RRE and a second sequence that acts here to suppress the expression of CAT. Many such repressive sequences have been identified throughout the HIV-1 provirus and are known, variously, as *cis*-acting repressor sequences (CRSs) (10), inhibitory sequences (INSs) (11), and *cis*-acting inhibitory regions (IRs) (12); their mechanism(s) of action, however, have yet to be elucidated. CAT expression is rescued from such vectors either by co-expression of Rev or by inactivation of the inhibitory sequences. This system therefore differs from reporter A in two major ways. First, the HIV-1 sequences are not necessarily derived from a contiguous region of the provirus and, secondly, splice sites (and introns) that are utilized in virally-infected cells do not have to be included. It is also worth noting that vectors that operate under similar principals but have had the *cat* gene replaced with reporters such as the HIV-1 *tat* gene have been described (11, 12).

An experiment using vectors of this type entails:

- co-transfection of the reporter and Rev expression plasmids
- preparation of a cell extract 48–72 h later
- analysis of the extract for CAT activity

Protocol 9. Assays for chloramphenicol acetyltransferase (diffusion-based)[a]

Equipment and reagents

- TEN buffer: 150 mM NaCl, 40 mM Tris–HCl pH 7.5, 1 mM EDTA
- Acetyl coenzyme A (acetyl-1-^{3}H) [Dupont (Sorvall Division) Co.]
- Econofluor [Dupont (Sorvall Division) Co.]

Method

1. Wash the cells once with PBS at room temperature.
2. Resuspend the cells in ~ 1 ml TEN buffer (scrape if using adherent cells).
3. Transfer to an Eppendorf tube, spin down the cells, and resuspend in 300–500 μl 100 mM Tris–HCl pH 7.8.

4. Lyse cells by freeze–thawing three times; 15 min in a dry ice/ethanol bath followed by 5 min at 37°C is ideal.
5. Centrifuge at full speed for 10 min at 4°C.
6. Transfer the supernatant to a fresh tube and heat inactivate the cellular CAT inhibitors by incubating at 70°C for 10 min; the extract may be stored at −70°C, however, some problems with protein insolubility may be encountered.
7. In a scintillation vial, add 100 μl of extract, or an appropriate fraction thereof diluted in 100 mM Tris pH 7.8, to 100 μl 100 mM Tris pH 7.8 supplemented with 2 mg/ml chloramphenicol and 2 μl/ml acetyl-1-^{3}H coenzyme A.
8. Carefully overlay the reaction mixture with 3 ml Econofluor, cap the vial, and monitor the progression of chloramphenicol acetylation by counting in a scintillation counter at ~60 min intervals.

[a] There are currently a number of methods available for determining levels of CAT activity in cell extracts. The one described here is convenient as it yields kinetic data and does not involve thin-layer liquid chromatography and autoradiography. The assays are usually performed between 48–72 h after transient transfection. The results can be normalized for the concentration of protein in the extracts by the method of Bradford (21).

This strategy can be improved by the inclusion of a constant amount of an internal control plasmid in each transfection. An ideal plasmid for this purpose is an efficient expressor of β-galactosidase (pCH110, available from Pharmacia Biotech Inc. is commonly used) as the same protein extract can then be used for both assays (13). The increase in CAT activity observed in the presence of Rev, relative to the absence of Rev, represents *trans*-activation and is typically between 5- and 20-fold when the plasmids express the wild-type RRE and wild-type Rev.

CAT reporters are attractive for two main reasons:

(a) CAT assays themselves are very sensitive and so a wide variety of cell types, including those with relatively poor transfection efficiencies, can be utilized.
(b) The assays are inexpensive to perform and require no specialized reagents, this is convenient if only a small number of experiments are envisioned.

One disadvantage to this approach, however, is the relatively high background of CAT expression obtained in cells lacking Rev. This system is therefore less suitable for studies in which low levels of Rev function are being examined.

Although vectors of this type have been used extensively for the identification of the negatively acting sequence elements and for the definition of the critical regions of both the RRE and Rev, their usefulness in experiments that aim to study the mechanism of Rev function by analysing patterns of RNA expression is less clear.

(a) Vectors frequently lack authentic viral splice sites and, as a result, the transcripts they express can be very different from those observed in virally-infected cells.

(b) The absence of splice sites presumably contributes to the apparent presence of Re-responsive mRNAs in the cytoplasms of transfected cells lacking Rev even when CAT expression is low (14). This finding is different to what is observed in infected cells (as well as with reporter A) and may, therefore, only be of indirect relevance to Revs mechanism of action.

2.4 CAT as a reporter gene—introns present

Reporter C (*Figure 1*; the original was termed pDM128, refer to ref. 13) combines attractive features from both of the above systems. This vector contains both 5′ and 3′ viral splice sites as well as the RRE in its natural location. However, instead of using a viral gene as the indicator, the *cat* gene has been inserted into the intron. As a result, the expression of CAT is dependent upon a Rev-responsive unspliced transcript exiting the nucleus and entering the cytoplasm. This reporter has several advantages:

(a) Rev function can be monitored by the induction of CAT activity.

(b) The effect of Rev on the fate of an unspliced mRNA that contains authentic HIV-1 splice sites and is confined to the nucleus in the absence of Rev can be examined.

(c) *Trans*-activations of 50- to 100-fold have been recorded with this system in cell lines such as CV-1 and HeLa. Inductions of CAT activity with this magnitude make this a very useful approach to be considered for experiments addressing Rev function and particularly for studies in cells that may be relatively difficult to transfect.

As with reporter B, experiments that measure CAT activity are improved by the addition of an internal control plasmid in the transfections.

2.5 Subcellular localization

In addition to assessing function, it is often interesting to determine the subcellular localization of either wild-type or mutant Rev proteins. As mentioned earlier, Rev is ordinarily found in the nucleus, and predominantly the nucleoli, of expressing cells. A straightforward way of examining subcellular localization is to perform indirect immunofluorescence (*Protocol 10*) on fixed and permeabilized cultures transiently transfected with appropriate plasmids. *Figure 3* shows typical immunofluorescence and phase-contrast photographs of HeLa cells transfected with a wild-type HIV-1 Rev expression vector.

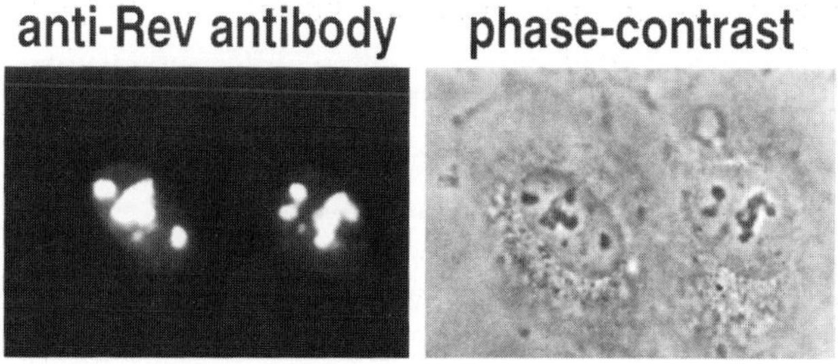

Figure 3. Subcellular localization of the wild-type Rev protein in transiently transfected HeLa cells. A subconfluent monolayer was transfected with 1.25 μg pcREV on a Lab-Tek 4 chamber slide using calcium phosphate. At ~ 48 h, the cells were fixed, permeabilized, and hybridized initially with an anti-Rev monoclonal antibody (Repligen Corp.) at 1.25 μg/ml, and finally with a goat anti-mouse antibody conjugated to Texas red (FisherBiotech). The slide was visualized through a Nikon Microphot-FXA microscope at a magnification of × 400.

Protocol 10. Indirect immunofluorescence[a]

Equipment and reagents

- 3% paraformaldehyde in PBS, for 100 ml, also add 250 μl 1 M NaOH and gently warm in a microwave for 15–30 sec
- PBS supplemented with 10 mM glycine
- Permeabilization buffer: PBS containing 10% goat serum, 1% Triton X-100
- Immunofluorescence hybridization buffer: PBS containing 250 mM NaCl, 10% goat serum, 0.1% Tween-20
- Detector antibody with texas red as the fluorochrome (FisherBiotech)
- Citifluor mounting medium (University of Kent Chem. Lab.)
- Cells growing in Lab-Tek 4 chamber slides (Nunc Inc.)—monolayer cultures can be transfected directly in the chambers; for example, use ~ 4 × 10^4 HeLa cells per sample and analyse ~ 48 h later

Method

1. Remove the medium and wash the cells twice with PBS.
2. Fix the cells by incubating with the paraformaldehyde solution for 30 min at room temperature.
3. Wash extensively with the PBS/glycine solution.
4. Incubate the cells in permeabilization buffer for 10 min at room temperature and wash with the PBS/glycine solution.
5. Prepare an appropriate dilution of the primary antibody or antiserum in immunofluorescence hybridization buffer (for a polyclonal serum raised in rabbits, this is usually between 1/100 and 1/1000) and incubate with the cells for 2 h at room temperature.
6. Wash extensively with the PBS/glycine solution.
7. Dilute the fluorochrome-conjugated species-specific second antibody

Protocol 10. *Continued*

(detector antibody) 1/100 in hybridization buffer and incubate with the cells for 30 min in the dark at room temperature.

8. Wash thoroughly with the PBS/glycine solution. Mount with Citifluor and a glass coverslip. Visualize by epifluorescence.

[a] The following protocol works well for adherent cells such as COS or HeLa 48–72 h after transient transfection and was used to generate *Figure 3*.

3. Analysis of dissected domains of Rev—experimental approaches

Many mutations have been introduced into the HIV-1 *rev* gene by site-directed mutagenesis. These studies have identified four attributes of Rev that, to date, appear to be essential for function (reviewed in refs 1 and 2):

- nuclear/nucleolar localization (refer to section 2.5)
- direct binding to the RRE (see Chapter 9)
- ability to assemble into homomultimers (or -oligomers) are all conferred by a relatively large N-terminal domain of ~ 45 aa
- a distinct leucine-rich C-terminal region of ~ 10 aa known as the activation (or effector) domain

The activation domain is believed to interact with the host cell factor(s) that are required for Revs activation of unspliced viral mRNA expression in the cytoplasm. Thus, Rev appears to possess a modular structure that is somewhat reminiscent of a number of transcription factors. Recently, assay systems in which segments of Rev are expressed as fusions with heterologous protein sequences have been developed that may facilitate the analysis of Revs domains in isolation. These are discussed briefly below and are diagrammed in *Figure 4*; in each case, the reporter gene itself is *cat* and the assays are performed following co-transfection as described above.

3.1 RNA binding

In the system diagrammed in *Figure 4A*, transcriptional *trans*-activation of the HIV-1 LTR promoter element is used to monitor the binding of Rev to its RNA target sequence. Under normal circumstances, the virally encoded Tat protein interacts with an RNA stem–loop structure, termed TAR, found at the 5′ terminus of all viral mRNAs and dramatically up-regulates LTR-directed transcription (see Chapter 10). Here, the binding site for Rev (derived from the RRE) is inserted into TAR and the *trans*-activating domain of an established transcription factor (TF) is fused to Rev. Thus, when the

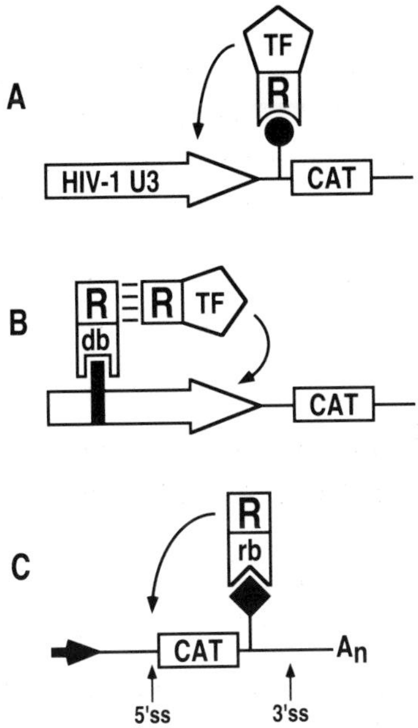

Figure 4. Reporter systems for addressing the RRE binding (A), multimerization (B), and activation domain (C) phenotypes of Rev. Open arrows, responsive promoter sequences; R, Rev sequences; TF, transcription factor, solid circle, RRE sequences; db, DNA binding domain; solid rectangle, DNA sequence cognate for db; rb, RNA binding domain; solid diamond, RNA sequence cognate for rb; otherwise refer to the legend for *Figure 1* and the text for further details.

Rev–TF fusion protein binds to nascent transcripts via the inserted RRE sequences, transcription is stimulated. To date, both the HIV-1 Tat protein itself as well as the extensively characterized activation domain of the herpes simplex virus 1 (HSV-1) activator of immediate early viral gene expression, VP16, have been used to supply the sequences that activate transcription (15, 16). This system can be used for mapping either Rev or RRE-derived determinants of binding.

3.2 Multimerization

In the assay shown in *Figure 4B*, domains of Rev are used to bridge the gap between the DNA binding and activation domains of a transcription factor that would normally function via a DNA target site. Systems of this type, often referred to as two hybrid systems (17), are becoming increasingly popular both for analysing interactions between known proteins and as tools used in the attempted identification of proteins that interact with a given protein of interest. Thus, in this scenario, there are three plasmids (in addition to the internal controls) to transfect:

- the reporter gene (*cat*) driven by an intrinsically weak (or inactive) promoter that contains the binding site(s) for the chosen transcription factor (GAL4 from bakers' yeast is often used)
- Rev expressed as a fusion to the DNA binding domain of the relevant transcription factor
- Rev expressed as a hybrid with the activation domain of a transcription factor (this can be derived from VP16 of HSV-1)

In the event of a Rev–Rev interaction occurring, transcription is activated and the determinants of multimerization may be mapped (18). It should be appreciated, however, that a positive result can also ensue when an additional molecule supplied by the transfected cell acts as a bridge to juxtaposition the two hybrid Rev molecules and, as a consequence, bypasses the requirement for a direct Rev–Rev interaction.

3.3 Activation domain function

To assay for activation domain function (*Figure 4C*), the primary binding site for Rev is replaced by the binding site for a heterologous RNA binding protein and Rev is expressed as a fusion to the RNA binding domain cognate for the introduced site. To date, this approach has utilized the coat protein of bacteriophage MS2 and its binding site, the 21 nucleotide operator (19, 20). Experiments utilizing this system have shown not only that mutations in the arginine-rich region of Rev do not affect Rev activation domain function but also that the binding of multiple Rev–MS2 fusion proteins are required for optimum *trans*-activation.

References

1. Cullen, B. R. and Malim, M. H. (1991). *TIBS,* **16**, 346.
2. Pavlakis, G. N. and Felber, B. K. (1990). *The New Biol.,* **2**, 20.
3. Mellon, P., Parker, V., Gluzman, Y., and Maniatis, T. (1981). *Cell,* **27**, 279.
4. Malim, M. H., Hauber, J., Fenrick, R., and Cullen, B. R. (1988). *Nature,* **335**, 181.
5. Malim, M. H., Böhnlein, S., Hauber, J., and Cullen, B. R. (1989). *Cell,* **58**, 205.
6. Emerman, M., Vazeux, R., and Peden, K. (1989). *Cell,* **57**, 1155.
7. Hammarskjöld, M.-L., Heimer, J., Hammarskjöld, B., Sangwan, I., Albert, L., and Rekosh, D. (1989). *J. Virol.,* **63**, 1959.
8. Knight, D. M., Flomerfelt, F. A., and Ghrayeb, J. (1987). *Science,* **236**, 837.
9. Malim, M. H. and Cullen, B. R. (1993). *Mol. Cell. Biol.,* **13**, 6180.
10. Rosen, C. A., Terwilliger, E., Dayton, A., Sodroski, J. G., and Haseltine, W. A. (1988). *Proc. Natl Acad. Sci. USA,* **85**, 2071.
11. Schwartz, S., Felber, B. K., and Pavlakis, G. N. (1992). *J. Virol.,* **66**, 150.
12. Maldarelli, F., Martin, M. A., and Strebel, K. (1991). *J. Virol.,* **65**, 5732.
13. Hope, T. J., Xiaojian, H., McDonald, D., and Parslow, T. G. (1990). *Proc. Natl Acad. Sci. USA,* **87**, 7787.

14. Cochrane, A. W., Jones, K. S., Beidas, S., Dillon, P. J., Skalka, A. M., and Rosen, C. A. (1991). *J. Virol.,* **65**, 5305.
15. Southgate, C., Zapp, M. L., and Green, M. R. (1990). *Nature,* **345**, 640.
16. Tiley, L. S., Madore, S. J., Malim, M. H., and Cullen, B. R. (1992). *Genes Dev.,* **6**, 2077.
17. Fields, S. and Song, O. (1989). *Nature,* **340**, 245.
18. Bogerd, H. and Greene, W. C. (1993). *J. Virol.,* **67**, 2496.
19. McDonald, D., Hope, T. J., and Parslow, T. G. (1992). *J. Virol.,* **66**, 7232.
20. Venkatesan, S., Gerstberger, S. M., Park, H., Holland, S. M., and Nam, Y.-S. (1992). *J. Virol.,* **66**, 7469.
21. Bradford, M. M. (1976). *Anal. Biochem.,* **72**, 248.

12

Analysis of HIV-1 LTR DNA binding proteins

S.-H. I. OU, F. WU, and R. GAYNOR

1. Introduction

The HIV-1 long terminal repeat (LTR) acts as a promoter and contains numerous regulatory elements that are critical for activation of gene expression in both the presence and absence of the *trans*-activator protein, *Tat* (1). Mutagenesis of the HIV-1 LTR has been performed and these mutants were used to direct the expression of a variety of reporter genes such as chloramphenicol acetyltransferase or luciferase. These constructs have been used to define critical regulatory domains which modulate HIV-1 gene expression either in the presence or absence of *Tat*. Some of these regulatory elements are involved in the control of gene expression in a variety of different cells while others appear to be tissue-specific. In addition to DNA regulatory elements, HIV-1 contains an additional element known as TAR (*trans*-activating region), which is capable of forming a double-strand RNA structure (see Chapter 9) (2, 3). This RNA element, which is critical for *Tat* activation, serves as the binding site for both *Tat* and cellular proteins (2, 3). This review will be on cellular factors that bind to DNA regulatory elements in the HIV-1 LTR.

The HIV-1 LTR can be divided into three functional regions designated as modulatory, core, and TAR as shown in *Figure 1* (1). The modulatory region which extends from −453 to −78 in the HIV-1 LTR is involved in regulating HIV-1 gene expression in a tissue-specific manner. This region includes two NF-κB binding sites which are essential for expression in activated T cells. The modulatory region is also involved in the negative regulation of HIV-1 gene expression. The core region, which extends from −78 to +1 contains three SP-1 binding sites and the TATA element is critical for HIV-1 gene expression in all cell lines tested. TAR extends from +1 to +60 in the HIV-1 LTR. In addition to encoding TAR RNA, this region contains multiple sequence motifs which serve as the binding sites for a variety of DNA binding proteins (4). Because different DNA binding proteins can often bind to the same DNA sequence, different effects on gene expression can result depending on the relative abundance of factors in the cell.

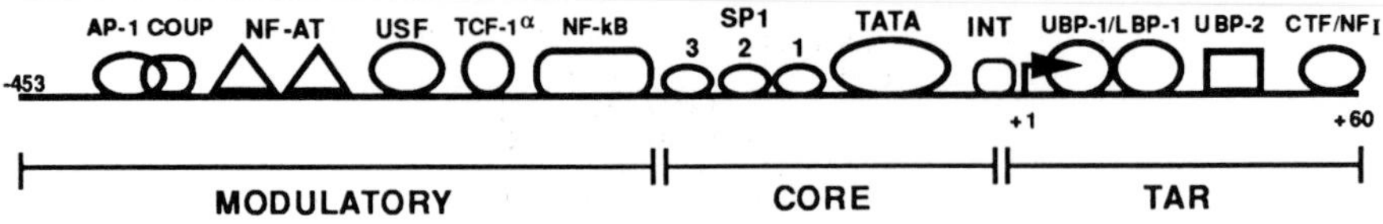

Figure 1. HIV-1 long terminal repeat (LTR) DNA. A schematic of the HIV-1 LTR, indicating potential sites of interaction for DNA and RNA binding proteins, is shown. These include binding sites for chicken ovalbumin upstream promoter (COUP), activator protein (AP-1), nuclear factor of activated T cells (NF-AT), upstream stimulatory factor (USF), T cell factor-1α (TCF-1α), nuclear factor (NF-κB), Spl, TATA, initiator, untranslated binding protein (UBP)-1, or binding protein (LBP)-1, UBP-2, and CTF/NF1.

Recent technological advances have made it possible to characterize DNA binding proteins prepared from cellular extracts, purify and clone the genes encoding these proteins, and analyse the domains in these cDNAs which mediate DNA binding. As a result, starting with only the regulatory sequence motif, it is possible to isolate the genes encoding these proteins that bind to these sequences. In this chapter we shall discuss the procedures for characterizing cellular factors that bind to the HIV-1 LTR. *Figure 2* shows a flow diagram outlining the steps needed to characterize, purify, and clone cellular factors that bind to the HIV-1 LTR.

2. Characterization of cellular proteins binding to HIV-1 regulatory elements

To determine the presence of distinct regulatory elements in the HIV-1 LTR wild-type and mutated HIV-1 LTR constructs are assayed in the presence or absence of *Tat*. Once the sequence of the regulatory element is identified, it then becomes critical to characterize the cellular proteins involved in this regulation. The binding sites for cellular proteins can be determined by using nuclear extract as a source of DNA binding proteins. The sequences that are bound can be identified using DNase I footprinting and methylation interference assays. The number of proteins binding to these sites and the molecular weight of these proteins can be characterized by both gel retardation and UV cross-linking studies.

2.1 DNase I footprinting assays

DNase I footprinting (*Protocol 1*) is a useful technique to determine the binding site of a protein in the HIV-1 LTR (5, 6). The binding of a protein to a specific DNA site protects the phosphodiester backbone of DNA from DNase I hydrolysis. An end-labelled DNA probe is digested with DNase I either in the presence or absence of nuclear extract. The products are separated by electrophoresis on denaturing DNA sequencing gels. Autoradiography is then used to visualize protected regions or 'footprints' in the DNA (5).

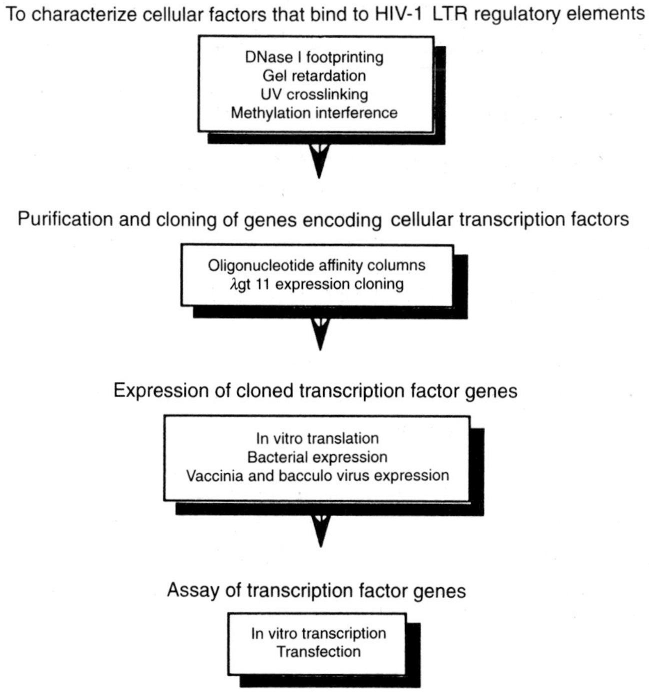
To characterize cellular factors that bind to HIV-1 LTR regulatory elements
DNase I footprinting
Gel retardation
UV crosslinking
Methylation interference
Purification and cloning of genes encoding cellular transcription factors
Oligonucleotide affinity columns
λgt 11 expression cloning
Expression of cloned transcription factor genes
In vitro translation
Bacterial expression
Vaccinia and bacculo virus expression
Assay of transcription factor genes
In vitro transcription
Transfection

Figure 2. Analysis of HIV-1 LTR DNA binding proteins. Techniques outlined in this proposal to characterize cellular factors that bind to the HIV-1 LTR, to purify and clone these genes, to express cloned transcription factor genes, and to assay these genes are listed.

The following technical aspects should be considered when performing DNase I footprinting using crude fractions of cellular proteins.

(a) Develop conditions that allow specific binding of proteins to the defined regulatory elements and preventing significant amounts of non-specific binding.

(b) To prevent non-specific binding include non-specific competitor molecules such as (poly dI–dC) (poly dI–dC), (poly dG–dC) (poly dG–dC), or (poly dA–dT) (poly dA–dT) (6).

(c) Titrate the amounts of DNase I and cellular extract that are added to the footprinting reaction to obtain a uniform ladder of labelled DNA fragments.

(d) Both large scale and microscale nuclear extract procedures have been used successfully to prepare cellular extracts for DNase I footprinting and gel retardation assays (7).

(e) The extracts can be tested using *in vitro* transcription assays with the HIV-1 LTR to determine their activity (see Chapter 10).

(f) DNase I footprinting of HIV-1 LTR DNA fragments can be performed using DNA labelled at either end of the DNA fragment and then incubated with various amounts of nuclear extract ranging from 10 μg to 100 μg.

(g) For most DNA binding proteins, we have also found that fractionation of HeLa nuclear extract on a heparin agarose column and elution with 0.4 M KCl enables the detection of low abundance DNA binding proteins by DNase I footprinting assays (6).

Protocol 1. DNase I footprinting

Equipment and reagents

- Sequencing gel electrophoresis system (Gibco–BRL Model S2)
- Buffer A: 10 mM Tris pH 7.4, 0.1 mM EDTA, 1 mM DTT, 5% glycerol

Method

1. The HIV-1 LTR *Mae*I/*Hin*dIII fragment was cloned into the *Hin*cII/*Hin*dIII site of pUC19, digested with *Eco*RI, and treated with 1–5 U of calf intestinal alkaline phosphatase.
2. The fragment was end-labelled with [γ-^{32}P]ATP with 10 U of T4 oligonucleotide kinase to generate the labelled fragment.
3. Fragments were gel isolated (*Eco*RI–*Hae*II for the coding strand), electroeluted, and used for DNase I footprinting assays.
4. Add 1–200 μg of HeLa nuclear extract with 1–5 ng of end-labelled probe in 50 μl of buffer A.
5. Incubate reaction for 20 min at room temperature.
6. Increase to 100 μl and final concentrations of DNase I (0.4–2.0 μg/ml) 5 mM $MgCl_2$, 2.5 mM $CaCl_2$.
7. The reaction was stopped after 30 min, and extracted with phenol/chloroform.
8. Ethanol precipitate and load on a 10% polyacrylamide, 8 M urea sequencing gel.
9. G + A and C + T Maxam–Gilbert sequencing reactions were performed for each.
10. All gels were then dried on Whatman 3MM paper and subjected to autoradiography.

2.2 Gel retardation assays

In gel retardation assays, the protein–DNA complexes formed are separated from unbound DNA by non-denaturing PAGE. The following technical points should be noted:

(a) Either unfractionated or fractionated nuclear extract can also be used in gel retardation analysis (8, 9).

(b) Either labelled double-stranded oligonucleotides or DNA fragments corresponding to the specific binding site can be used as probes in gel retardation analysis (6).

(c) Oligonucleotide probes ranging from 15 to 50 bp are preferable to DNA fragment probes due to the ability to label them to higher specific activity.

(d) After annealing, double-stranded oligonucleotides should be labelled with T4 polynucleotide kinase using [γ-^{32}P]ATP of high specific activity (6000 Ci/mmole) and gel purified or purified by NucTrap® probe purification columns (Stratagene).

(e) Incubate 1–10 μg of nuclear extract with 10 000–30 000 c.p.m. of probe at room temperature for 10–30 min.

(f) Include non-specific competitor such as (poly dI–dC) (poly dI–dC) (10).

(g) Electrophoresis is performed on native polyacrylamide gels (ranging from 4–6%) at either 4°C or room temperature. It is preferable to pre-run the gel at 20 mA constant current for half an hour.

(h) The number of retarded complexes can be used as an indication of the number of the state of proteins that bind to each regulatory element.

(i) Competition analysis with increasing concentrations of unlabelled specific and non-specific oligonucleotides can be used to define the binding specificity of each complex.

Protocol 2. Gel retardation

Equipment and reagents

- Vertical gel electrophoresis apparatus (Gibco–BRL model V16–2)
- 1 × DNA binding buffer: 10 mM Tris pH 7.9, 50 mM KCl, 0.1 mM EDTA, 1 mM DTT, 10% glycerol
- (Poly dI–dC) (poly dI–dC) [1 μg/μl] (Pharmacia)
- 50 × DNA loading dye: 0.5 g xylene cyanol, 0.5 g bromophenol blue, 0.25 ml 1 mM EDTA, 50 ml glycerol, H_2O to 10 ml

Method

1. Set-up binding reaction by mixing 1–10 μg of nuclear extract in a 20–50 μl reaction containing 100 ng–1 μg of (poly dI–dC) (poly dI–dC) in a final concentration of 1 × binding buffer.
2. Incubate at room temperature for 10 min.
3. To show specific competition, add increasing amounts of specific or non-specific annealed oligonucleotides or double-stranded DNA.
4. Incubate at room temperature for 10 min.
5. Add 10 000–30 000 c.p.m. of labelled DNA probe.

Protocol 2. *Continued*

6. Incubate at room temperature for 20 min.
7. Add DNA loading dye to the samples. Electrophorese the sample on a 4–6% native polyacrylamide gels in 0.5 × TBE at constant 125 V at 4°C or room temperature for 2–4 h.
8. Dry the gel on Whatman 3MM paper and expose to film with intensifying screens overnight at −70°C as in *Protocol 1*.

2.3 Analysis of DNA binding properties using methylation interference

Methylation interference (*Protocol 3*) studies are useful to determine the specific nucleotides which are in contact with the DNA binding site (11, 12). The following points should be noted:

(a) The DMS (dimethylsulfate) reagent utilized in these studies methylates guanine residues and to a lesser degree adenine residues (12).
(b) Each DNA molecule should be methylated on average of one guanine residue per DNA molecule.
(c) Both the bound and free probe are separated by gel electrophoresis.
(d) The DNA from both bound and free pools is cleaved and then fractionated on DNA sequencing gels (11).
(e) Absent bands in the bound probe pattern indicate guanine, to a lesser degree adenine residues, where methylation prevents protein binding.

Protocol 3. Methylation interference

Equipment and reagents

- Equipment (see *Protocol 1*)
- G buffer: 50 μl 1 M cacodylic acid, 10 μl 1 M $MgCl_2$, 2 μl 0.5 M EDTA, 936 μl H_2O
- G stop: 500 μl 3 M NaOAc pH 7.0, 71 μl 14 M (β-mercaptoethanol), 25 μl 4 mg/ml poly I/C, 404 μl H_2O

Method

1. Prepare a reaction mix containing: 30 μl end-labelled DNA, 200 μl G buffer, 1 μl of DMS.
2. Incubate at room temperature for 6 min. (Note time, only partially methylates DNA.)
3. Add 50 μl of G stop and 750 μl of ethanol, −20°C, 1 h.
4. Retrieve the sample by spinning in microcentrifuge.
5. Scale up gel retardation reaction tenfold and fractionate according to *Protocol 2*.

6. Perform autoradiography for 1–2 h at 4°C with 'wet' gel resting on a glass plate wrapped by Saran Wrap.
7. Cut bands containing the free and bound labelled DNA species from the gel.
8. Electroelute DNA. Purify by Qiagen™ Tip 5. Phenol/chlorform extract, add tRNA as carrier, and ethanol precipitate at −20°C for 30 min.
9. Resuspend the pellet in 100 μl of 1 M piperidine. Heat to 90–95°C for 30 min, cool on ice, and transfer to new tube. Add Na acetate, ethanol precipitate, and resuspend in G dye.
10. Add formamide loading dye, heat to 95°C for 5 min, and electrophorese on an 8 M urea, 6–8% polyacrylamide sequencing gel with (G) and (G + A) sequence lanes, and then perform autoradiography.

2.4 Determination of molecular weight of DNA binding proteins by UV cross-linking

It is important to determine the molecular weight of the cellular proteins which bind to specific regulatory sites. This can be accomplished by UV cross-linking of proteins to labelled DNA probes containing the binding site of interest (*Protocol 4*). Incubate nuclear extracts with either ^{32}P-labelled oligonucleotides or DNA fragment probes under conditions used for gel retardation (11, 13). Expose to a UV transilluminator at 254 nm and an intensity of 7000 μW/cm^2 for 5–30 min. Digest the cross-linked products with low concentrations of DNase I and separate proteins on an SDS–polyacrylamide gel (2).

Following autoradiography, the molecular weight of cellular proteins that bind to the specific regulatory elements can be determined. Modifications of this technique can also be made to ascertain the molecular weight of specific gel retarded species. After performing gel retardation expose the wet gel to X-ray film, and elute retarded DNA–protein complex from the gel. Expose the gel slice to UV light, soak gel slice in Laemli binding buffer, heat to 100°C for 5 min, and electrophorese on an SDS–PAGE.

Protocol 4. UV cross-linking

Equipment and reagents

- UV transilluminator (Fotodyne)
- Uniformly labelled probe of between 50–200 bp
- Vertical gel running apparatus (see *Protocol 2*)
- Reagents (see *Protocol 2*)

Method

1. Set-up the binding reaction as described in *Protocol 2*.
2. Place the sample about 4.5 cm from a Fotodyne UV transilluminator (maximum emission wavelength 7000 μW/cm^2) and irradiate the sample for 10–30 min at 4°C or room temperature.

Protocol 4. *Continued*

3. Bring up the reaction mixture to 10 mM $CaCl_2$.
4. Add 3 μg of DNase I.
5. Incubate the reaction for 30 min at 37 °C.
6. Add 5 × protein loading dye buffer, heat to 95 °C for 5 min, and electrophorese on an 8–10% SDS–polyacrylamide gel. Include ^{14}C-labelled protein molecular weight markers to help in determining protein size.
7. Dry the gel and expose it for autoradiography using an intensifying screen.

3. Purification and cloning of cellular transcription factors

To determine the regulatory properties of the different transcription factors that bind to the HIV-1 LTR, it is necessary to purify and clone these factors. Two techniques have been developed which have greatly facilitated the cloning of cDNAs encoding cellular transcription factors.

(a) Oligonucleotide affinity chromatography allows purification of cellular factors based on their binding affinity to a specific regulatory sequence (14).

(b) cDNA libraries can be synthesized and expressed as λgt11 β-galactosidase fusion proteins and the ability of the fusion proteins to bind to specific DNA regulatory motifs can then be assayed (15, 16).

3.1 Oligonucleotide affinity column purification

The purification of cellular factors that bind to important DNA regulatory regions has been markedly improved by the use of oligonucleotide affinity column chromatography (*Protocol 5*). This technique is based on the fact that DNA binding proteins will bind to oligonucleotides corresponding to specific binding sequences (14). After multiple passages of cellular extract through these columns, moderate to high affinity DNA binding proteins can be purified and silver staining of SDS–polyacrylamide gels can be used to visualize the protein of interest. This technique has been used to purify a variety of sequence-specific proteins that bind to the HIV-1 LTR including SP 1, NF-κB, and UBP-1 (2, 17). It is important to note:

(a) Because of the low abundance of most DNA binding proteins, the process of oligonucleotide affinity chromatography can not be performed with crude nuclear extract since the high protein concentration of nuclear extract limits the ability to load enough material on the oligonucleotide column. Usually, several conventional chromatography columns must first be used to partially purify the proteins of interest. An initial purification step will both improve the flow rate of the column and allow increased quantities of partially purified cellular proteins to be applied to the column.

(b) Oligonucleotide affinity columns containing mutated binding sequences that prevent the binding of the DNA binding protein of interest can be used to remove non-specific binding proteins prior to the use of the specific oligonucleotide column.

(c) 60–100 mg of nuclear extract is usually required to obtain 50–100 pmol of the desired protein.

(d) Aliquots of the proteins should be assayed for DNA binding by gel retardation and DNase I footprinting, and subjected to SDS–polyacrylamide gel electrophoreses and silver staining to quantify the amounts of protein (3). The purified protein can then be transferred to nitrocellulose or PVP paper, proteolytically cleaved, applied to HPLC for purification, and subjected to peptide sequence analysis (17). The peptide sequence can then be used to design oligonucleotides which can be used to clone the cDNA of interest.

Protocol 5. Oligonucleotide affinity column

Equipment and reagents

- Oligonucleotides 15–50 bp
- CNBr-activated Sepharose CL-2B
- Sialized Eppendorf tubes
- Econo Column™ chromatography column (Bio-Rad)
- Buffer A: 67 mM Tris–HCl pH 7.6, 10 mM $MgCl_2$, 1 mM EDTA, 5 mM DTT
- Buffer B: 10 mM Tris–HCl pH 7.5, 10 mM $MgCl_2$, 5 mM DTT, 5 mM ATP
- Buffer C: 10 mM Tris–HCl pH 7.6, 300 mM NaCl, 1 mM EDTA, 0.02% sodium azide
- Buffer Z: 25 mM Hepes pH 7.8, 0.1% NP-40, 100 mM KCl, 1 mM DTT, 20% glycerol

Method

1. Anneal approx. 200 μg of each strand of complementary oligonucleotides in 200 μl of buffer A which contains the binding site of interest. Incubate at 95°C for 2 min; at 65°C for 15 min; at 37°C for 15 min; at room temperature for 5 min; and on ice for 2 min.
2. Add (100 U) T4 polynucleotide kinase and (20 mM) ATP.
3. Incubate the reaction at 37°C for 2 h.
4. Heat the mixture at 65°C for 15 min to inactivate the kinase and then ethanol precipitate the sample.
5. Dissolve the DNA in 200 μl of buffer B. Add 30 U of T4 DNA ligase, and incubate at 4°C overnight. Phenol extract the sample, ethanol precipitate, and then dissolve in 100 μl of water.

Protocol 5. *Continued*

6. Check a small aliquot of the oligonucleotide by agarose gel electrophoresis to demonstrate that ligation has occurred.
7. Activate 10 ml of Sepharose CL-2B by treatment with cyanogen bromide. Couple the ligated DNA to the resin at room temperature for 16 h. Wash the resin with H_2O extensively, and store at 4°C in 20 ml of buffer C. The concentration of covalently bound DNA in the resin is ~ 20 μg/ml of resin.
8. Prepare nuclear extract from the equivalent of 60–100 litres of HeLa cells. Fractionate on one or more conventional columns such as heparin–agarose. Assay column fractions by DNA binding assays including gel retardation (*Protocol 2*) and DNase I footprinting (*Protocol 1*).
9. Dialyse cellular extract into buffer Z.
10. Incubate the cellular extract with 100–500 μg (poly dI–dC)·(poly dI–dC) (25–30 μg) on ice for 30 min.
11. Load cellular extract on to a 5 ml oligonucleotide column equilibrated with buffer Z at 4°C. Always load the column by gravity flow, over a 1–2 h period. Wash the column with at least 20 column volumes of cold buffer Z. Remove the buffer at the top of the column, block column flow, add one column volume of elution buffer (buffer Z containing 1.0 M KCl), and let sit for 10 min at 4°C. Collect 200 μl fractions in sialinized Eppendorf tubes. Wash column with 0.6 column volume of buffer Z and collect 200 μl fractions.
12. Dialyse active samples for DNA binding against buffer Z.
13. Re-load on to additional oligonucleotide columns if further purification is desired, or freeze samples in 100 μl aliquots in liquid nitrogen.
14. Regenerate oligonucleotide columns with 5 ml of buffer Z containing 1.0 M KCl and then place into 10 ml of storage buffer.

3.2 Bacteriophage lambda gt11 expression cloning

Though oligonucleotide affinity chromatography has markedly improved our ability to purify sequence-specific DNA binding proteins, this technique works best to purify high affinity and moderately abundant transcription factors. If the boundaries of the binding sequence of interest are not clearly defined, or if the binding affinity of the protein is weak, then lambda gt11 expression cloning is an attractive alternative method for cloning cellular transcription factors (*Protocol 6*) (15, 16).

(a) Oligomerized oligonucleotides corresponding to the binding site of interest are used as probes (18).

(b) Labelled oligonucleotides with mutations in critical nucleotides in the binding domain are used as negative controls in these screens.

(c) Phage are plated at between 20 000–50 000 plaques/150 mm plate, and 4 h later nitrocellulose filters incubated in IPTG are used to overlay the plates.

(d) The filters are incubated for 4–6 h to induce proteins containing β-galactosidase fused to cDNA libraries.

(e) The filters are blocked with milk, and then may either be incubated directly with probe or first be subjected to renaturation and denaturation with guanidine hydrochloride prior to incubation with probe.

(f) The filters are then washed, subjected to autoradiography, and positive plaques are aligned with plates containing the phage.

(g) The phages are isolated and then subjected to secondary and tertiary screening to purify and isolate a single positive phage.

(h) At each stage, both wild-type and mutant oligonucleotides are used in the screening procedure to determine the specificity of fusion protein binding.

(i) Stocks of purified phages are then obtained for preparation of DNA and sequence analysis.

Protocol 6. Screening of lambda gt11 expression library for DNA binding proteins

Equipment and reagents

- 150 mm and 60 mm Petri plates (Falcon)
- Nitrocellulose filter (0.45 μm pore size) (Millipore)
- India ink
- 23 gauge needle
- Tweezers
- Water-bath incubator with thermometer
- Glogos™ II autoradiogram markers (Stratagene)
- 5 × PB (litre): 25 g NaCl, 25 g yeast extract, 50 g bactotryptone, 50 ml 1 M Tris pH 7.5, 50 ml 1 M $MgSO_4$
- 1 × PB plates (litre): 15 g agar, 5 g NaCl, 5 g yeast extract, 10 g bactotryptone, 10 ml 1 M Tris pH 7.5, 10 ml 1 M $MgSO_4$
- Top agar: same as plates except 7 g of agarose in place of 15 g agar
- 1 × SM (litre): 5.8 g NaCl, 2 g $MgSO_4.7H_2O$, 50 ml 1 M Tris pH 7.5, 5 ml 2% gelatin
- 1 M IPTG: 1 g IPTG + 3.8 ml sterile H_2O
- 20% milk
- 10 × binding buffer: 0.2 M Tris pH 7.5, 0.5 M KCl, 2 mM EDTA

Method

1. Start an overnight of *E. coli* strain Y1088 with 10 ml PB + 0.2% maltose + 50 μg/ml ampicillin.
2. Pre-warm 150 mm plates to 37°C, melt top agar, and place in 50°C water-bath.
3. Dilute λgt11 phage stock to appropriate titre for primary screen (approx. 30 000 plaques/150 mm plate). For secondary and later

Protocol 6. *Continued*

screens, dilute phage stock 1 : 1000 and plate 10–30 ml/100 mm plate. Add 100 μl to 0.3 ml Y1088 overnight in a 15 ml Falcon tube. Incubate at room temperature for 15 min. Add 8 ml 1 × PB top agar. Mix, and plate on 150 mm plates. Incubate at 42°C for 3–4 h or until plaques are visible.

4. Dip numbered nitrocellulose filters into a large Petri dish containing 50 ml of 10 mM IPTG.
5. Remove plates from incubator (five plates at a time). Place IPTG saturated filters on plates, and mark filters with a 23 gauge needle dipped in India ink for orientation. Incubate filters and plates in 37°C incubator for 4–6 h.
6. Cool plates for 5 min at 4°C. Remove IPTG-induced filters from the plates. Block the filters with 5% milk in 1 × binding buffer at room temperature. Place each filter in 25 ml of milk solution in large Petri dishes with protein bound side facing up.
7. Take 10 μg of annealed, double-stranded oligonucleotides corresponding to the binding site of interest and treat with DNA ligase (see *Protocol 4*).
8. Nick translate 50 ng of ligated oligonucleotides using [α-^{32}P]dNTPs, and purify the labelled oligonucleotides by gel electrophoresis.
9. Rinse filters with binding buffer + 1 mM DTT two times.
10. Probe filters at room temperature for 1 h with 10^6 c.p.m./ml of probe in binding buffer + 1 mM DTT containing 5–10 μg of sheared salmon sperm DNA/ml of probe. Rock gently and continuously.
11. Wash filters with binding buffer and 1 mM DTT in trays. Use 600 ml/ 8–10 filters in each tray. Repeat the wash three times.
12. Dry filters on Whatman 3 MM paper. Wrap in between Saran Wrap on boards and expose overnight with intensifying screen at −70°C. Orient the filter and the film with Glogos™ II (Stratagene) autoradiogram markers.
13. Align phages with autoradiographs and isolate positive plaques and surrounding agarose.
14. Elute phages in 1 ml of 1 X sterile SM at 4°C overnight or room temperature for 4–6 h.
15. Plate secondary and tertiary phages on to 100 mm plates and repeat the protocol as described for primary screening.

4. Expression systems for transcription factors

After the identification of cDNAs which encode HIV-1 DNA binding proteins, it is necessary to determine if the cloned cDNA encodes a full-length protein. In-frame stop codons should be present at both the 5′ and 3′ portions of the cDNAs. The position of these stop codons should be confirmed by reverse transcriptase–polymerase chain reaction (RT–PCR) on RNA prepared from the same cells as the cDNA library was prepared using 5′ and 3′ untranslated region primers and internal cDNA primers. Nuclease S1 mapping, primer extension, and Northern analysis should also be used to confirm the 5′ start and the 3′ end of each cDNA. Once the full-length protein is identified, a detailed characterization of the cDNA should be performed. This should include:

(a) Northern analysis to determine the expression pattern.

(b) Southern analysis to determine if the gene is single copy.

(c) Locate homologies with other DNA binding proteins using a database such as Swiss Prot. This homology search will aid in the construction of cDNA constructs that mutate potential DNA binding and/or other domains which can then be assayed using both *in vivo* and *in vitro* assays.

(d) Use expression systems to produce adequate quantities and re-confirm its DNA binding properties of the protein of interest.

There are three main approaches to the expression of the cDNA clones:

(a) *In vitro* transcription of the cDNA of interest and subsequent translation using rabbit reticulocyte lysate can be used to express many proteins (19). The quantities of proteins produced using reticulocyte lysate translation are sometimes sufficient to allow analysis of proteins using gel retardation assays.

(b) Expression of the protein in bacterial systems is useful to generate sufficient quantities of protein for DNA binding assays, *in vitro* transcription, and the preparation of monoclonal and polyclonal antibodies (18, 20).

(c) Many proteins that are very large, or require specific post-translational modifications can be expressed in either vaccinia (21, 22) (see Chapter 7) or baculovirus (23) systems (see Chapter 6).

4.1 *In vitro* transcription and translation using cloned transcription factor genes

In vitro transcription and translation in rabbit reticulocyte lysate will allow the production of enough protein to assay some transcription factor cDNAs by both DNA binding and *in vitro* transcription (19).

(a) The addition of [^{35}S]methionine allows the labelling and visualization of the translated protein by autoradiography.

(b) A number of expression vectors can be used for *in vitro* transcription and translation of transcription factor cDNAs such as the pGEM series which contains both the T7 or SP6 promoters. However, we have found that the pTM1 expression vector which contains the encephalomyocarditis virus virus untranslated leader sequences downstream of the T7 promoter markedly increases the ability to transcribe and translate cloned cDNAs (21, 22).

(c) Since the vector also contains a T7 transcription terminator sequence located downstream of the cloned cDNA, the pTM1 vector does not require linearization by restriction digestion to terminate the RNA (21, 22).

The translated proteins have numerous applications: 1–3 μl of the lysate can be used in gel retardation assays. A variety of truncations of the protein can be synthesized by using unique restriction enzymes found in the transcription factor cDNA. These truncated proteins can then be used to map the DNA binding and/or dimerization domains of proteins using gel retardation assays. Translated proteins have also been used in *in vitro* transcription assays with the HIV-1 LTR to determine the transcriptional activating or repressing abilities of these proteins (24). However, the problems with *in vitro* translation include:

- difficulty in synthesizing full-length proteins of high molecular weight
- low yields of translation product
- partial products which interfere with binding and transcription assays may also be produced.
- Many important post-translational modifications such as phosphorylation are not introduced

4.2 Bacterial expression systems

To produce large quantities of bacterial fusion proteins, we have found that two systems are extremely useful:

- proteins fused to glutathione-S-transferase
- proteins tagged with a histidine tail

Proteins containing a portion of the *S. japonica* glutathione-S-transferase protein are fused to the cDNA of interest. These fusion constructs are cloned downstream of the T7 promoter in a variety of vector backgrounds such as pGEX-2T (20), and can be expressed using a variety of *E. coli* strains carrying T7 polymerase such as the protease negative strain BL21 DE3 lys E.

(a) The bacteria are grown to an OD at 600 nm of 0.6, induced with IPTG, and harvested 2–4 h later.

(b) After lysis of the *E. coli,* the fusion proteins are bound to Sepharose agarose beads containing glutathione, washed extensively, and eluted with buffer containing glutathione.

(c) The glutathione-S-transferase moiety can be removed by cleavage with either thrombin or factor X because these recognition motifs are present in the fusion protein (2).

(d) Cleavage can be performed while the fusion protein is bound to the glutathione beads, allowing the purification of native protein.

(e) Fusion proteins can be obtained in high yield (up to 500 μg–1 mg per litre of culture) and with purities ranging from 50–95%.

(f) Following dialysis, these proteins can be used for DNA binding, *in vitro* transcription assays, and for immunization of rabbits and mice for antibody production.

Protocol 7. Bacterial expression using pGEX vector protocol

Equipment and reagents

- 5 ml chromatography column (Bio-Rad)
- 1 M IPTG (Sigma)
- 0.5 M glutathione (Sigma)
- Lysis buffer 1 × PBS: 150 mM NaCl, 16 mM Na_2HPO_4, 4 mM Na_2HPO_4 pH 7.4, 1% Triton X-100, 1 mM EDTA, 1 mM DTT, 0.5 mM PMSF, 0.5 mg/litre leupeptin, 0.7 mg/litre pepstatin
- Thrombin cleavage buffer: 50 mM Tris pH 8.0, 150 mM NaCl, 2.5 mM $CaCl_2$, 1 M M DTT
- Bradford assay solution (Bio-Rad)
- Sonicator (Heat Systems, Inc. Model XL2020)
- Glutathione–agarose (Sigma)

Method

1. Inoculate an overnight culture of BL21 DE3 lys E transformed with the recombinant pGEX plasmid in a 1/200-fold dilution into 500 ml of LB media.
2. Grow the cells at 37 °C to mid-log phase OD_{600}-~0.6. Induce expression with IPTG to a final concentration of 0.1 mM. Grow for an additional 2–3 h at 37 °C.
3. Pellet the cells by centrifugation and resuspend in 1/50 vol. of lysis buffer.
4. Lyse cells by sonication five times, 20% power, 30 sec each.
5. Centrifuge the bacterial extract at 5000 *g* for 30 min at 4 °C. Apply the supernatant to 1.0 ml of glutathione–agarose beads in columns. Wash with 5–10 ml of cold 1 × PBS. Elute the bound proteins with five column volumes of 5 mM glutathione in 50 mM Tris–HCl pH 8.0. Collect in 0.5 ml fractions. Peak protein concentration usually resides in the second fraction.
6. Alternatively, the fusion protein can be cleaved while remaining bound

Protocol 7. *Continued*

to the glutathione–agarose beads. Equilibrate the column with thrombin cleavage buffer after PBS wash. Add one column volume of thrombin cleavage buffer + 6 μg (60 μl) of thrombin (Sigma). Seal the column with Parafilm, and rotate at room temperature for 40 min. Collect the 1 ml in one fraction, wash with 2–3 ml of thrombin cleavage buffer, and collect. Add PMSF to 0.5 mM final concentration to stop the cleavage reaction.

7. Assay protein concentrations using a Bradford assay.
8. Dialyse the peak protein fractions into buffer containing 100 mM KCl, 20% glycerol, 0.2 mM EDTA, 20 mM Tris–HCl pH 7.9, 1 mM DTT, 0.5 mM PMSF (2 × gel retardation solution).

The use of histidine-tagged proteins can also facilitate the purification of bacterial proteins (25) (see Chapters 4 and 5).

(a) The cDNA of interest can be modified using the polymerase chain reaction to add six histidine residues to either the amino- or carboxy-terminus of the protein. This modified cDNA is then cloned downstream of a T7 promoter and ribosome binding site using several commercially available plasmid vectors.
(b) After expression of the histidine-tagged proteins in suitable hosts such as M15 pREP4, the bacteria are lysed and the extract is applied to Sepharose beads which are coupled to nickel (25).
(c) Following extensive washing of the column, the histidine-tagged protein is eluted with increasing concentrations of imidazole.
(d) Following dialysis the protein can be used in gel retardation or DNase I footprinting assays.

Bacterial produced proteins are very useful for studying the DNA binding properties of transcription factors. The ability to obtain relatively pure preparations of proteins and the ease of manipulation of wild-type and mutated plasmid constructs in bacteria provides a means to determine the functional domains required for DNA binding. The bacterially-expressed proteins can also be used to generate monoclonal and polyclonal antibodies. These antibodies can be used to identify the nature of the endogenous transcription factor with regard to its nuclear localization, state of phosphorylation, relative abundance, and binding and dimerization properties. Such antibodies are critical in characterizing the properties of cellular transcription factors.

Potential problems with bacterial expression of transcription factors include:

- the inability to synthesize proteins of high molecular weight
- toxicity of these proteins to bacteria

- protein degradation
- failure to post-translationally modify proteins

For many proteins up to 100 kDa, the full-length proteins can be synthesized in bacterial expression vectors. Though it is frequently difficult to synthesize larger proteins in bacteria due to degradation of the full-length protein, portions of these proteins containing the DNA binding domain can be synthesized and used in DNA binding assays. Some proteins regardless of their sizes are extremely toxic and often only degraded proteins are obtained. Finally, post-translational modifications do not occur in bacteria making it difficult to analyse factors that require modifications for their activity in *in vitro* transcription or DNA binding.

4.3 Eukaryotic expression systems

Both baculovirus and vaccinia expression systems have recently been used to synthesize transcription factors. The vaccinia system requires the cloning of the transcription factor gene downstream of the T7 promoter in the pTM1 vector (21, 22, 26).

(a) HeLa cells are transfected with the pTM1 vector and then infected with a recombinant vaccinia virus which constitutively produces T7 RNA polymerase.
(b) Cells are harvested at 24–48 hours post-transfection and nuclear extract is prepared.
(c) A histidine tag or the influenza haemagglutinin epitope tag can be inserted into the transcription factor gene at either the amino- or carboxy-terminus to allow purification by affinity chromatography using the nickel chromatography (28) or 12CA5 monoclonal antibody respectively (27).
(d) About 50 μg of protein can be purified from 20–150 mm plates of HeLa cells.

Similar experiments can be performed using co-infection protocols with recombinant vaccinia virus containing the gene of choice and a second vaccinia virus containing T7 RNA polymerase (26). However, the recombinant vaccinia virus containing the gene of choice must first be generated and this process can be laborious. Vaccinia expression systems have several advantages in that they allow high levels of expression proteins including proteins with 200 kDa in molecular weight which are correctly post-translationally modified.

Other systems (23, 29) which can produce high levels of correctly modified proteins are the baculovirus expression systems. The baculovirus expression system has been used extensively to over-produce cellular transcription factors (23, 29). Baculovirus has a number of advantages over other expression systems:

- high level expression of proteins regardless of their molecular weight
- solubility of the majority of these proteins
- lack of requirement for helper virus

The gene of interest is first cloned into a baculovirus vector downstream of the polyhedrin gene promoter and this DNA is co-transfected with wild-type viral DNA into *Sf*9 cells. Recombinant virus from the supernatant is plaque purified by several rounds of growth—the recombinants can be visually screened due to their unique morphology. High titre viral stocks are obtained and maximum protein production is determined by varying the multiplicity and time of infection. Baculovirus proteins can be purified by histidine tagging or influenza epitope insertion.

5. Assays of eukaryotic transcription factors

The ability to assay the effects of cellular transcription on the activity of the HIV-1 LTR is critical in understanding the control of both basal and Tat-induced gene expression. Two approaches for assaying the activity of transcription factor cDNAs are possible. Each has significant advantages as well as potential problems.

(a) Transcription factor cDNA is cloned into a eukaryotic expression vector and co-transfected into cells with an HIV-1 LTR reporter construct. The transcription factor gene can be cloned either in its native form or as a fusion protein with GAL4 (30).

(b) Transcription factor cDNAs are over-produced in bacterial or eukaryotic expression vectors and assayed for their ability to stimulate gene expression from the HIV-1 LTR using *in vitro* transcription assays.

5.1 Transfection assays

To determine the function of transcription factors that bind to the HIV-1 LTR, both *in vivo* and *in vitro* assays should be performed. The ability to assay transcription factors by transfection is complicated by the fact that most transcription factors are constitutively expressed in a variety of cell lines. This fact may prevent marked activation of the HIV-1 reporter constructs by most transcription factor genes in co-transfection assays. However, in a number of cell lines, certain transcription factors are expressed at very low levels. If the level of gene expression of certain transcription factors is elevated sufficiently following transfection, the factor may then be capable of activating HIV-1 gene expression in transfection experiments (31). If no effects are seen in transfection experiments using the full-length cDNA, then fusion of the transcription factor cDNA with the DNA binding domain of the yeast transcription factor GAL4 should be performed (30).

To assay the transcription factor cDNA, it should be cloned downstream of either a strong eukaryotic promoter such as SV40, CMV, or RSV. Transfection into a variety of cell lines such as HeLa and Jurkat with an HIV-1 reporter construct should then be performed (see Chapters 10 and 11). Similar experiments can be performed using GAL4–cDNA fusions.

(a) Reporter constructs containing multimerized GAL4 binding sites upstream of a TATA element and either luciferase or chloramphenicol acetyltransferase reporter genes can be used to assay the activating potential of the GAL4–cDNA (30).
(b) At 36–48 hours post-transfection, the cells are harvested and assayed for CAT or luciferase activity. The transfection assays test the ability of different transcription factors to activate or repress HIV-1 gene expression. Once the results from these assays are established, domains in the cDNAs which are involved in regulating HIV-1 gene expression can be defined by mutagenesis experiments.

5.2 *In vitro* transcription

It is also important to determine the role of cloned transcription factor genes on *in vitro* transcription from the HIV-1 LTR (see Chapter 10).

(a) Proteins assayed in this system can be synthesized using *in vitro* transcription and translation, bacterial, vaccinia, or baculovirus expression systems (24).
(b) The proteins can be purified using affinity chromatography with either antibodies or nickel chromatography.
(c) The proteins can be added directly to unfractionated or fractionated nuclear extract and together with a HIV-1 reporter template, run-off transcription can then be performed (32).
(d) The role of these proteins on *Tat* activation *in vitro* can be assayed.
(e) If the level of an endogenous transcription factor is high in nuclear extract, the factor to be assayed can be removed from the nuclear extract using either conventional or oligonucleotide affinity chromatography or immunodepleted by antibody affinity chromatography. Thus, it is usually possible to assay the activity of transcription factors using *in vitro* transcription assays.

Acknowledgements

We thank Ty Lawrence and Brian Finley for the preparation of this manuscript. In addition we thank David Harrich, Ajay Nirula, Ben McLemore, and Jacob Seeler for scientific input and suggestions incorporated into this manuscript.

References

1. Gaynor, R. B. (1992). *AIDS*, **6**, 347.
2. Wu, F., Garcia, J., and Sigman, D. (1991). *Genes Dev.*, **5**, 2128.
3. Dingwall, C., Ernberg, I., and Gait, M. J. (1990). *EMBO J.*, **9**, 4145.

4. Jones, D. A., Luciw, P. A., and Duchange, N. (1988). *Genes Dev.*, **2**, 1101.
5. Galas, D. and Schmitz, A. (1978). *Nucleic Acids Res.*, **5**, 3153.
6. Gaynor, R. B., Kuwabara, M. D., and Wu, F. K. (1988). *Proc. Natl Acad. Sci. USA*, **85**, 9406.
7. Dignam, J. D., Martin, P. L., Shastry, B. S., and Roeder, R. G. (1983). In *Methods in enzymology* (eds Wu, R., Grossman, L., and Moldave, K.), Vol. 101, pp. 582–9.
8. Fried, M. and Crothers, D. M. (1984). *J. Mol. Biol.*, **172**, 241.
9. Garner, M. M. and Revzin, A. (1981). *Nucleic Acids Res.*, **9**, 3047.
10. Staudt, L. M., Singh, H., Sen, R., Wirth, T., Sharp, P. A., and Baltimore, D. (1986). *Nature*, **323**, 640.
11. Baldwin, A. and Sharp, P. (1988). *Proc. Natl Acad. Sci. USA*, **85**, 723.
12. Maxam, A. and Gilbert, W. (1980). In *Methods in enzymology* (eds Grossman, L. and Moldave, K.), Vol. 65, pp. 499–560.
13. Chodosh, L. A., Carthew, R. W., and Sharp, P. A. (1986). *Mol. Cell. Biol.*, **6**, 4723.
14. Kadonga, J. T. and Tjian, R. (1986). *Proc. Natl Acad. Sci. USA*, **83**, 5889.
15. Singh, H., LeBowitz, J., Baldwin, A., and Sharp, P. A. (1988). *Cell*, **52**, 415.
16. Vinson, C. R., LaMarco, K. L., Johnson, P. F., Landshulz, W. H., and McKnight, S. L. (1988). *Genes Dev.*, **2**, 801.
17. Kadonaga, J. T., Carner, K. R., Masiarz, F. R., and Tjian, R. (1987). *Cell*, **51**, 1079.
18. Li, C., Lai, C., Sigman, D. S., and Gaynor, R. B. (1991). *Proc. Natl Acad. Sci. USA*, **88**, 7739.
19. Hope, I. A. and Atruhl, K. (1986). *Cell*, **46**, 885.
20. Smith, D. B. and Johnson, K. S. (1988). *Gene*, **67**, 31.
21. Moss, B. (1990) *Annu. Rev. Biochem.*, **59**, 661.
22. Moss, B., Elroy-Stein, O., Mizukami, T., Alexander, W. A., and Fuerst, T. R. (1990). *Nature*, **348**, 91.
23. Summers, M. D. and Smith, G. E. (1987). *Texas Agriculture Experiment Station Bulletin No. 1555*, College Station, Texas.
24. Garcia, J. A., Ou, S.-H. I., Wu, F., Lusis, A. J., Sparkes, R. S., and Gaynor, R. B. (1992). *Proc. Natl Acad. Sci. USA*, **89**, 9372.
25. Gentz, R., Chen, C. H., and Rosen, C. A. (1989). *Proc. Natl Acad. Sci. USA*, **83**, 821.
26. Fuerst, T. R., Niles, E. G., Studier, F. W., and Moss, B. (1986) *Proc. Natl Acad. Sci. USA*, **83**, 8122.
27. Field, J., Nikawa, J., Broek, D., MacDonald, B., Rodgers, L., Wilson, I. A., *et al.* (1988). *Mol. Cell Biol.*, **8**, 2159.
28. Janknecht, R., de Martynoff, G., Lou, J., Hipskind, R. A., Nordheim, A., and Stunnenberg, H. G. (1991). *Proc. Natl Acad. Sci. USA*, **88**, 8972.
29. Carbonell, L. F., Klowden, M. J., and Miller, L. A. (1985). *J. Virol.*, **56**, 153.
30. Southgate, D. D. and Green, M. R. (1991). *Genes Dev.*, **5**, 2496.
31. Schmid, R. M., Perkins, N. D., and Duckett, C. S. (1991). *Nature*, **352**, 733.
32. Marciniak, R. A., Calnan, B. J., Frankel, A. D., and Sharp, P. A. (1990). *Cell*, **63**, 791.

13

Transient assay for Nef-induced down-regulation of CD4 antigen expression on the cell surface

J. SKOWRONSKI and R. MARIANI

1. Introduction

The *nef* genes of the human and simian immunodeficiency viruses encode closely related proteins of 25 kDa to 27 kDa which are expressed in the cytoplasm of virally-infected cells. *Nef* is essential for SIV pathogenesis in the rhesus monkey model of human AIDS (1) but does not appear to be required for viral growth *in vitro*. As a result, *nef* genes of many commonly studied viral isolates are truncated prematurely, or contain loss-of-function alleles. Recent observations from several laboratories indicate that a large fraction of *nef* alleles derived from either laboratory HIV and SIV isolates, or cloned directly from peripheral blood leucocytes of HIV-infected patients, are able to decrease CD4 antigen expression on the cell surface. This effect of *nef* on CD4 expression is neither species-, nor cell type-specific. CD4 down-regulation has been documented in both human and murine CD4-positive T cells, and in non-lymphoid cells expressing endogenous, or transfected, CD4 genes (2–6). The CD4 antigen is required both for the normal development and function of helper T cells, and to mediate immunodeficiency virus infection (7). Thus, the interaction of *nef* with CD4 is likely to have profound, yet mostly unexplored, consequences for the HIV-1-infected CD4-positive T cell.

In this chapter we describe a transient assay which allows the rapid identification of *nef* alleles that decrease CD4 antigen expression on the cell surface. This assay has been used in our laboratory to characterize a large number of *nef* alleles isolated directly from clinical samples (6), and of mutant *nef* genes generated for structure–function studies of the Nef protein.

2. Assays to measure CD4 down-regulation by Nef

CD4 down-regulation by *nef* requires co-expression of both Nef and CD4 in the same cell. Two assays can be used to detect the effect of *nef* on CD4 antigen on the cell surface:

- stable expression assay
- transient expression assay

The stable expression assay involves generation of cell lines that express both CD4 and Nef. This is achieved by transduction of CD4-positive cell lines with retroviral Nef expression vectors. CD4 expression on the cell surface is detected following selection and expansion of productively infected cells. By contrast, the transient expression assay does not require construction of Nef expressing cell lines. Instead, CD4-positive cell lines are transfected with vectors designed to direct rapid, high level expression of Nef. Level of CD4 antigen on the cell surface is then measured 24 hours following transfection.

The transient assay has several advantages:

(a) The transient assay can be completed within 24 hours. In contrast, selection and expansion of retrovirally transduced cells requires at least 7–14 days.

(b) The transient assay avoids selection against cells that express Nef. Cells transduced with Nef expression vectors grow more slowly than those transduced with control vectors. During prolonged culture, cells with measurably lower Nef expression, and an increased CD4 expression on the cell surface, are selected.

(c) The transient assay permits quantitative measurements of CD4 down-regulation as a function of the amount of Nef protein expressed in transfected cells. Such dose–response experiments can be used to measure the strength of mutant and variant *nef* alleles (see section 3 below).

2.1 Transient assay for CD4 down-regulation by Nef

The design of a transient assay for CD4 down-regulation by Nef, as used in our laboratory, is as follows:

(a) CD4-positive cells are co-transfected with a mixture of Nef expression vector and a reporter plasmid that directs expression of the *E. coli lacZ* gene (see *Figure 1* and *Protocol 1*). A fraction of transfected cells both express Nef and show elevated β-galactosidase activity. The *lacZ* reporter gene provides a positive control for transfection efficiency and serves as a marker for productively transfected cells.

(b) 12–24 hours after transfection, cells are harvested and simultaneously tested for CD4 antigen and β-galactosidase activity. Cells are loaded with fluorescein di-β-D-galactopyranoside (FDG), a fluorogenic substrate for β-galactosidase that becomes fluorescent only after enzymatic cleavage and is detected by green fluorescence (see *Protocols 2* and 3). The FDG-

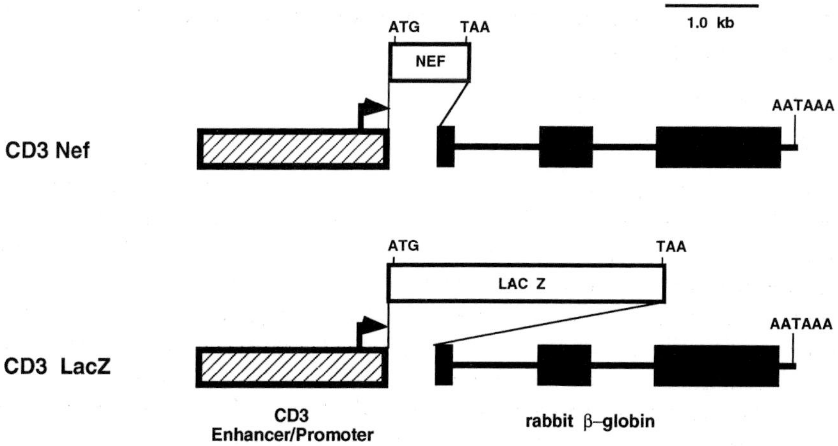

Figure 1. CD3 β expression vectors. The generic CD3 β expression vector contains the enhancer and promoter elements derived from the murine gene encoding the δ subunit of the CD3 complex expressed in all T cells. RNA processing signals are derived from the rabbit β-globin gene. *Lac Z, nef,* or other open reading frames of interest, are cloned between the unique *Xba*I (5′) and *Mlu*I (3′) sites (not shown). The transcription initiation site within the CD3 promoter, the first initiation codon, and in-frame termination codon in the *nef* and *lacZ* open reading frames are indicated.

loaded cells are then reacted with a monoclonal anti-CD4 antibody coupled to phycoerythrin which is revealed by red fluorescence (see *Protocol 4*).

(c) CD4 expression on the cell surface and β-galactosidase activity are detected in live cells on a cell-by-cell basis by two-colour flow cytometry (see *Figure 2*).

Induction of a readily detectable decrease in CD4 antigen on the cell surface requires a relatively high level Nef expression in a large fraction of transfected cells. This will vary depending on particular combinations of cell lines, expression vectors and transfection protocols used in the assay. These variables are briefly discussed below.

2.1.1 Cell lines

In our experience, the best results are obtained with CD4-positive T cell lines. The two cell lines of choice, routinely used in our laboratory, are CEM E5 T cells and Jurkat T cells, both selected for high CD4 expression (8).

- both cell lines are easy to maintain and can be efficiently transfected by electroporation (see *Protocol 1*)
- CD4 expression is relatively uniform
- a relatively small decrease in CD4 on the cell surface (10–20%) can be readily detected

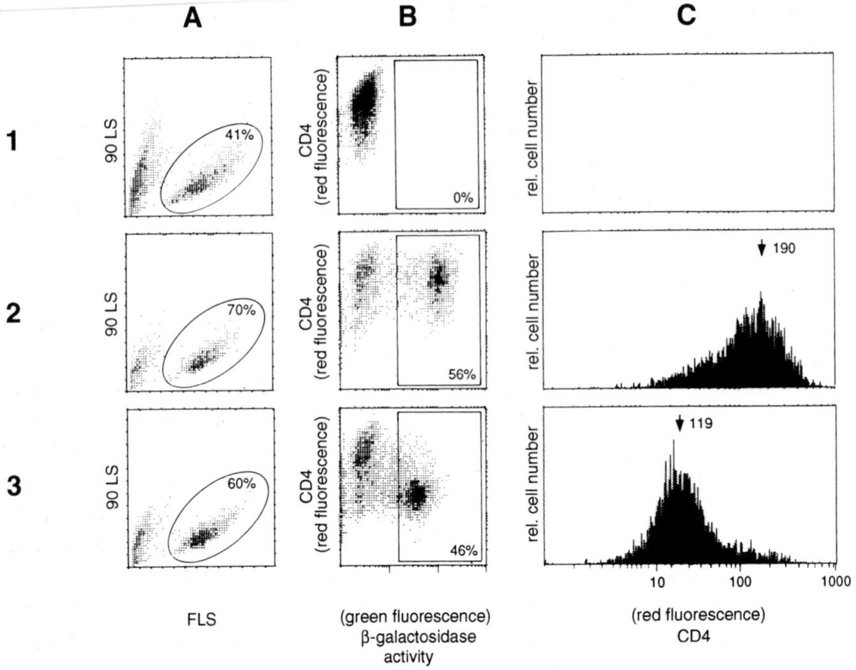

Figure 2. Flow cytometry analysis of CD4 antigen expression on Jurkat T cells transiently expressing HIV-1 *nef*. Exponentially growing cultures of Jurkat T cells were electroporated with control expression vectors containing: (1) HIV-1 *tat* cDNA (15 μg of CD3 Tex plasmid, *upper* panels); (2) a mixture of CD3 Tex and CD3 LacZ reporter plasmids (14 μg and 1 μg, respectively, *middle* panels); (3) a mixture of CD3 NA7, containing the wild-type HIV-1 *nef* allele, together with CD3 LacZ and CD3 Tex (3 μg, 1 μg, and 11 μg, respectively, *bottom*). Cells were cultured overnight, loaded with FDG (Molecular Probes), and reacted with α-CD4 monoclonal antibody conjugated to phycoerythrin (α-Leu3A, Becton Dickinson). Stained cells were analysed on an Epics C flow cytometer equipped with an argon laser. A. 90° light scatter (90 LS) versus forward light scatter (FLS) analysis of electroporated cells revealed that 41%, 70%, and 60% of events analysed in samples (1), (2), and (3), respectively, correspond to live cells (boxed). The remaining events reflected cellular debris (not boxed, characterized by low FLS values). Live cells (circled) were selected by bitmap gating for further analysis. B. β-galactosidase activity (green fluorescence) and CD4 antigen expression of the surface (red fluorescence) of live cells, selected by bitmap gating as shown in (A), are shown on the abscissa and ordinate in the logarithmic scale, respectively. Cell populations expressing high levels of β-galactosidase activity are boxed and the percentage fraction of these cells are shown in the boxes. C. Cells with above background β-galactosidase activity were selected by bitmap gating (see boxes in the panel B) and histograms of CD4 antigen expression on the surface of these β-galactosidase positive cells are shown. *Arrows* reflect the median level of CD4 expression on the surface of β-galactosidase positive cells and the numbers shown at the arrows are the channel numbers corresponding to the position of the peak of CD4 fluorescence recorded in the logarithmic scale (full scale comprises 256 channels).

Although epithelial or fibroblastic adherent cells that express CD4 can be used instead of CD4-positive T cell lines, this approach is not recommended for the following reasons:

(a) Levels of CD4 antigen expression on adherent cells are usually lower than those on CD4-positive T cells and this results in lower sensitivity of the assay.

(b) Preparation of cell suspensions and processing them for flow cytometry analysis is more laborious than for the non-adherent T cells.

2.1.2 Expression vectors

Transient assays of CD4 down-regulation require a relatively high level Nef expression and many of the commonly used vectors do not direct sufficiently high Nef expression in T cells. Specifically, those employing the transcription regulatory sequences from SV40 or CMV early region genes, or from the Moloney MLV long terminal repeat perform relatively poorly in T cells. The best results in T cells have been obtained with T cell-specific CD3 β expression vectors. As shown in *Figure 1*, these CD3 β plasmids take advantage of transcription regulatory sequences from the murine δ CD3 gene, which is expressed in all T cells. RNA processing signals are provided by a segment derived from the rabbit β-globin gene. These CD3 β vectors have been found to direct high level expression of heterologous genes both in transient and stable experiments (5, 6, 9).

2.1.3 Transfections

The two transfection protocols most frequently used with lymphoid cells are:

- electroporation
- DEAE–dextran transfection (see Chapter 11)

Electroporation is more efficient and results in at least 10- to 100-fold higher expression of transfected genes when performed under optimal conditions. The critical parameter, which needs to be established empirically, is the voltage of the pulse at which electroporation is performed.

(a) Increasing the voltage results in a larger fraction of productively transfected cells, but cell survival is inversely related to the voltage (10).

(b) Voltages in the range of 200 V to 240 V provide an acceptable compromise between the expression level and the survival of electroporated cells, but the exact conditions will vary for different cell lines.

(c) Optimization of electroporation conditions can be conveniently performed using CD3 LacZ reporter plasmid and the procedure described in *Protocol 1*.

(d) Under optimal conditions of expression, electroporation of Jurkat or CEM T cells with CD3 Nef or CD3 LacZ vectors results in readily detectable expression in 40–80% of the surviving cells (*Figure 2*).

The following points should be borne in mind when performing *Protocol 1*:

(a) Cells from exponentially growing cultures survive electroporation better than those from a stationary phase.

(b) All manipulations are at room temperature (20°C to 25°C).

(c) Both Jurkat and CEM E5 T cells are relatively unaffected by 1 h to 1.5 h incubation at room temperature in the complete RPMI 1640 medium which permits processing of up to 25–30 samples per experiment.

(d) Gene Pulser cuvettes can be reused several times. For this purpose, immediately after electroporation, cuvettes should be cleaned by several rinses with deionized water, air dried, and sterilized for 1 h (or longer) under a UV lamp (for example in the tissue culture hood).

(e) All plasmid DNAs used for electroporation are purified by two sequential CsCl density gradients.

(f) The fraction of productively transfected cells that co-express both markers depends on the initial ratio between the Nef and LacZ expression vectors. Also, the doses of the expression vectors and LacZ reporter plasmids that confer detectable phenotypes differ for different T cell lines. Thus, in the initial experiments both the LacZ and Nef expression vectors should be titrated to determine the conditions, where more than 80% of β-galactosidase expressing cells will also express Nef, as revealed by a decrease in CD4 antigen expression on the cell surface.

(g) The total amount of the CD3 vector DNA in transfected DNA should be kept constant in order to minimize possible promoter competition effects.

Protocol 1. Electroporation of Jurkat and CEM T cells

Equipment and reagents

- Gene Pulser apparatus with Capacitance Extender (Bio-Rad, or equivalent)
- Low speed table-top centrifuge
- Tissue culture incubator
- 50 μl aliquots containing DNA for electroporation (CD3 Nef expression vector, CD3 LacZ reporter plasmid, and carrier CD3 Tex DNA: 40–70 μg total
- 1.5 M NaCl
- 1.5 ml microcentrifuge tubes
- 'Complete RPMI 1640 medium': RPMI 1640 medium supplemented with 10% fetal calf serum, 20 mM Hepes pH 7.4, and 2 mM glutamine
- Gene Pulser cuvettes with a 0.4 cm electrode gap

Method

1. Combine in a sterile disposable 1.5 ml microcentrifuge tube:
   - CD3 Nef expression vector 0.1–20 μg
   - CD3 LacZ reporter plasmid 0.1–10 μg
   - CD3 vector up to 30 μg

- carrier DNA — up to the total of 30–70 μg
- 1.5 M NaCl — 5 μl
- deionized sterile water — up to the total volume of 50 μl

2. Grow Jurkat and CEM T cells in 10 cm tissue culture dishes in complete RPMI 1640 medium.
   (a) Maintain the Jurkat T cells by diluting cultures 1:15 every three to four days, so that the cell density does not exceed 5×10^5 cells/ml.
   (b) Maintain the CEM T cells by diluting cultures 1:3 to 1:10 every three to five days.
3. Pool cells from several dishes of exponentially growing cultures[a] into 50 ml Falcon tubes and determine the cell density by counting in a haemocytometer.
4. Sediment cells by low speed centrifugation at 800 *g* to 1200 *g* (1200 r.p.m. in a table-top centrifuge), at room temperature.
5. Resuspend cell pellets in complete RPMI 1640 at 5×10^7 cells/ml.
6. Transfer aseptically 200 μl aliquots of the cell suspension to microcentrifuge tubes containing the 50 μl aliquots of DNA for transfection.[c]
7. Transfer the cells and DNA mixtures (250 μl), with an automatic pipettor, into 0.4 cm electroporation cuvettes.[d]
8. Electroporate with a single 200 V pulse with the Capacitance Extender set at 960 μF. Let the cells rest for 5–10 min at room temperature.
9. Add 3 ml of the complete RPMI 1640 medium to 15 ml conical Falcon tubes.
10. Transfer each aliquot of electroporated cells (including clumps of dead cells) into the tubes containing medium (using a Pasteur pipette).
11. Sediment the cells by a low speed centrifugation for 10 min at room temperature.
12. Discard the supernatant and floating debris.
13. Resuspend the cell pellet in 10 ml of complete RPMI 1640 medium and transfer into 10 cm tissue culture dishes.
14. Incubate for 12–24 h at 37°C prior to flow cytometry or immunoblot analysis.

2.2 Simultaneous detection of CD4 antigen and β-galactosidase activity in live cells

2.2.1 FDG detection of β-galactosidase activity

A straightforward method to detect and quantitate β-galactosidase activity in live mammalian cells by flow cytometry has been developed by L. Herzenberg and collegues at Stanford University (11):

(a) Cells are loaded with the β-galactoside analogue fluorescein di-β-D-galactopyranoside (FDG, see *Protocols 2* and *3*).
(b) Cleavage of FDG releases free fluorescein (FITC), which can then be detected by its green fluorescence.
(c) FDG and FITC is retained in cells at temperatures of 4°C and below, but can cross the cell membrane at higher temperatures. Thus, it is critical to keep the FDG-loaded cells at temperatures below 4°C at all times throughout the procedure.
(d) At 4°C the rate of enzymatic reaction is only tenfold lower than that at 37°C and therefore carrying out the reaction on ice results only in a modest loss of sensitivity. Usually, 1–2 h incubation on ice is sufficient to reveal the LacZ reporter activity in productively transfected cells.

Protocol 2. Preparation of FDG stock solution [a,b]

Equipment and reagents

- FDG (Molecular Probes)
- DMSO (Fluka)
- 1.5 ml microcentrifuge tubes
- Circulating water-bath

Method

1. Warm a vial containing 5 mg of FDG for 1–2 min at 37°C.
2. Add to the vial 38 μl of 50% DMSO in water.
3. Place the vial, protected from direct light, in a 37°C water-bath and incubate for 5 min.
4. Add gradually 152 μl of water.
5. Dispense the resulting 40 mM FDG solution in 10% DMSO into microcentrifuge tubes and store at −20°C.

[a] FDG is available from several vendors. In our experience FDG from Molecular Probes has been consistently of very good quality.
[b] FDG is most stable when stored in solution.

Protocol 3. FDG loading of electroporated T cells

Equipment and reagents

- Phosphate-buffered saline (PBS) supplemented with 3% fetal calf serum and 0.1% sodium azide (PBS-FA)
- 40 mM FDG stock (*Protocol 2*)
- Monofilament cloth—nylon mesh: 35 μm (Small Parts Inc.)
- Haemocytometer
- 15 ml conical disposable tubes
- Low speed refrigerated centrifuge

Method

1. Determine the cell density in electroporated cultures by haemocytometer counting.
2. Filter the cultures through a nylon cloth into 15 ml conical Falcon tubes. This removes aggregates of cells and debris.
3. Sediment cells by low speed centrifugation in a table-top centrifuge.
4. Resuspend the cell pellet in ice-cold PBS-FA at 2×10^7 cells/ml.
5. Transfer 30 μl of cell suspension into 1.5 ml microcentrifuge tubes and place them in a wet-ice water-bath.
6. Equilibrate the 2 mM FDG stock at 37°C (or 42°C) for 5 min.
7. Equilibrate the 30 μl aliquots of cells to be loaded with FDG in a 37°C water-bath for 5 min.
8. Add 30 μl aliquots of 2 mM FDG to cell suspensions. Swirl the tubes to mix and incubate at 37°C for exactly 90 sec.
9. Transfer the tubes to the wet-ice water-bath and immediately add 700 μl of ice-cold PBS-FA.
10. Incubate the FDG-loaded cells on ice for 1–2 h. Protect from direct light.
11. Pellet the cells by low speed centrifugation at 4°C.[a]
12. Remove the supernatant completely.
13. Resuspend the cell pellet in 100 μl of ice-cold PBS-FA.
14. Analyse β-galactosidase activity by flow cytometry, or stain the FDG-loaded cells to reveal CD4 antigen (*Protocol 4*).

[a] It is critical that cells are maintained at 0–5°C during all operations. If a reliable refrigerated low speed centrifuge is not available, place the microcentrifuge tubes containing FDG-loaded cells inside larger tubes filled with wet-ice (e.g. 15 ml Falcon tubes), and spin them together.

2.2.2 Detection of the CD4 antigen

The preferred way to detect CD4 expression is by direct immunofluorescence (see also Chapter 13, Volume 1):

(a) The FDG-loaded cells are reacted with saturating amounts of an anti-CD4 monoclonal antibody conjugated with phycoerythrin (see *Protocol 4*).

(b) Phycoerythrin is then detected by its red fluorescence.

(c) Alternatively, CD4 can be detected by indirect immunofluorescence. This approach requires additional operations in order to remove an excess of primary antibody, etc. These steps increase the probability of FDG leakage from loaded cells, and is therefore not recommended.

Protocol 4. Immunofluorescent detection of CD4 antigen on the surface of FDG-loaded cells

Equipment and reagents

- FDG-loaded cells (*Protocol 3*)
- Leu3A phycoerythrin-conjugated anti-CD4 antibody (Becton Dickinson or equivalent)
- PBS-FA (*Protocol 3*)
- Low speed refrigerated microcentrifuge

Method

1. Mix:
   - FDG-loaded cells in 100 μl of PBS-FA
   - 100 μl of ice-cold PBS-FA containing a saturating amount of phycoerythrin-conjugated anti-CD4 monoclonal antibody
2. Incubate on ice for 20 min, protected from direct light.
3. Add 800 μl of ice-cold PBS-FA.
4. Pellet cells by low speed centrifugation at 4°C.[a]
5. Remove the supernatant and resuspend the cell pellet in 300 μl of PBS-FA.
6. Analyse the stained cells by flow cytometry (see section 2.3).

[a] To prevent leakage of FDG or FITC all operations must be carried out at 4°C.

2.3 Flow cytometry

Figure 2 shows results obtained from a typical flow cytometry analysis (see Chapter 13, Volume 1) of a functional HIV-1 *nef* allele. Usually, preparations of FDG-loaded cells stained for CD4 contain variable amounts of dead cells and cellular debris. This debris shows dramatically different physical properties than the live cells and can be easily revealed and eliminated electronically during flow cytometry analysis. As shown in *Figure 2* (panel A), the forward versus 90° light scatter analysis provides a criterion to electronically select live cells for further analysis of CD4 and β-galactosidase expression. In this particular experiment live cells, shown within the bitmaps, corresponded to 40% and 70% of events analysed. Under the optimal experimental conditions approximately 40–80% of cells that survived electroporation show a high level β-galactosidase activity.

Nef is expressed in 80–90% of β-galactosidase positive cells as judged from a sharp decrease in CD4 on the surface of β-galactosidase positive cells (see panel B3 in *Figure 2*). As shown in panels C (*Figure 2*), further electronic selection of cells with high β-galactosidase activity (boxed in panel B) is used to display distribution of CD4 expression on the productively transfected cells. This allows one to quantitate the level of CD4 antigen on productively

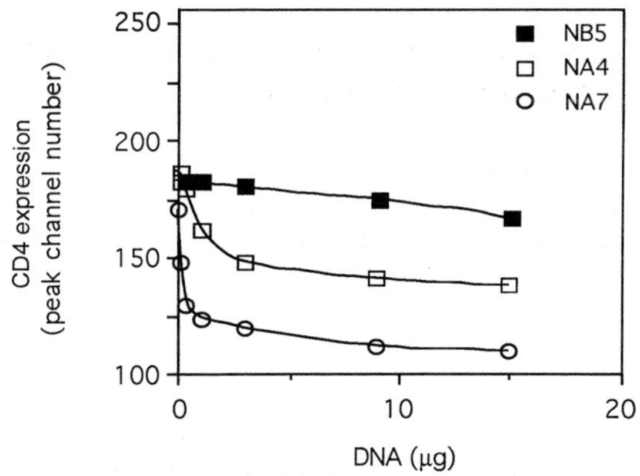

Figure 3. Dose–response analysis of HIV-1 *nef* alleles. Jurkat T cells were co-electroporated with CD3 LacZ reporter plasmid (0.1 μg–1 μg) and different amounts of CD3 NA7, CD3 NA4, or CD3 NB5 *nef* expression vectors (0.1 μg–15 μg, shown in the abscissa), which contain HIV-1 *nef* alleles cloned directly from PBLs of HIV-1-infected patients. CD3 Tex plasmid was included in all electroporations to keep the total amount of CD3 vectors at 20 μg. Transfected cells were analysed as illustrated in *Figure 2*. The level of CD4 antigen on the surface of electroporated cells, represented by the peak channel number of CD4 fluorescence recorded in the logarithmic scale, and determined as shown in *Figure 2C*, is shown in the ordinate.

transfected cells and express it numerically as a median value of the CD4 fluorescence. This transformation is useful for presentation of data from dose–response experiments, as shown in *Figure 3*.

3. Identification of weak and loss-of-function *nef* alleles

The extent of a decrease in CD4 antigen expression on the cell surface correlates with the amount of Nef protein expressed in the cell. The absence of a detectable effect of Nef on CD4 antigen on the cell surface reflects either the expression of a non-functional protein or the inefficient expression of a wild-type allele. These possibilities can be distinguished by titration experiments using increasing amounts of Nef expression vector to transfect cells. An example of this type of experiment is shown in *Figure 3*.

(a) Jurkat T cells were electroporated with increasing amounts of CD3 vectors containing three independent *nef* alleles isolated by PCR from PBLs of HIV-1-infected, seropositive patients.
(b) Data from live cells, recorded and processed as described in the legend to *Figure 2*, present CD4 antigen expression on the cell surface, as a function of the amount of CD3 Nef expression vector used for transfection.

(c) For the NA7 *nef* DNA there is little effect on CD4 antigen on the cell surface, when cells are transfected with less than 0.1 μg of DNA.

(d) DNA doses of more than 1 μg saturate the response. Such a dose–response curve is characteristic of strong HIV-1 *nef* alleles.

(e) With NA4 *nef* the overall decrease in CD4 expression observed over a range of DNA doses is much less dramatic. Such a dose–response curve reveals that NA4 is a relatively weak *nef* allele.

(f) The NB5 allele has almost no effect on CD4 antigen expression, even at DNA doses 20 times higher than the saturating doses for the active NA7 allele. These results indicate that NB5 allele is inactive for CD4 down-regulation.

(g) Further immunoblot analysis of Nef protein expression in transfected cells can be used to assess whether this reflects expression of unstable, or stable loss-of-function Nef protein.

Acknowledgements

We thank Dan Littman for providing us with CD4-positive Jurkat T cells and Pat Burfeind for expert and creative assistance in flow cytometry analysis. We thank Winship Herr for critical reading of the manuscript and editorial help. This work was supported by a grant from NIH and Johnson & Johnson (to J. S.) and by Cold Spring Harbor Laboratory funds.

References

1. Kestler, H. W., Ringler, D. J., Mori, K., Panicali, D. L., Sehgal, P. K., Daniel, M. D., *et al.* (1991). *Cell*, **65**, 651.
2. Garcia, V. and Miller, A. D. (1991). *Nature*, **350**, 508.
3. Garcia, J. V., Alfano, J., and Miller, A. D. (1993). *J. Virol.*, **67**, 1511.
4. Anderson, S., Shugars, D. C., Swanstrom, R., and Garcia, J. V. (1993). *J. Virol.*, **67**, 4923.
5. Skowronski, J., Parks, D., and Mariani, R. (1993). *EMBO J.*, **12**, 703.
6. Mariani, R. and Skowronski, J. (1993). *Proc. Natl Acad. Sci. USA*, **90**, 5549.
7. Koga, Y., Sasaki, M., Yoshida, H., Wigzell, H., Kimura, G., and Nomoto, K. (1990). *J. Immunol.*, **144**, 94.
8. Scheppler, J. A., Nicholson, J. K. A., Swan, D. C., Ahmed-Ansari, A., and McDougal, J. S. (1989). *J. Immunol.*, **143**, 2858.
9. Lee, N. A., Loh, D. Y., and Lacy, E. (1992). *J. Exp. Med.*, **175**, 1013.
10. Chu, G., Hayakawa, H., and Berg, P. (1987). *Nucleic Acids Res.*, **15**, 1311.
11. Nolan, G. P., Fiering, S., Nicolas, J.-P., and Herzenberg, L. A. (1988). *Proc. Natl Acad. Sci. USA*, **85**, 2603.

14

Vpu

E. A. COHEN and H. G. GÖTTLINGER

1. Introduction

The *vpu* gene in the central region of the HIV-1 genome partially overlaps the *env* reading frame and is predicted to encode a protein of 81 amino acids with a hydrophobic N-terminus and a hydrophilic C-terminus. Vpu is unique to HIV-1 and is absent from HIV-2 and SIV (1). Biochemical studies have shown that Vpu is a phosphorylated type I integral membrane protein and is capable of forming homo-oligomers (2, 3). Immunofluorescence and immune electron microscopy studies have demonstrated that Vpu is concentrated in the Golgi apparatus and in endosomal vesicles (4, 5). Vpu can also be localized to the cell membrane, but does not accumulate there, suggesting that it may be recycled between different cellular compartments (5). Vpu is not incorporated into the virion.

Although Vpu affects different steps of the virus replication cycle, it is not essential for virus replication in tissue culture. In the absence of a functional *vpu* gene, the production of viral particles is reduced by about five- to tenfold in infected T cell lines (6, 7). The magnitude of the effect of Vpu on particle production is cell type-dependent and is particularly pronounced in the epithelial cell line HeLa. Vpu does not act at the level of expression or processing of the viral internal structural (Gag) proteins. Rather, Vpu appears to facilitate the final release of nascent viral capsids, since budding structures accumulate at the surface of cells expressing vpu^- mutant proviruses (48). Vpu can significantly enhance particle production by Gag proteins from widely divergent retroviruses, implying that Vpu acts indirectly through modification of the cellular environment rather than by directly interacting with viral proteins (9).

Apart from its effect at a late stage of capsid morphogenesis, Vpu also modulates processing and transport of the viral envelope glycoprotein. Vpu can reduce the intracellular formation of complexes between CD4, the receptor for HIV-1, and the envelope glycoprotein precursor, apparently by inducing the rapid and specific degradation of CD4 molecules that are retained in the endoplasmic reticulum by binding to envelope glycoprotein (10, 11). Nevertheless, Vpu appears to induce a two- to threefold decrease in the concentration of envelope protein on the surface of infected T cells, an observation that potentially explains the decreased cytopathicity of Vpu^+ strains (8).

2. Assays for Vpu function

2.1 Viral particle release

In human T lymphocytic cell lines such as Jurkat cells and the CEM cell derivative A3.01, Vpu has a five- to tenfold effect on the efficiency of HIV-1 particle production. However, the use of T cell lines is restricted by the need for virus replication to yield measurable amounts of virus particles. For an analysis of the effect of Vpu on particle production independent of virus replication, the epithelial cell line HeLa has proven particularly suitable (9). HeLa cells do not support replication of HIV-1 as they lack the CD4 receptor. However, they can be transfected with high efficiency, resulting in the transient expression of sufficient amounts of viral protein for analysis:

(a) The effect of Vpu on particle production in HeLa cells (CCL2, American Type Culture Collection) is similar, but more pronounced, than in lymphoid cells permissive for HIV-1 replication.
(b) HIV-1 particle production in HeLa cells is very inefficient in the absence of Vpu. The block appears to be at the level of particle release, since particles are assembled and accumulate at the cell surface in large numbers.
(c) Upon expression of Vpu, up to 100-fold increase in particle release can be observed. The HeLa CCL2 line therefore provides a very sensitive model system to study the requirements for Vpu-facilitated particle release (9).
(d) COS-7 cells are not suitable for this purpose, since HIV-1 particle production is high even in the absence of Vpu and is not significantly enhanced in the presence of Vpu.

Transient expression after transfection of HeLa cells by a calcium phosphate precipitation technique (12) permits an analysis of the effect of Vpu on the production of capsids by replication defective viral genomes (see Chapter 2, Volume 1). Expression and release of HIV-1 Gag proteins can be analysed by metabolic labelling of the transfected cultures, followed by immunoprecipitation from the cell lysate and supernatant fractions (see *Protocol 1*).

The effect of Vpu on the levels of particulate Gag protein released into the supernatant fractions can be determined without the need to rely on immunological methods (see *Protocol 2*).

(a) The transfected cells are metabolically labelled.
(b) Particulate material released into the supernatant is pelleted through a 20% sucrose cushion.

(c) The viral proteins in the pellet are analysed by SDS–PAGE. Occasional background bands representing cellular proteins can easily be discerned by their appearance in mock-transfected samples.

(d) HIV-1 virion production can also be monitored by measuring reverse transcriptase or HIV-1 p24 antigen levels in the pelleted material (see Chapters 2 and 7, Volume 1).

Protocol 1. Transfection of HeLa cells, metabolic labelling, and immunoprecipitation

Reagents

- HeLa cells (CCL2; American Type Culture Collection)
- DMEM
- HIV-1 proviral plasmid
- 1 mM Tris–HCl pH 8.0
- 2.5 M $CaCl_2$
- 2 × HBS buffer: 50 mM Hepes, 1.5 mM Na_2HPO_4, 280 mM NaCl, pH 7.1
- Cysteine- or methionine-free RPMI 1640/5% FCS
- $[^{35}S]$cysteine or $[^{35}S]$methionine
- RIPA buffer: 140 mM NaCl, 8 mM Na_2HPO_4, 2 mM NaH_2PO_4, 1% Nonidet P-40, 0.5% sodium deoxycholate, 0.05% SDS
- Protein A-Sepharose beads (Pharmacia)
- SDS sample buffer: 125 mM Tris–HCl pH 6.8, 2% SDS, 4.8% 2-mercaptoethanol, 20% glycerol

Method

1. Seed 10^6 HeLa cells/80 cm^2 tissue culture flask in 12 ml DMEM/10% FCS 24 h prior to transfection.
2. Add 25 μg proviral plasmid and 50 μl 2.5 M $CaCl_2$ to 1 mM Tris–HCl pH 8.0 (500 μl final volume) and mix.
3. Add dropwise, mix slowly to 0.5 ml 2 × HBS buffer. Avoid extra mixing or shaking of the solution! Leave for 30 min at room temperature to allow precipitate to form.
4. Add precipitate to cells, moving flasks as little as possible.
5. Incubate for 15 h.
6. Wash cells twice with DMEM; continue incubation in 12 ml DMEM/10% FCS.
7. Remove culture medium 48 h post-transfection. Replace medium with 10 ml cysteine- or methionine-free RPMI 1640/5% FCS containing 50 μCi/ml $[^{35}S]$cysteine or methionine. Incubate for 12 h.
8. Harvest labelling medium. Centrifuge for 20 min at 500 *g* and 4°C. Add one-fourth volume of 5 × RIPA buffer to lyse virions. Wash cells once with PBS. Lyse in 700 μl of ice-cold 1 × RIPA buffer.
9. Clarify cell lysates by centrifugation at 35 000 r.p.m. for 1 h at 4°C in a Beckman 45Ti rotor.

Protocol 1. *Continued*

10. Prepare immunosorbents as follows. Mix in an Eppendorf tube:
- 500 μl of a 10% suspension of protein A–Sepharose beads per sample in 1 × RIPA buffer
- 5–10 μl of appropriate antiserum or serum from an HIV-1-infected individual

Incubate 1 h at room temperature. Wash beads twice with 1 × RIPA buffer.

11. Add supernatant fractions to 50 μl of packed beads in a 15 ml conical tube and incubate overnight on a rotating platform at 4°C. For immunoprecipitation from cell lysates, pre-incubate 50 μl of packed immunosorbents for 1 h in an Eppendorf tube with 500 μl of unlabelled lysate derived from 5–10 × 10^6 untransfected cells to minimize unspecific binding of cellular proteins. Add the labelled cell lysate on top of the unlabelled lysate. Incubate overnight at 4°C.

12. Wash beads six times with 1 ml of ice-cold RIPA buffer.

13. Boil 5 min in SDS sample buffer containing 2-mercaptoethanol. Separate proteins by SDS–PAGE. Visualize protein bands by autoradiography of the dried gel at −70°C.

To study the effect of Vpu on capsid release in the HeLa cell system, it is not necessary that Vpu is expressed in the context of the viral genome. A similar enhancement of particle release can be obtained when Vpu is provided in *trans* by co-transfection of a Vpu expressor plasmid (*Protocol 2*). As a control, a plasmid can be used that is identical to the Vpu expression vector, except for the absence of a *vpu* initiation codon (9, 13).

Protocol 2. Quantitation of Vpu effect on viral particle production

Equipment and reagents

- Beckman SW41 rotor or similar model
- Ultracentrifuge
- HeLa cells (CCL2; American Type Culture Collection)
- *gag* and *vpu* expression vectors
- Cysteine-free RPMI 1640
- [^{35}S]cysteine
- 20% sucrose in PBS, pH 7.2
- 0.45 μm pore size filters
- SDS sample buffer: 125 mM Tris pH 6.8, 2% SDS, 4.8% 2-mercaptoethanol, 20% glycerol

Method

1. Co-transfect HeLa cells in parallel with a 1:1 molar ratio of a *gag* expressor construct and a *vpu* expression vector or an appropriate *vpu*$^-$ control vector. Up to 50 μg plasmid DNA can be used to transfect 10^6 HeLa cells in a 80 cm^2 tissue culture flask.

2. Label the cultures with 50 μCi/ml of [^{35}S]cysteine from 48–60 h after transfection, using 11 ml of cysteine-free RPMI 1640 with 5% FCS.
3. Harvest the supernatant. Clarify from cell debris by centrifugation for 20 min at 500 *g* and 4°C. Pass supernatant through 0.45 μm pore size filters.
4. Layer 10 ml of clarified supernatant on top of 2 ml of a 20% sucrose cushion in a centrifuge tube for the Beckman SW41 rotor.
5. Centrifuge sample for 2 h at 27 000 r.p.m. and 4°C.
6. Carefully aspirate medium and resuspend the invisible pellets in 50 μl of SDS sample buffer.
7. Heat in boiling water for 5 min and load the entire sample into a single well of a 12% SDS–polyacrylamide slab gel.

2.2 Viral cytopathicity

HIV-1 infection of $CD4^+$ human T cell lines in culture results in marked cytopathic effects (see Chapters 2 and 3, Volume 1). Cell killing is characterized by initial formation of large, multinucleated giant cells and syncytia, followed later by destruction of single cells that swell and die. The extent of cell killing of human $CD4^+$ T cells in culture by different isolates of HIV-1 has been shown to be dependent on several independent factors, including the rate of virus replication, the origin of the viral envelope glycoprotein, as well as upon the expression of non-structural viral proteins including Vif, Vpr, Vpu, and Nef (14).

Vpu expression was shown to delay the cytopathic effect of HIV-1 in Jurkat, MT-4, and A3.01 CD4-positive T cell lines by reducing the rate of syncytium formation (4, 7, 11). Infection of T cell lines with Vpu defective (vpu^-) viruses results in a more rapid onset of syncytium formation in comparison to vpu^+ virus, even though significantly less progeny virus is produced.

To examine the effect of Vpu on the rate of syncytium formation, cell killing kinetic studies are performed using infectious HIV molecular clones isogenic except for the expression of Vpu (*Protocol 3*).

(a) Jurkat cells support long-term (up to 12 days) HIV-1 replication and therefore represent a good system to study the effect of Vpu on virus-induced cytopathic effects (7).
(b) Infected or transfected cells are monitored daily from day 1 to day 12 for total viable cell number, percentage of dead cells, and appearance of syncytia.
(c) The spread of infection and virus release can be determined periodically by measuring cell-associated $p24^{gag}$-specific immunofluorescence and supernatant reverse transcriptase activity.

Protocol 3. Transfection of Jurkat cells and quantitation of cytopathic effects

Equipment and reagents

- Immunofluorescence microscope
- Jurkat cells
- RPMI 1640
- HIV-1 proviral plasmid
- DEAE–dextran solution (5 mg/ml in 1 M Tris–HCl pH 7.4)
- DMEM
- Trypan blue stain (0.4% in 0.85% saline) (Gibco)
- Methanol/acetone mixture (1 : 1)
- Mouse anti-p24^{gag} monoclonal antibody (American Bio-Technologies)
- Goat anti-mouse IgG–FITC conjugate (Gibco–BRL)

Method

1. Maintain Jurkat cells in RPMI 1640 medium supplemented with 10% heat inactivated FCS prior to transfection.
2. Mix gently in a 15 ml conical tube:
   - 10 μg of proviral plasmid
   - 150 μl DEAE–dextran solution
   - 2 ml DMEM
3. Wash Jurkat cells (10^7) once with DMEM and resuspend cells in 1 ml of DMEM.
4. Add gently cells (10^7) to the DNA–DEAE–dextran mix. Incubate for 1 h at 37 °C. Gently swirl the conical tube after 30 min. Wash cells once with DMEM/10% FCS.
5. Resuspend cells with 10 ml RPMI/10% FCS in a T25 tissue culture flask and incubate at 37 °C.
6. Monitor daily the total number of viable cells and the percentage of trypan blue-stained cells by light microscopy. Resuspend cells in an appropriate volume of fresh RPMI 1640/10% FCS in order to maintain 10^6 cells/ml of trypan blue-excluding viable cells in the culture.
7. Evaluate periodically the spread of infection by cell-associated p24^{gag} immunofluorescence. Dispense cells on to microscopic 15-well glass slides. Fix with methanol and acetone (1 : 1) for 30 min. Label cells with an HIV-1 anti-p24^{gag} monoclonal antibody for 1 h. Wash twice with PBS. Incubate cells with goat anti-mouse IgG–FITC conjugate for 1 h. Wash twice with PBS. Determine the percentage of stained cells by fluorescence microscopy.
8. Score daily for syncytia by light microscopy as follows:
   +, 1–2 syncytia per (× 100) field in a culture of 10^6 cells/ml;
   ++, 3–5 syncytia per field;
   +++, 5–10 syncytia per field;

++++ 10–20 syncytia per field;
+++++, > 20 syncytia per field.
The final counts should average the analysis of at least ten fields.

11. Harvest aliquots for assay of virus-specified reverse transcriptase activity (see Chapter 2, Volume 1).

2.3 Envelope glycoprotein expression at the cell surface

Syncytium formation is the result of cell-to-cell fusion resulting from the interaction of envelope glycoprotein gp120 expressed at the surface of infected cells with the CD4 receptor of uninfected target cells (see Chapter 8). It has been shown by flow cytofluorometric analysis that the relative amounts of gp120 on the surface of cells infected with vpu^- virus was enhanced two- to threefold as compared to cells infected with vpu^+ virus (8). Mutational analysis of the Vpu protein demonstrated that this reduction in cell surface expression of envelope glycoproteins in the presence of Vpu was not solely the result of facilitated release (15). Data from this study suggested there was a distinct effect of Vpu on cell surface envelope glycoprotein expression. It is unclear at this point whether this effect of Vpu on surface envelope glycoprotein expression and syncytium formation is related to the Vpu-induced degradation of CD4 (10, 11, 16).

The effect of Vpu on envelope glycoprotein levels at the cell surface is measured as follows (*Protocol 4*):

(a) Jurkat or MT-4 cells are transfected or infected at a high m.o.i. with infectious molecular clones of HIV-1 that are isogenic except for the expression of Vpu.
(b) Infected cells are reacted with an anti-gp120 monoclonal antibody and stained with a FITC-conjugated goat anti-mouse IgG.
(c) The percentage of positively stained cells as well as the relative amount of envelope glycoprotein gp120 at the cell surface are evaluated periodically by flow cytofluorometry.

Protocol 4. Analysis of Vpu effect on Env cell surface expression

Equipment and reagents

- Fluorescence activated cell sorter
- Mouse anti-gp120 monoclonal antibody (American Bio-Technologies)
- Goat anti-mouse IgG–FITC conjugate (Gibco–BRL)
- Paraformaldehyde (1% in PBS)

Method

1. Wash transfected Jurkat cells (2 × 10^5) (*Protocol 3*) twice with cold PBS/5% FCS.

Protocol 4. *Continued*

2. Pellet cells by centrifugation at 500 *g* for 5 min.
3. Incubate cells on ice with 10 μl of 1:10 diluted anti-gp120 monoclonal antibody for 1 h.
4. Wash cells three times with cold PBS/5% FCS.
5. Incubate cells on ice with goat anti-mouse IgG–FITC conjugate for 1 h.
6. Wash cells three times with cold PBS/5% FCS.
7. Fix cells with 1% paraformaldehyde in PBS.
8. Analyse stained cells by flow cytofluorometry (see Chapter 13 and Chapter 13, Volume 1).

2.4 Degradation of the CD4 receptor

HIV-1 infection of CD4-positive T lymphocytes leads to reduced cell surface expression of the CD4 receptor. Several mechanisms involving either the virus encoded gp160 envelope precursor protein (16) or the Nef protein (17) have been reported to cause down-modulation of the CD4 receptor (see Chapter 13). The formation of intracellular complexes between CD4 and gp160 retains CD4 and gp160 in the endoplasmic reticulum and prevents their maturation and transport to the cell surface. Vpu was shown recently to disrupt the intracellular gp160–CD4 complexes as a result of a rapid degradation of CD4 in the endoplasmic reticulum (10, 11).

The effect of Vpu on gp160–CD4 complex formation and envelope glycoprotein processing can be measured in HeLa cells co-transfected with a CD4 expressor plasmid and Vpu/envelope glycoprotein expressor plasmids isogenic except for the expression of Vpu. Vpu was also shown to induce destabilization of gp160–CD4 complexes in *trans*.

(a) Transfected Hela cells are pulse chase labelled and immunoprecipitates are prepared from cell lysates at various time points, employing either antisera against HIV-1 envelope glycoprotein or CD4-specific monoclonal antibodies.

(b) The extent of envelope glycoprotein processing and levels of gp160–CD4 co-precipitation can be determined after separation by SDS–PAGE and autoradiography (*Protocol 5*).

The consistently lower amounts of CD4 recovered from cells expressing Vpu suggested that reduced complex formation between gp160 and CD4 was the result of Vpu-induced destabilization of CD4. Experimental systems have been developed by Willey *et al.* and by Chen *et al.* (11, 18) to examine the effect of Vpu on CD4 stability. Using an *in vivo* experimental system, Willey *et al.* showed that Vpu induces rapid degradation of CD4, reducing the half-life of CD4 from 6 h to 12 min (11). A prerequisite for Vpu-induced CD4 degradation is that CD4 be prevented from leaving the endoplasmic reticulum, which is accomplished through its binding to gp160. The involvement of gp160 in the induction of CD4 degradation is restricted to its function as a CD4 'trap'. In the absence of envelope glycoprotein, an endoplasmic reticulum retention mutant of CD4 as well as wild-type CD4 in cultures treated with brefeldin A (BFA), a fungal metabolite that blocks the transport of protein from the endoplasmic reticulum, is degraded in the presence of Vpu (11).

CD4 stability can be evaluated in the presence of Vpu by co-transfecting HeLa or COS cells with expressor plasmids for CD4 and Vpu (*Protocol 5*).

(a) Vpu does not have to be expressed in the context of a provirus or from an envelope expressor plasmid.

(b) HeLa and COS cell lines provide a very sensitive model system to study Vpu-induced degradation of CD4, since both cell lines are easily transfectable and yield large amounts of proteins from the transfected plasmids.

(c) Following transfection, cells are pulse chase labelled in the presence of BFA and immunoprecipitated with a CD4-specific antibody.

(d) The effect of Vpu on CD4 in the absence of BFA constitutes an appropriate negative control.

(e) The extent of CD4 degradation can be determined quantitatively by densitometric scanning of the CD4 protein band.

Protocol 5. Effect of Vpu on intracellular Env–CD4 complexes and stability of CD4

Equipment and reagents

- Ultracentrifuge (Beckman T45)
- HeLa or COS-7 cells
- Expressor plasmids for CD4, Vpu, and Env
- Brefeldin A (BFA; Sigma)
- Methionine-free DMEM
- [^{35}S]methionine
- Methionine
- Lysis buffer: 50 mM Tris–HCl pH 7.4, 300 mM NaCl, 0.5% Triton X-100, 10 mM iodoacetamide, 0.5 mM leupeptin, 0.2 mM phenylmethylsulfonyl fluoride

Method

1. HeLa (1 × 10^6) or COS-7 (1.4 × 10^6) cells are seeded in 80 cm^2 flasks in 12 ml of DMEM/10% FCS 24 h prior to transfection.
2. Transfect cells with:
   (a) A CD4 expression plasmid and a construct that expresses both Vpu and Env.
   (b) A CD4 expression plasmid and separate Vpu and Env expression plasmids.

Protocol 5. *Continued*

(c) A CD4 expression plasmid and an expression plasmid for Vpu alone. In this case cells are treated with BFA to trap CD4 in the endoplasmic reticulum. The molar ratio between the CD4 plasmid and other plasmids should approximately be 1:3.

3. Remove culture medium between 20–48 h post-transfection. Wash cells once with warmed PBS. Add 5 ml of warmed methionine-free DMEM/10% FCS and incubate 30 min at 37°C. Starting from this step, media of cells co-transfected with a CD4 expression plasmid and a Vpu expression plasmid should contain 10 μM of BFA.
4. Scrape cells off the flask in 10 ml of the same medium. Collect cells in 15 ml conical tube. Centrifuge at 500 *g* for 5 min. Discard the supernatant.
5. Add 250 μl of methionine-free DMEM containing 500 μCi of [^{35}S]methionine.
6. Incubate for 30 min at 37°C with gentle shaking at 5 min intervals.
7. Add 10 ml warmed DMEM/10% FCS containing an excess of cold methionine, and centrifuge cells at 500 *g* for 5 min to remove medium.
8. Resuspend cells in 4 ml of DMEM/10% FCS with excess of cold methionine. Aliquot into four tubes and incubate at 37°C.
9. Collect aliquots of cells at 0, 1, 2, and 3 h. Wash cells in cold PBS. Lyse cells by freezing and thawing in lysis buffer.
10. Centrifuge cell lysates at 35 000 r.p.m. for 1 h at 4°C in a Beckman Ti45 rotor.
11. Immunoprecipitate Env and CD4 (*Protocol 1*) using PBS instead of RIPA buffer for the preparation of immunosorbents and washing steps.

2.5 *In vitro* assay for CD4 degradation

Recently, an *in vitro* assay which bypasses the complexity of a whole cell system and permits experimental manipulations not possible in intact cells has been developed (18).

(a) Expression of CD4 and Vpu was achieved by *in vitro* transcription and translation using a rabbit reticulocyte lysate supplemented with canine microsomal membranes and resulted in Vpu-dependent decay of CD4, thus duplicating previous observations made in whole cells.

(b) Degradation of CD4 in this system requires membrane association and expression of Vpu and CD4 in the same membrane compartment.

(c) The rate of CD4 decay in the rabbit reticulocyte lysate is approximately one-third of that observed previously in tissue culture (40 min versus 12 min).

This *in vitro* degradation assay (*Protocol 6*) constitutes a valuable system to determine the domains of CD4 and Vpu which are involved in this process. The ability to study CD4 decay outside a whole cell system should significantly advance studies of the biochemical mechanism and allow the definition of the parameters which govern Vpu-induced degradation of CD4.

Protocol 6. *In vitro* CD4 degradation assay

Reagents

- Template DNAs containing the genes for CD4 or Vpu downstream of a T7 or SP6 RNA polymerase promoter
- 5 × transcription buffer: 200 mM Tris–HCl pH 7.5, 30 mM $MgCl_2$, 10 mM spermidine, 50 mM NaCl
- rNTP solution containing 2.5 mM each of ATP, CTP, and UTP (prepared by mixing equal volumes of the three 10 mM rNTP stocks)
- GTP (10 mM)
- m7 G(5′)ppp(5′)G (10 mM) (Pharmacia)
- DTT (100 mM)
- RNasin ribonuclease inhibitor (Promega)
- T7 or SP6 RNA polymerase
- Cytidine 5′-triphosphate, tetrasodium salt, (5-^{3}H)-(NEN)
- RQ1 RNase-free DNase I (Promega)
- Amino acid mixture minus methionine (Promega)
- [^{35}S]methionine
- Nuclease treated rabbit reticulocyte lysate (Promega)
- Canine pancreatic microsomal membranes (Promega)
- RNase A
- TCA

Method

1. Plasmids containing the genes for CD4 or Vpu downstream of either a bacteriophage SP6 or T7 RNA polymerase promoter are linearized 3′ to the genes using restriction enzymes that generate 5′ protruding or blunt ends.
2. Assemble in a 100 μl reaction mixture (at room temperature):
   - 20 μl of 5 × transcription buffer
   - 10 μl of 100 mM DTT
   - 15 μl of rNTP solution
   - 1 μl of GTP (10 mM)
   - 10 μl of 10 mM m7 G(5′)ppp(5′)G
   - 100 U RNasin
   - Linearized template DNAs (2–5 μg)
   - Cytidine 5′-triphosphate, tetrasodium salt, (5-^{3}H)-(10 μCi)
   - SP6 or T7 RNA polymerase (30–50 U)

   Complete to a final volume of 100 μl with nuclease-free water.
3. Incubate the reaction mixture for 120 min at 37–40°C.

Protocol 6. *Continued*

4. Add 4 μl of RQ1 RNase-free DNase I and incubate 15 min at 37°C to remove the DNA template.
5. Purify synthesized RNA. Extract twice with phenol/chloroform. Ethanol precipitate. Resuspend the RNA sample in 30 μl of nuclease-free water.
6. Mix the following reagents in a mini-Eppendorf tube at 4°C:
   - 1 μl of RNasin
   - 1 μl of 1 mM amino acid mixture minus methionine
   - 5 μl of CD4 RNA transcript (250 ng)
   - 10 μl of [^{35}S]methionine (100 μCi)
   - 35 μl reticulocyte lysate
   - 5 μl microsomal membranes (1.2 equivalents/μl)
7. Incubate for 30 min at 30°C.
8. Mix:
   - 5 μl of Vpu RNA
   - 20 μl of fresh reticulocyte lysate
   - 25 μl of pre-translated lysate

   For a negative control, follow the same procedure except that Vpu RNA is replaced by 5 μl of nuclease-free water.
9. Incubate for an additional 15 min at 30°C to allow for Vpu synthesis.
10. Stop the reaction by adding RNase A to a final concentration of 1 mg/ml.
11. Re-incubate the RNase A treated lysate at 30°C to test for CD4 degradation.
12. Remove equal aliquots (5 μl) at various times (0, 30, 60, 90, 120 min).
13. Precipitate *in vitro* translated proteins with TCA.
14. Boil samples in SDS sample buffer (*Protocol 1*). Separate proteins by SDS–PAGE. Visualize bands by autoradiography of the dried gels (−70°C).

Acknowledgements

We thank Jacques Friborg and Xiao J. Yao for helpful discussions. E. A. C. is the recipient of a career award from the National Health Research and Development Program (NHRDP) of Canada and is supported by grants from NHRDP and MRC. H. G. G. is supported by National Institutes of Health Grants AI29873 and AI34267.

References

1. Cohen, E. A., Terwilliger, E. F., Sodroski, J. G., and Haseltine, W. A. (1988). *Nature*, **334**, 532.
2. Strebel, K., Klimkait, T., Maldarelli, F., and Martin, M. A. (1989). *J. Virol.*, **63**, 3784.
3. Maldarelli, F., Chen, M.-Y., Willey, R. L., and Strebel, K. (1993). *J. Virol.*, **67**, 5056.
4. Klimkait, T., Strebel, K., Hoggan, M. D., Martin, M. A., and Orenstein, J. M. (1990). *J. Virol.*, **64**, 621.
5. Yao, X. J., Boisvert, F., Garzon, S., and Cohen, E. A. Manuscript in preparation.
6. Strebel, K., Klimkait, T., and Martin, M. A. (1988). *Science*, **241**, 1221.
7. Terwilliger, E. F., Cohen, E. A., Lu, Y., Sodroski, J. G., and Haseltine, W. A. (1989). *Proc. Natl Acad. Sci. USA*, **86**, 5163.
8. Yao, X. J., Garzon, S., Boisvert, F., Haseltine, W. A., and Cohen, E. A. (1993). *J. Acq. Immune Defic. Syndr.*, **6**, 135.
9. Göttlinger, H. G., Dorfman, T., Cohen, E. A., and Haseltine, W. A. (1993). *Proc. Natl Acad. Sci. USA*, **90**, 7381.
10. Willey, R. L., Maldarelli, F., Martin, M. A., and Strebel, K. (1992). *J. Virol.*, **66**, 226.
11. Willey, R. L., Maldarelli, F., Martin, M. A., and Strebel, K. (1992). *J. Virol.*, **66**, 7193.
12. Cullen, B. R. (1987). In *Methods in enzymology* (ed. R. Wu), Vol. 152, pp. 684–704. Academic Press, London.
13. Yao, X. J., Göttlinger, H., Haseltine, W. A., and Cohen, E. A. (1992). *J. Virol.*, **66**, 5119.
14. Haseltine, W. A. (1992). In *AIDS: Etiology, diagnosis, treatment and prevention* (ed. S. Hellman and S. A. Rosenberg), pp. 39–59. J. B. Lippincott, Philadelphia.
15. Friborg, J. A., Ladha, A., Göttlinger, H., Haseltine, W. A., and Cohen, E. A. (1995). *J. Acq. Immune Defic. Synd. and Human Retroviral.* **8**, 10–22.
16. Crise, B. L., Buonocore, L., and Rose, J. K. (1990). *J. Virol.*, **64**, 5585.
17. Garcia, J. V. and Miller, A. D. (1991). *Nature*, **350**, 508.
18. Chen, M.-Y., Maldarelli, F., Karczewski, M. K., Willey, R. L., and Strebel, K. (1993). *J. Virol.*, **67**, 3877.

15

Vif

DIDIER TRONO

1. Introduction

The HIV-1 *vif* open reading frame (previously called '*sor*' or '*A*') encodes a protein of 192 amino acids, with an approximate molecular weight of 23 kDa. Vif is highly conserved among different HIV-1 isolates. Vif is also present in HIV-2 and simian immunodeficiency virus, and in non-primate lentiviruses, such as feline and bovine immunodeficiency viruses, caprine arthritis encephalitis virus, and visna virus, a cytopathic lentivirus of sheep only distantly related to HIV. HIV-1 Vif is translated from a singly spliced viral mRNA, together with other late viral proteins whose expression is induced by Rev. Initial studies showed that HIV-1 strains defective in *vif* generated virions that infected cells approximately a 1000 times less efficiently than wild-type viruses, hence the acronym Vif, for *v*irion *i*nfectivity *f*actor.

Vif is required for the formation of HIV-1 particles competent for cell-to-cell as well as cell-free spread in a number of cell types, including those targets of the infection *in vivo* (1, 2). Although Vif is required at the stage of viral particle formation, its function is manifested only after the virus enters target cells. *Vif* defective viruses are fully competent for entry. For example, HIV(MLV) pseudotypes, which should bypass any requirement for Vif at the entry step were it defective, do not rescue the *vif* defect (2). The functional role of Vif was identified by monitoring the various intermediates and products of reverse transcription by PCR (2) in cells freshly infected with WT and ΔVif viruses:

(a) *Vif* defective viruses were dramatically affected in their ability to complete the synthesis of the proviral DNA after viral internalization.

(b) The reverse transcription process *per se* appears to be normal in *vif* mutated viruses. Wild-type amounts of partial reverse transcripts are seen in ΔVif virions and in cells infected with mutant viruses.

(c) The defect in DNA synthesis is most likely secondary to a block in the uncoating and/or in the transport of the internalized viral particle. Vif function could for instance result in the partial disruption of the core, making it permeable for cellular components essential for reverse transcription, such as cations or nucleotides. Alternatively, Vif might be required because, in lentiviruses, the ribonucleoprotein complex is routed to a specific subcellular compartment, where DNA synthesis proceeds. In that respect, the discovery that the matrix protein of lentiviruses directs incoming viral cores to the nucleus of newly infected cells might be of great relevance to understand Vif function (3).

It remains unclear whether Vif exerts its role directly or through some intermediate. Vif has recently been detected in HIV-1 virions; however, Vif could act indirectly by modifying another virion-associated factor involved in the early steps of the virus life cycle, such as one of the *gag/pol* gene products. Although analyses of wild-type and *vif* defective particles, produced from cells restrictive for the mutant virus, have so far failed to detect a difference in the amount or in the proteolytic processing of these various proteins, it is possible that subtle differences or post-translational modifications not addressed by the techniques used have been missed. The essential role played by Vif in the virus life cycle is a strong incentive to further examine this point.

2. Growth of ΔVif HIV-1 viral stocks

Vif is required for the infectivity of HIV-1 virions only when these are produced from a few specific cell types. Thus, viruses genotypically defective in *vif* may either show a mutant or a wild-type (WT) phenotype, depending on their cells of origin (1, 2). Cells revealing of the *vif* defective (ΔVif) phenotype are:

- peripheral blood lymphocytes (PBL)
- primary macrophages
- established T cell lines, such as H9 and CEM

ΔVif viruses produced from the T cell lines SupT1, C8166, and Jurkat, or from the adherent cell lines COS and 293 (a human fibroblastic cell line) are phenotypically normal. The production of viral stocks showing a Vif phenotype is complicated by this particularity. Cells that are readily transfectable with proviral DNA constructs, such as COS or 293 cells, generate HIV-1 viruses that are normally infectious even in the absence of *vif*. For practical reasons, the two most convenient sources of functionally relevant ΔVif virions are CEM and H9 cells. CEM cells, although they reveal of the ΔVif phenotype, are partially permissive for the growth of the *vif* defective virus. Therefore, high titre stocks of ΔVif are usually obtained from these cells. H9 cells are fully restrictive for ΔVif. H9 cultures which are 100% infected with *vif* viruses can therefore be obtained only with proviral constructs carrying dominant selectable markers.

2.1 Generating ΔVif from CEM cells

The following points should be noted when preparing stocks of *vif* defective viruses in CEM cells:

(a) WT virus should spread in the whole culture in approximately eight to ten days, whereas ΔVif virus should lag by one to three weeks.

(b) At the peak of infection, 80–90% of the cells will be killed by either virus.

(c) The cultures can be maintained by changing medium once a week, to obtain chronic survivors. On average, these will produce between 200–350 ng p24/ml of supernatant, for one to two months, after which low producers will outgrow other cells.

(d) Stocks of chronic survivors should be frozen as soon as they emerge.

As an alternative to transfection, CEM cells can be electroporated with WT and ΔVif proviral DNAs (*Protocol 2* and Chapter 13). This significantly shortens the procedure, the peak of infection being reached within five to six days with WT. Unfortunately following electroporation, the ΔVif phenotype is less pronounced, which may be a problem when assaying the properties of various *vif* mutants.

Protocol 1 describes the preparation of *vif* viral stocks in CEM cells. Work with infectious viruses must be performed using appropriate biological Containment Facilities (see Chapters 1, 2, and 3, Volume 1).

Protocol 1. Infection of CEM cells with cell-free virus produced from ΔVif permissive cells

Equipment and reagents

- COS or 293 cells transfected with WT and ΔVif proviral DNAs (see Chapter 11 and Chapter 2, Volume 1)
- 0.45 μm nitrocellulose filter membrane
- p24 antigen
- Tissue culture medium with serum
- p24 antigen ELISA kit (see Chapter 7, Volume 1)
- 37°C CO_2 incubator

Method

1. Transfect WT and ΔVif proviral DNAs into COS or 293 cells, using the DEAE–dextran or calcium phosphate technique (see Chapter 11 and Chapter 2, Volume 1) (4).
2. Harvest cell supernatant at two days post-transfection.
3. Filter supernatant through 0.45 μm nitrocellulose filter membrane, and measure viral content by p24 antigen ELISA (see Chapter 7, Volume 1).
4. Incubate 2×10^6 CEM cells with supernatant containing approx. 100 ng p24, in total volume of 5 ml, from 6–16 h at 37°C in CO_2 incubator.

Protocol 1. *Continued*

5. Wash cells three times in tissue culture medium with serum.
6. Resuspend cells in 10 ml culture medium. Incubate at 37°C in CO_2 incubator. Monitor viral growth by examining culture for signs of cytopathicity (syncytia), and by measuring p24 antigen activity in the supernatant, by ELISA.

Protocol 2. Infection of CEM cells by electroporation[a]

Equipment and reagents

- Bio-Rad Gene Pulsor
- CEM cells
- DNA
- PBS
- Electroporation cuvette
- Tissue culture medium with serum
- 37°C CO_2 incubator

Method

1. Wash 1×10^7 CEM cells[b] with ice-cold PBS.
2. Resuspend in 0.75 ml PBS in electroporation cuvette.
3. Add 20–40 μg DNA, in PBS.
4. Leave on ice for 10 min.
5. Electroporate cells at a voltage of 250 V, and a capacitance of 960 μF.
6. Incubate cells on ice for 5–10 min.
7. Resuspend cells in 20 ml tissue culture medium with serum, and place in CO_2 incubator at 37°C.

   Note: For best results, cells should be in logarithmic phase of growth, at a density of 2.4×10^5 cells/ml.

[a] Conditions are given for the Bio-Rad Gene Pulser electroporator.
[b] For best results cells should be in logarithmic growth.

2.2 Generating ΔVif stocks from H9 cells

Because H9 cells do not allow the spread of the ΔVif virus, the techniques used to grow *vif*-minus viruses in CEM cells are inefficient and allow only a fraction of the H9 population to become infected. H9 cells can be more efficiently infected by co-cultivation of H9 cells with transfected ΔVif permissive cells, or proviral constructs containing dominant selectable markers (*Protocol 3*). 20–50% of the H9 cell population becomes infected by this method, as assessed by indirect immunofluorescence. A major limitation of this approach is that, with time, ΔVif-infected cells are counterselected (probably due to the virus cytopathicity). This results in an overgrowth of uninfected cells.

The use of proviral constructs with dominant selectable markers limits the overgrowth of cultures by uninfected cells:

(a) Transfect COS or 293 cells with WT or ΔVif proviral constructs containing a dominant selectable marker in the *nef* location (as described in ref. 5). Suitable markers are the Salmonella typhimurium His D gene, or the neomycin phosphotransferase gene, conferring resistance to histidinol and G418, respectively (5). The dihydrofolate reductase gene, adequate in other cell types, is problematic in H9 cells, because a significant proportion of these cells is spontaneously resistant to methotrexate.

(b) Infect H9 cells with supernatant of transfected cells, as described for CEM cells in *Protocol 1*.

(c) Two days post-infection, place H9 cells in selection medium, in 24-well dishes. For proviruses containing the His D gene, use 1 mM histidinol (Sigma) in His^- tissue culture medium. For proviruses containing the neomycin phosphotransferase gene, use 1 mg/ml G418 (Geneticin, Gibco) in regular medium.

(d) Feed cells every three to four days. Resistant populations are obtained within eight to ten days with WT and within three to four weeks with ΔVif.

Protocol 3. Infection of H9 cells by co-cultivation

Equipment and reagents

- COS cells transfected with ΔVif proviral DNA (*Protocol 1*)
- H9 cells

Method

1. Add 2×10^6 H9 cells two days after transfection.
2. Co-cultivate for 6–12 h.
3. Transfer suspension cells (H9) to new flask. This may have to be repeated two or three times, at daily intervals, as some adherent cells are also transferred.

3. Measurement of *vif*-minus HIV-1 growth

3.1 RNA content of HIV-1 particles

Protocol 4 is an extremely simple method for measuring virus RNA levels by hybridization. Perform hybridization using a ^{32}P-labelled RNA probe generated with T7 polymerase. We commonly use a riboprobe complementary to nucleotides 8475 to 8900 of the HIV-1 genome (6, 7). Hybridization is performed overnight at 60°C, and membranes are washed in 0.2 × SSC three times 20 min at 68°C before exposure to X-ray film. It can also be used to titre the supernatant of packaging cell lines which produce MLV-based retroviral vectors.

Protocol 4. Measurement of virion RNA by slot blot analysis

Equipment and reagents

- Virus-containing culture medium
- 10 mg/ml proteinase K
- 10 mg/ml tRNA
- 0.5 M EDTA
- 20% SDS
- Phenol
- 0.4 M NaCl
- 70% ethanol
- 2 mM EDTA
- 50% formamide/17% formaldehyde
- 15 × SSC: 1 × SSC is 0.15 M NaCl, 0.015 M sodium citrate
- Nitrocellulose
- ^{32}P-labelled DNA probe
- Stratalinker (Stratagene)
- Vortex
- Microcentrifuge
- Slot blot apparatus
- 37°C water-bath

Method

1. Remove 700 µl virus-containing supernatant for p24 antigen measurement.
2. Mix the following:
   - 700 µl virus-containing culture medium
   - 35 µl 10 mg/ml proteinase K
   - 7 µl 10 mg/ml tRNA
   - 3.5 µl 0.5 M EDTA
   - 17.5 µl 20% SDS
3. Incubate at 37°C for 45 min.
4. Add 700 µl of phenol, vortex, and microcentrifuge for 5 min.
5. Ethanol precipitate the aqueous phase after addition of 0.4 M NaCl. Rinse the RNA pellet with 70% ethanol. Resuspend in 20 µl 2 mM EDTA.
6. Prepare fivefold serial dilutions of RNA.
7. Denature RNA in 50% formamide/17% formaldehyde. Heat to 50°C for 20 min. Mix with 120 µl 15 × SSC. Bind to nitrocellulose by aspiration through a slot blot apparatus.
8. Expose to UV to cross-link RNA to membrane, using a Stratalinker (Stratagene), according to manufacturer's instructions.
9. Hybridize to ^{32}P-labelled RNA probe.

3.2 PCR analysis of cells freshly infected with HIV

Due to its extreme sensitivity, the polymerase chain reaction (see Chapters [ ], Volume 1) allows the study of the early steps of HIV-1 infection by monitoring the accumulation of viral DNA in freshly infected cells. Because of the risk of contamination associated with this technique, precautions should be taken in preparing the virions, and later in handling the samples to be analysed:

- a set of dedicated pipettors, filtered pipette tips, and screw-cap microtubes are essential
- contamination is much more frequent with snap-cap microtubes

Protocol 5 describes the preparation of virions suitable for PCR analysis. We have observed that DNase digestion is much less efficient if performed on straight supernatant rather than on concentrated virions. After DNase treatment, some PCR detectable DNA can still be found associated with HIV-1 or MLV virions. This corresponds to partial reverse transcripts, contained within the viral particles (if virions are first lysed with NP-40 and then treated with DNase, these products are no longer detectable) (8, 9).

Protocol 5. Preparation of 'PCR suitable' virions

Equipment and reagents

- Viral-infected cells
- PBS or RPMI
- DNase I
- 10 mM $MgCl_2$
- Ultracentrifuge with SW28 rotor (Beckman) (23 000 r.p.m./4 °C)

Method

1. Concentrate virus from filtered supernatants of infected cells by ultracentrifugation at 23 000 r.p.m. in an SW28 rotor (Beckman) for 2.5 h at 4 °C.
2. Resuspend viral pellet in PBS or RPMI (in a volume between 0.2–2.0 ml).
3. To remove residual cellular or viral DNA, released into the culture supernatant by lysed cells, digest with 20 μg/ml DNase I at 37 °C for 30 min in the presence of 10 mM $MgCl_2$.
4. Take an aliquot for p24 antigen measurement, and store the viral stock in 200 μl aliquots at −70 °C.

The sequences of HIV-specific primers commonly used in our laboratory to monitor infection are as follows (positions of nucleotides in the HIV-1

HXB2D sequence, according to Ratner *et al.* (10), are indicated in parentheses):
U3, CACACACAAGGCTACTTCCCT (57 to 77);
R, GGCTAACTAGGGAACCCACTGCTT (496 to 516);
U5, CTGCTAGAGATTTTCCACACTGAC (635 to 612);
5NC, CCGAGTCCTGCGTCGAGAGAGC (698 to 677, 5′-non-coding region);
Vif1, GGGAAAGCTAGGGGATGGTTTTAT (5136 to 5159);
Vif2, CAGATGAATT AGTTGGTCTG (5366 to 5347).
The U3–U5 pair amplifies all viral DNAs, including minus-stand strong stop DNA, the earliest reverse transcription product; the Vif1–Vif2 pair amplifies elongated minus-strand, synthesized after the first template switch, as well as later products; the R–5NC pair is specific for completed reverse transcripts, made after the second template switch. *Protocol 6* gives a method for measuring HIV yield by PCR.

Protocol 6. PCR analysis of cells freshly infected with HIV

Equipment and reagents

- Target cells
- p24 from concentrated ΔVif or WT virions
- PBS
- PCR lysis buffer: 50 mM KCl, 10 mM Tris pH 8.3, 1.5 mM $MgCl_2$, 0.01% gelatin, 0.45% NP-40, 0.45% Tween-20, 60 μg/ml proteinase K
- Primer
- Deoxynucleoside triphosphates (all four bases)
- 50 mM KCl, 10 mM Tris pH 8.3
- 1.5 mM $MgCl_2$
- 0.01% gelatin
- *Taq* DNA polymerase (Stratagene)
- Crude lysate of 1200–60 000 cells or purified virions (template)
- Mineral oil
- 1.2% agarose gel
- 0.4 N NaOH, 2 mM EDTA
- 50% formamide
- 4 × SSPE: 1 × SSPE is 0.18 M NaCl, 10 mM Na_2PO_4, 1 mM EDTA pH 7.7
- 0.1% SDS, 1 × Denhardt's solution
- 100 μg/ml herring sperm DNA
- Buffer containing 1 × 10^6 c.p.m./ml ^{32}P-labelled probe[a]
- 2 × SSC 1% SDS
- 1 × SSC
- 0.5 × SSC
- Nylon membrane (Zetabind, Cuno)
- X-ray film
- filters
- Perkin-Elmer thermocycler
- Electrophoresis apparatus
- 4°C water-bath
- 55°C water-bath
- 42°C water-bath

Method

1. Mix 3 × 10^6 target cells (for instance H9) with 250 ng p24 of concentrated ΔVif or WT virions.
2. Incubate at 4°C for 30 min.
3. Harvest aliquots of the cultures corresponding to 6 × 10^5 cells (0 h samples). Wash cells three times in culture medium to remove virus non-specifically bound to the cell surface. Lyse in PCR lysis buffer as described in steps 5–8.

4. Incubate the remaining cells at 37°C for 4 h to allow entry of the viruses. Wash and resuspend cells in fresh medium. At regular time points (for instance 4 h, 8 h, 12 h, and 24 h) harvest and lyse aliquots of 6×10^5 cells.
5. Wash the cells in PBS, and then disrupt them in 100 μl PCR lysis buffer.
6. Incubate the lysate at 55°C for 2 h to overnight.
7. Inactivate the proteinase by heating to 100°C for 10 min.
8. Store samples at −20°C.
9. Assemble 50 μl PCR reactions containing:
   - 50 pmol of each primer
   - 0.2 mM each of the four deoxynucleoside triphosphates
   - 50 mM KCl, 10 mM Tris pH 8.3
   - 1.5 mM $MgCl_2$
   - 0.01% gelatin
   - 2.5 U of *Taq* DNA polymerase (Stratagene)
   - either crude lysate of 1200–60 000 cells, or purified virions as template
10. Overlay PCR reaction mixtures with a drop of mineral oil and subject to 30 cycles in a Perkin-Elmer thermocycler:
    - 1 min at 94°C
    - 1 min at 55°C
    - 2 min at 72°C
    - 7 min at 72°C
11. Analyse amplified products by electrophoresis through a 1.2% agarose gel.
12. Transfer to nylon membrane (Zetabind, Cuno) in 0.4 M NaOH/2 mM EDTA.
13. Pre-hybridize for 2–3 h at 42°C in a solution consisting of:
    - 50% formamide
    - 4 × SSPE
    - 0.1% SDS, 1 × Denhardt's solution
    - 100 μg/ml herring sperm DNA
14. Hybridize for 6 h to overnight at 42°C, in the same buffer containing 1×10^6 c.p.m./ml ^{32}P-labelled probe.[a]
15. Wash filters, once each, in the following conditions: room temperature in 2 × SSC, 1% SDS; 65°C for 20 min in 2 × SSC, 1% SDS; 65°C for 20 min in 1 × SSC; 65°C for 20 min in 0.5 × SSC.

Protocol 6. *Continued*

16. Expose to X-ray film.

[a] Probes are generated by random oligonucleotide labelling with a random primer kit (Stratagene), following the manufacturer's instructions, and purified through Nuctrap push columns (Stratagene). The template commonly used in our laboratory to generate the HIV-1-specific probe is a *Hpa*I–*Xba*I fragment from plasmid R7-HXB2 (2), containing the entire proviral sequence.

3.3 Detecting Vif protein in HIV-1-infected cells

Detection of Vif in cells can be difficult, due to the low abundance of this protein, its sensitivity to contaminating proteases, and its apparent unstability. However, proper methods allow these problems to be circumvented. *Protocol 7* has been used with success in our laboratory to detect Vif.

Protocol 7. Immunoblot detection of HIV-1 Vif protein

Equipment and reagents

- HIV-1-infected cells
- PBS
- Lysis buffer: 10 mM NaCl, 10 mM Tris pH 7.5, 0.5% Nonidet P-40, 1% SDS, 1 mM phenylmethylsulfonyl fluoride (PMSF), 2 μg/ml aprotinin
- BCA protein assay (Pierce)
- Buffer A: 25 mM Tris, 192 mM glycine, and 20% methanol, pH 8.3
- TBST: 10 mM Tris–HCl pH 8.0, 150 mM NaCl, 0.05% Tween-20
- Horse-radish peroxidase-conjugated secondary antibody
- Non-fat dry milk
- 12% SDS–polyacrylamide gel
- Tuberculin syringe
- 23 G needle
- Polyvinilidene diflouride (PVDF) membranes (Micron Separations, Inc.)
- Electrophoresis apparatus
- ECL Western Blotting Kit (Amersham)

Method

1. Wash approx. 5×10^6 HIV-1-infected cells once in PBS.
2. Disrupt cells in 100 μl lysis buffer, containing:
   - 10 mM NaCl
   - 10 mM Tris pH 7.5
   - 0.5% Nonidet P-40
   - 1% SDS
   - 1 mM phenylmethylsulfonyl fluoride (PMSF)
   - 2 μg/ml aprotinin
3. Shear liberated cellular DNA by several passages through a 23 G needle, using a tuberculin syringe.
4. Measure protein concentration of lysate, using the BCA protein assay (Pierce).

5. Aliquot and store samples frozen at −20 or −70°C.[a]
6. Separate proteins by SDS–polyacrylamide gel electrophoresis, loading approx. 50 μg of proteins on a 12% gel.
7. Transfer proteins to polyvinilidene difluoride (PVDF) membranes (Micron Separations, Inc.) in buffer A at 500 mA for 2 h.
8. Block membranes by incubating for 2–12 h in TBST containing 5% non-fat dry milk, at room temperature with gentle rocking.
9. Incubate membranes in the presence of anti-Vif antibody (we use a 1:500 dilution of a monoclonal antibody recognizing the carboxy-terminal end of HIV-1 Vif), in TBST.
10. Wash membranes with TBST.
11. Perform detection with horse-radish peroxidase-conjugated secondary antibody, by enhanced chemoluminescence (ECL Western Blotting Kit, Amersham).

[a] Avoid freezing and thawing repeatedly the extracts.

Acknowledgements

Work on Vif in the author's laboratory is supported by a grant from NIH. D. T. is also the recipient of a Pew Scholarship in the Biomedical Sciences, and as such receives support from the Pew Charitable Trust.

References

1. Gabuzda, D. H., Lawrence, K., Langhoff, E., Terwilliger, E., Drofman, T., Haseltine, W. A., *et al.* (1992). *J. Virol.*, **66**, 6489.
2. von Schwedler, U., Song, J., Aiken, C., and Trono, D. (1993). *J. Virol.*, **67**, 4945.
3. Bukrinsky, M. I., Haggerty, S., Dempsey, M. P., Sharova, N., Adzhubei, A., Spits, L., *et al.* (1993). *Nature*, **365**, 666.
4. Ausubel, F., Brent, R., Kingston, R. E., Moore, D. D., Smith, J. A., Seidman, J. E., *et al.* (ed.) (1987). *Current protocols in molecular biology*. John Wiley & Sons, New York.
5. Aldovini, A. and Feinberg, M. B. (1990). In *Techniques in HIV research* (ed. A. Aldovini and B. D. Walker), pp. 147–75. Macmillan, London.
6. Trono, D., Andino, R., and Baltimore, D. (1988). *J. Virol.*, **62**, 2291.
7. Trono, D. and Baltimore, D. (1990). *EMBO J.*, **9**, 4155.
8. Lori, R. F., di Marzo Veronese, F., De Vico, A. L., Lusso, P., Reitz, M. S., and Gallo, R. C. (1992). *J. Virol.*, **66**, 5067.
9. Trono, D. (1992). *J. Virol.*, **66**, 4893.
10. Ratner, L. W., Haseltine, W. A., Patarca, R., Livak, K. J., Starcich, B., Josephs, S. F., *et al.* (1985). *Nature*, **313**, 277.

Part II
Drug discovery

16

Cellular assays for antiviral drugs

N. MAHMOOD

1. Introduction

The detailed understanding of the growth of HIV, reflected by the chapters in this book has enabled chemists to produce compounds targeted at specific sites in the HIV replicative cycle. HIV infects human T lymphocytes, monocytes/macrophages, and nerve cells both in AIDS patients and *in vitro*. Compounds have been shown to exhibit differences in antiviral activity and cytotoxicity depending on the cell type studied. It is therefore necessary to evaluate prospective anti-HIV drugs in multiple cell lines of different origins.

Assay systems using specific cell/virus combinations are now available for the assessment of compounds with various mechanisms of anti-HIV action.

In the study of structure–activity relationships of chemically synthesized compounds small differences in structure can markedly affect the efficacy of any compound.

- assay methods should be sensitive enough to detect low levels of antiviral activity and cytotoxicity
- methods should be standardized so that results are comparable between experiments

2. Primary screening

2.1 The XTT–Formazan assay

The ideal assay system for preliminary screening of a broad spectrum of compounds of different modes of anti-HIV action has two characteristics:

- the assay should be automated and simple to perform
- the assay should provide a quantitative measure of drug toxicity

A compound showing some efficacy and selectivity in primary screening should be evaluated further using specific and more sensitive tests measuring various end-points in HIV replication.

Measurement of cell viability of HIV-infected and uninfected cells by the soluble Formazan assay (1) is a reliable and convenient way of assessing the antiviral effect of a large number of compounds (*Protocol 1*).

(a) This assay is based on the metabolic reduction of XTT (2,3-bis[2-methoxy-4-nitro-5-sulfophenyl]-5-[(phenylamino)carbonyl]-2H-tetrazoliumhydroxide) to a soluble brown product (XTT–Formazan) in surviving cells; it replaces 3-(4,5-dimethylthiazol-2-yl)-2,5-di-phenyl-tetrazolium bromide (MTT) which has been used routinely in our laboratory (2).

(b) The MTT–Formazan product is insoluble and requires vigorous and time-consuming mixing with isopropanol and Triton X-100 prior to colorimetric determination. Particularly in the case of adherent cell lines, the XTT–Formazan assay is easier to perform.

(c) It is extremely important to use cells which are sensitive to HIV infection and its lytic effect. C8166 cells are human T lymphoblastoid cells which have been reported to carry but not express the HTLV-1 gene expressing the Tat sequence (3). These are our first choice. Infected C8166 cells produce large visible syncytia with a ballooning effect resulting in cell death within four to six days of infection. MT4 cells have also been used successfully in the MTT–Formazan assay (2). MT4 are human T cells transformed by co-cultivating with leukaemia lymphocytes harbouring HTLV-1. They may shed HTLV-1 in the culture supernatant. HIV infection of these cells is lytic, but the cytopathic effect (syncytium formation and ballooning) is difficult to visualize. Other T cell lines such as H9, JM, CEM, and U937 are less suitable for this assay as they have a tendency of becoming chronically infected and resistant to infection and its lytic effect. All cell cultures should be passaged twice a week in RPMI 10% (see *Protocol 1*).

(d) Although C8166 cells are sensitive to infection by most of HIV-1 strains, HIV-2 and SIV, HIV-1_{IIIB}, or HIV-1_{MN} are preferentially used for primary screening. The virus stocks are maintained in the culture supernatant of chronically-infected H9 cell line, the titre is determined in C8166 cells (see Chapter 2, Volume 1).

(e) Prior to initial screening of an anti-HIV candidate it is essential to acquire information about its solubility and stability in different solvents at different temperatures. For testing, water solutions are preferable which are diluted 1/10 in medium but a 1/100 dilution is necessary if the solution is prepared in DMSO or ethanol. It is always useful to test the toxicity of different solvents used for dissolving compounds. The stock solution should be stored at −20°C.

The data from the plate reader is collected via a remote link on a computer located outside the Containment Laboratory. The proportions of viable cells, both infected and uninfected are readily calculated using a spread sheet (e.g. Lotus123 or XL), in order to assess antiviral activity and cytotoxicity. The data is expressed as percentages of untreated uninfected control. The results of cell viability assay from one plate are presented in *Figure 1*.

Protocol 1. Primary screening of antiviral compounds using the XTT–Formazan method

Equipment and reagents

- XTT (Sigma Chemical)
- *N*-methylphenazonium methosulfate (BDH Ltd, UK)
- RPMI 1640 (ICN Flow, UK) containing 10% fetal calf serum, 2 mM glutamine, 100 U/ml penicillin, 100 μg/ml streptomycin and kanamycin (RPMI 10%)

Method

1. Make six serial fivefold dilutions of each compound in triplicate in 50 μl of medium in microtitre plate wells using a multichannel pipette. Use only the central 60 wells and fill the outer row wells with 50 μl of medium only.
2. Centrifuge C8166 cells at 170 *g* for 5 min and resuspend in fresh medium at a concentration of 1.6×10^6/ml.
3. Add 25 μl of the resuspended cells to each well.
4. Add 25 μl of a dilution of virus HIV-1_{111B} or HIV-1_{MN}, supernatant from chronically-infected H9s (clarified by centrifugation at 170 *g* for 5 min) to a duplicate set of dilutions immediately after the addition of cells. The dilution of virus which kills 70–80% infected cells in five days is predetermined.
5. Add 25 μl of medium to the third set of dilutions to measure cytotoxicity. Include untreated wells as controls for both infected and uninfected cells.
6. Incubate the plates at 37°C in humidified boxes with 5% CO_2.
7. Examine the cell cultures for evidence of virus-induced cytopathic effects two to three days after infection.
8. Add 25 μl of a mixture of XTT 1 mg/ml and 0.02 mM *N*-methylphenazonium methosulfate to infected and uninfected cultures five days post-infection.
9. Incubate plates for 3–4 h and replace plastic covers with adhesive plate sealers (Flow).
10. Agitate the mixture using a plate shaker (Vari-Shaker, Dynatech) for 1 h.
11. Read the optical densities at a test wavelength of 450 nm and a reference wavelength of 650 nm.
12. Include AZT as control in each run.

A: Readout from the XTT-Formazan cell viability assay.

	1	2	3	4	5	6	7	8	9	10	11	12
A	0.004	0.005	0.011	0.008	0.009	0.008	0.009	0.010	0.010	0.009	0.006	0.009
B	0.006	0.543	0.158	0.152	0.142	0.140	0.506	0.525	0.226	0.244	0.514	0.009
C	0.005	0.554	0.168	0.159	0.140	0.175	0.516	0.508	0.286	0.273	0.523	0.008
D	0.007	0.501	0.159	0.170	0.159	0.170	0.527	0.502	0.326	0.338	0.517	0.008
E	0.008	0.539	0.161	0.177	0.191	0.212	0.389	0.501	0.479	0.499	0.162	0.010
F	0.008	0.462	0.168	0.162	0.187	0.179	0.224	0.511	0.508	0.516	0.154	0.010
G	0.008	0.134	0.125	0.109	0.177	0.170	0.179	0.527	0.501	0.563	0.157	0.009
H	0.007	0.007	0.008	0.009	0.004	0.006	0.007	0.003	0.008	0.008	0.007	0.007

B: Analysis of XTT-Formazan cell viability assay

	Compound 1 (Inactive plant extract)			Compound 2 (Glycosidase inhibitor)			Compound 3 (AZT)		
Row	**Conc. (μM)**	**CT (%)**	**CV (%)**	**Conc. (μM)**	**CT (%)**	**CV (%)**	**Conc. (μM)**	**CT (%)**	**CV (%)**
	0	100	29	0	100	29	0	100	29
B	0.32	105	29	0.16	98	26	0.003	101	45
C	1.6	107	31	0.8	100	29	0.016	98	53
D	8	97	31	4	102	31	0.08	97	64
E	40	104	32	20	75	38	0.4	97	94
F	200	89	31	100	42	34	2	99	99
G	1000	25	21	500	34	32	10	102	103

Figure 1. Example of XTT-Formazan cell viability assay. Panel A: Readout from microtitre plate. The plate layout was as follows: Wells B2 to G2, compound 1 (inactive plant extract) and uninfected C8166 cells. Wells B3 to G4, compound 1, HIV-1 IIIB infected cells. Wells B5 to G5 and B6 to G6, compound 2 (glycosidase inhibitor) and HIV-1 IIIB infected cells. Wells B7 to G7, compound 2 and uninfected cells. Wells B8 to G8, compound 3 (AZT) and uninfected cells. Wells B9 to G9 and B10 to G10, compound 3 and HIV-1 IIIB infected cells. Wells B11 to D11, untreated uninfected control cells. Wells E11 to G11, untreated HIV-1 IIIB infected control cells. Wells A1 to A12, A1 to H1, A12 to H12 and H1 to H12 are background. The compounds were assayed at increasing concentrations in wells B to G. The concentration ranges were: 0.32 μM to 1000 μM (compound 1), 0.16 μM to 500 μM (compound 2) and 0.003 μM to 10 μM (compound 3). The average values for the background wells was 0.008. The uninfected control cells gave an average value of 0.510 and the infected control cells gave an average value of 0.150. The analysis was performed 5 days after infection of the cells. Panel B: Analysis of result of XTT viability assay. For each concentration of compound (Conc.) cell viability is expressed as a percentage of untreated uninfected control cells. CT: Uninfected control cells, CV: HIV-1 infected cells.

It is important that the following technical points are considered when performing the XTT–Formazan assay:

(a) Use a low m.o.i. (multiplicity of infection) to allow a few cycles of virus replication to occur so that compounds inhibiting at the late stage of virus replication will also show pronounced antiviral effects.
(b) For inhibitors of gp120/CD4 binding it is critical to treat the cells with compounds prior to the addition of virus.
(c) To avoid false positives, plates must be examined microscopically for evidence of syncytium formation and cytotoxicity.

3. Confirmatory assays

The XTT–Formazan method provides a useful preliminary screen for antiviral activity, but the method has some limitations:

- coloured compounds absorbing at 540 nm
- reduction of XTT by test compounds
- precipitation of calf serum proteins by test compound

gp120 assays are routinely used in our laboratory to confirm antiviral activity. Other assays may also be used depending on the mechanism of action of the compound being tested.

3.1 Antigen gp120 assay

At the end of the XTT–Formazan assay, the plates are stored at 4°C prior to measurement of virus-related antigens gp120 or p24 in the supernatant from selected wells in the presence of XTT. A previously used method, the MTT–Formazan employs isopropanol which interferes with antigen assays requiring fresh experiments for confirmatory tests.

The selective interaction of a lectin GNA (from *Galanthus nivalis*) with glycoproteins of HIV and SIV provides the basis for quantitative determination of these proteins in cell culture fluids of infected cells (*Protocol 2*) (4).

Protocol 2. gp120 ELISA using GNA for antigen capture

Equipment and reagents

- GNA (Vector, UK)
- Empigen BB, alkyl dimethyl amine betaine (Albright and Wilson Ltd, UK)
- Horse-radish peroxidase-conjugated sera (Amersham International)
- Human anti-HIV sera, pooled from three patients (a gift from Professor Karpas, Cambridge)
- *O*-phenylenediamine, OPD (Sigma Chemical)

Method

1. Coat microtitre plate (Falcon) wells with 0.5 μg GNA in 50 μl PBS at room temperature overnight.

Protocol 2. *Continued*

2. Wash three times with PBS.
3. Add 100 μl RPMI 10% to block at room temperature for 1–6 h until used.
4. Repeat wash step 2.
5. Add 25 μl of RPMI 10% containing 0.5% Empigen to all wells.
6. Add 25 μl of supernatant from drug treated wells and 25, 12.5, and 6.25 μl from untreated wells for standard curve (1:2, 1:4, and 1:8 dilutions are made by adding 12.5 μl and 18.75 μl of RPMI 10% to second and third wells).
7. Incubate at 37°C for 3 h or at 4°C overnight.
8. Remove unbound antigen by washing three times with PBS containing 0.1% Tween-20 (PBS T20).
9. Add 50 μl of human anti-HIV sera (1:200) diluted in RPMI 10% containing 0.1% Tween-20, incubate at 37°C for 4 h. Use antibodies to HIV-2 for gp120 assay from HIV-2. Use antibodies to SIV to detect gp120 from SIV.
10. Wash three times with PBS T20.
11. Add 50 μl of a 10^3 dilution of anti-human Ig peroxidase conjugate.
12. Incubate at 37°C for 90 min.
13. Repeat wash as in step 10.
14. Add 50 μl of OPD 1 mg/ml in 0.1 M sodium citrate buffer pH 5 containing 0.03% hydrogen peroxide.
15. Stop the reaction as soon as sufficient colour is produced (2–3 min) by the addition of 50 μl of 1 M sulfuric acid.
16. Measure the A_{492}.

This method is convenient, inexpensive, and it is also useful for quantitating large numbers of samples from 96-well plates. Using WIACALC (Pharmacia Biotech) the amount of antigen is calculated from linear logarithmic plots using three dilutions of untreated infected cultures as standards. The data obtained from the computer is presented in *Figure 2*. The values are calculated as a percentage of untreated infected control.

ELISAs can also be developed for the estimation of gp120 and p24 using specific antibodies for coating the plates, but this depends entirely on the availability of the correct antibodies specific to strains of HIV-1, HIV-2, and SIV (see Chapter 10, Volume 1).

3.2 Infectivity assay

Estimation of the yield of infectious virus is the only reliable method for the assessment of compounds affecting virus maturation and infectivity. For this assay 5–10 μl of supernatant is taken (prior to the addition of XTT to cultures, as it inhibits virus replication) and added to 100 μl of medium in the first row of 96-well plates, then serial twofold dilutions are made, and infectivity end-point is measured on C8166 cells by examining syncytia and by the XTT–Formazan assay. These are then compared with the untreated infected controls.

3.3 Reverse transcriptase assay

Determination of RT is an important indicator of virus replication where low levels of infection are monitored for example in peripheral blood lymphocytes and monocytes/macrophages, but this assay is not practical for the assessment of a large number of samples. Although p24 and gp120 assays are more sensitive, RT assays should be used as an additional confirmatory test. We use a modified RT assay (5) (see Chapter 2). The following points should be noted:

(a) Collect culture supernatant prior to the addition of XTT.
(b) Dilute 15 μl samples 1:2 in double strength RT reaction mixture. The PEG precipitation of samples from 96-well plates is impractical.
(c) Incubate samples at 37 °C for 4–5 h to get the best possible incorporation of label.
(d) Calculate results as the percentages of untreated infected controls.

3.4 Indirect immunofluorescence assay

This is also useful as an additional confirmatory assay, predominately when antigen levels are low in the culture supernatant. The cells from the 96-well plates (after the cell viability assay) are recovered, fixed with acetone, and treated with human anti-HIV and FITC-conjugated sheep anti-human IgG. Uninfected controls are included for comparison.

The initial screening programme outlined above, using C8166 cells infected with a laboratory adapted strain, HIV-1_{111B} or HIV-1_{MN} at a low m.o.i. and in the presence of compounds, is suitable for the assessment of antiviral agents with the following modes of action:

- inhibitors of the absorption of virus to cells
- inhibitors of virus replication
- inhibitors of virus maturation or infectivity

The results of tests made on three different compounds are presented in graph and table form (*Table 1, Figure 3.*)

- EC_{50} is the concentration of compound which inhibits virus antigen (gp120 or p24), RT, or infectious virus yield in infected cultures by 50%
- TC_{50} is the concentration of compound that inhibits XTT–Formazan production in uninfected cultures by 50% of that in untreated controls

Measurement of various end-points of virus replication are required for accurate and complete evaluation of antiviral activities of a compound, particularly where the potential target is unknown.

A Plate number : 1
Variables : ABSORB

	1	2	3	4	5	6	7	8	9	10	11	12
A	--	--	--	--	--	--	--	--	--	--	--	--
B	--	0.867	0.874	0.903	0.925	--	--	--	--	--	--	--
C	--	0.955	0.943	0.924	0.913	--	--	--	--	--	--	--
D	--	0.527	0.015	0.034	0.042	0.138	0.406	0.768	--	--	--	--
E	--	0.609	0.013	0.026	0.053	0.124	0.425	0.771	--	--	--	--
F	--	0.275	--	--	--	--	--	--	--	--	--	--
G	--	0.346	--	--	--	--	--	--	--	--	--	--
H	--	--	--	--	--	--	--	--	--	--	--	--

B

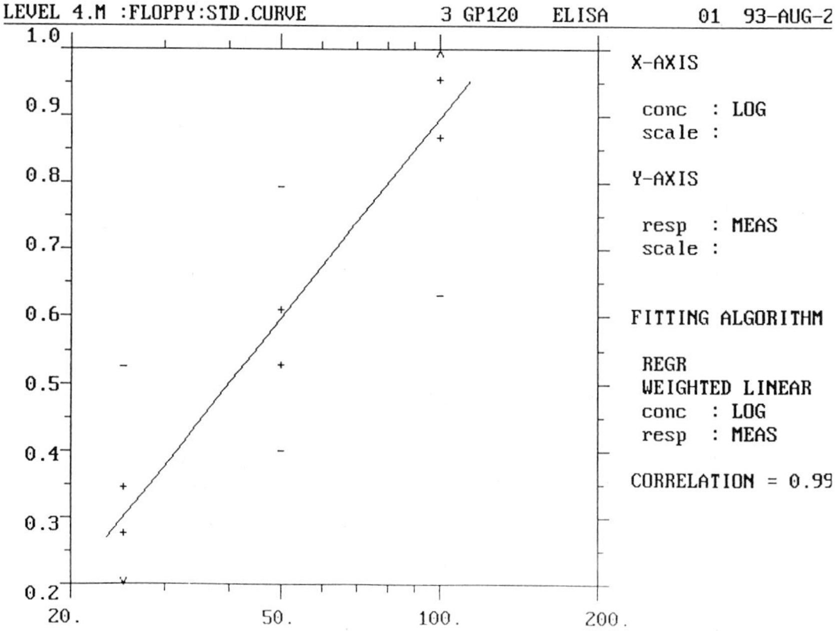

Figure 2. Example of output from 'WIACALC'. Panel A: Layout of microtitre plate. Values indicate levels of gp120 secretion. Panel B: Standard curve for gp120 secretion using three

4. Specific tests with different cell/virus systems, additional cellular assays

Using the procedures described above, potential compounds with high efficacy and selectivity are also tested against SIV in C8166 and HIV-2 in MT2 (HIV-2 produces better CPE in MT2 cells), and further evaluated in different virus/cell systems depending on the nature of the antiviral activity.

C

WELL	SEQ	CODE	PAT	CONC	%CV
B02	1	STD		100.000	
C02	2	STD		100.000	
	AVG.	**STD**		**100.000**	**6.8**
D02	3	STD		50.000	
E02	4	STD		50.000	
	AVG.	**STD**		**50.000**	**10.2**
F02	5	STD		25.000	
G02	6	STD		25.000	
	AVG.	**STD**		**25.000**	**16.2**
		Plant Extract			
B03	7		1	95.375	
C03	8		1	112.018	
	AVG.	**200μg**	**1**	**103.696**	**11.3**
B04	9		2	102.045	
C04	10		2	107.165	
	AVG.	**40μg**	**2**	**104.605**	**3.5**
B05	11		3	107.415	
C05	12		3	104.452	
	AVG.	**8μg**	**3**	**105.934**	**2.0**
		AZT			
D03	13		4	12.876	
E03	14		4	12.816	
	AVG.	**10μM**	**4**	**12.846**	**0.3**
D04	15		5	13.459	
E04	16		5	13.210	
	AVG.	**2μM**	**5**	**13.334**	**1.3**
D05	17		6	13.712	
E05	18		6	14.068	
	AVG.	**0.4μM**	**6**	**13.890**	**1.8**
D06	19		7	17.151	
E06	20		7	16.601	
	AVG.	**0.08μM**	**7**	**16.876**	**2.3**
D07	21		8	32.035	
E07	22		8	33.486	
	AVG.	**0.016μM**	**8**	**32.760**	**3.1**
D08	23		9	74.493	
E08	24		9	75.016	
	AVG.	**0.0032μM**	**9**	**74.755**	**0.5**

dilutions of untreated HIV-1 infected control cells. Panel C: Effects on gp120 secretion of treatment of HIV-1 IIIB infected C8166 cells with plant extract and AZT.

4.1 Peripheral blood lymphocytes (PBLs)

The active compounds tested in C8166 cells usually produce identical results in PBLs. Since there is little or no advantage over C8166 cells, the cost and labour involved in the isolation and maintenance of PBLs can not be justified for routine assays and only selected compounds are tested in these cells.

PBLs are isolated from the 'buffy coat' obtained from blood transfusion

Table 1. Anti-HIV activity

Virus—HIV-1 IIIB
Cells—C8166

Compound	**Conc (μM)**	**Syncytia (+/−)**	**Virus progeny produced % of control**	**Ag gp120 % of control**	**Estimated cell growth % of control**		**EC$_{50}$**[a]	**TC$_{50}$**[b]
					Infected	**Uninfected**		
Compound 1	1000 [c]				21	25	Inactive	500
(inactive	200	+	100	104	31	89		
plant extract)	40	+	100	105	32	104		
	8	+	100	106	31	97		
Compound 2	500				32	34	10	50
(glycosidase	100				34	42		
inhibitor)	20	+/−	6–12		38	75		
	4	+	100		31	102		
	0.8	+	100		29	100		
Compound 3	10	−		13	103	102	0.008	>1000
(AZT)	2	−/+		13	99	99		
	0.4	−/+		14	94	97		
	0.08	+/−		17	64	97		
	0.016	+/−		33	53	98		
	0.0032	+		75	45	101		
Control			100	100	29	10		

[a] EC$_{50}$ represents the concentration which reduces the Ag gp120 or virus progeny by 50% in infected cell cultures.
[b] TC$_{50}$ represents the concentration of drug which reduces cell growth by 50%.
[c] μg/ml.

centres, using Ficoll-hypaque gradient centrifugation (see Chapters 2, 3, and 7, Volume 1) (6).

The basic procedure used for testing compounds in PHA-activated PBLs is similar to that described in *Protocol 1*. It is important to note that PBLs isolated from volunteers differ greatly in their sensitivity to infection. The modifications are as follows:

(a) Use 100 μl of compound dilutions, 50 μl of cells, and 50 μl of virus HIV-1_{111B} at a m.o.i. 100 times higher than that used for C8166 cells.
(b) To test against clinical isolates (see Chapter 3, Volume 1), use undiluted supernatant from infected PBLs (PBLs are infected by co-culturing PBLs from a normal donor and from a patient for three weeks).
(c) Replace the supernatant containing virus and compounds with RPMI 10% containing fresh dilutions of compounds and IL-2, five to six days post-infection.

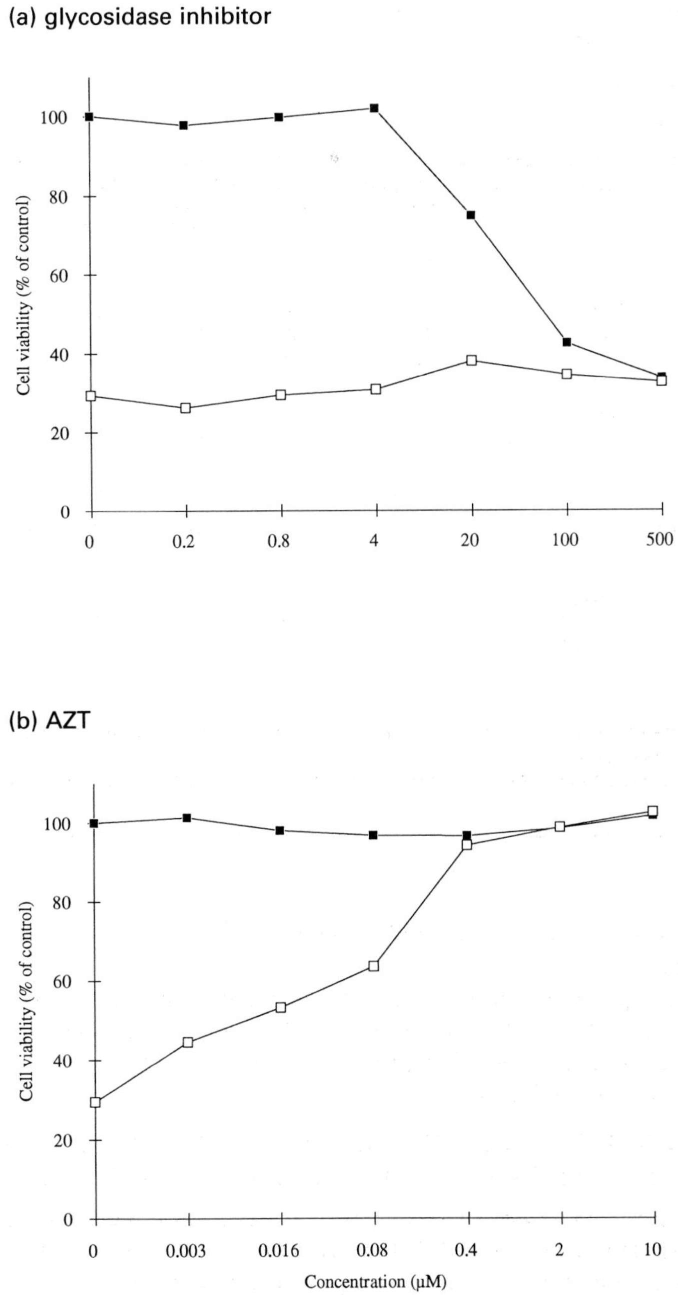

Figure 3. Cell viability after treatment with 0 to 500 μM glycosidase inhibitor (Panel a) or 0 to 10 μM AZT (Panel b). Data is expressed as a percentage of untreated control cell values. (■), Uninfected cells. (□), Cells infected by HIV-1 strain IIIB.

(d) 10–15 days post-infection, collect 100 μl supernatant for antigen and RT assays.
(e) Measure cytotoxicity by the XTT–Formazan method.
(f) The virus replication in primary cells is slow, CPE and cell death is minimal and can not be quantitated by XTT–Formazan method. The EC_{50} values are calculated from RT and gp120 estimations.

4.2 Peripheral blood monocytes/macrophages

Compared to T lymphocyte cells, monocytes have low levels of deoxynucleoside kinase activity and reduced ability to phosphorylate AZT, ddC, and other nucleoside analogues, which are only effective after phosphorylating to 5′-triphosphates inhibiting DNA chain elongation (7). Hence, a potential antiviral nucleoside analogue is evaluated in these cells. Monocytes from blood have been shown to be infected *in vivo* and *in vitro* (6).

PBMCs are isolated from gradients, washed with PBS, and seeded in flasks in RPMI containing 20% FCS. The non-adherent PBLs are removed by washing with PBS, 45 minutes after initiation of cultures. The adherent monocytes are maintained in RPMI 1640 supplemented with 15% pooled human AB serum (see Chapter 4, Volume 1).

The method for the evaluation of compounds in PBMCs is similar to that used for PBLs with following exceptions:

(a) Detach monocytes from the flasks by vigorous pipetting and agitation by vortexing.
(b) Add 50 μl containing around 10^5 cells to 96-well plates containing 100 μl of compound dilutions.
(c) Infect with 50 μl of undiluted supernatant from macrophages infected with HIV-$1_{BA\text{-}L}$. This strain is 10–100-fold more efficient in infecting macrophages than T cells (6).

The isolation, maintenance, and infection of macrophages is time-consuming and the assessment of compounds in these cells is problematic as HIV replication in these cells is more prolonged and less cytopathic. Only low levels of antigen and RT are detected after 12–15 days of infection of PBMCs. Hence, only selected compounds of high priority are tested in these cell lines.

4.3 JM and U937 cell lines

Cell lines, possessing many characteristics of monocytes are used routinely in our laboratory for the assessment of nucleoside analogues.

- JM is a semi-mature human T cell line from lymphoblastoid leukaemia (8)
- U937 is a human monocyte-like cell type from histiocytic lymphoma (9)

Like monocytes, these cells are insensitive to the antiviral effects of a number of nucleoside analogues because they may have low levels of kinase activities.

The basic procedure for testing compounds in these cell lines is similar to that described in *Protocol 1* with the following modifications:

(a) Instead of HIV-1_{IIIB}, use the monocytotropic virus strains HIV-1_{GB8} or HIV-1_{U455}. HIV-1_{U455} is preferred since high titre stocks are obtained from JM cells. The virus stock is titrated on C8166 cells.
(b) Examine syncytia formation microscopically.
(c) Use the XTT–Formazan assay for the measurement of the cytotoxicity.
(d) Use confirmatory assays (gp120 and infectious virus yield) for monitoring virus replication.

The assessment of compounds in these cells has proven to be a useful tool for selecting compounds likely to inhibit infection in monocytes/macrophages. The infection of JM cells with HIV-1_{U455} is highly cytopathic, giant syncytia are distinctly visible but the quantitation by cell viability assay is less reliable due to the emergence of chronically-infected cells.

4.4 Evaluation of compounds in chronically-infected H9 cells

The drugs currently used for the treatment of AIDS such as AZT and other nucleoside analogues are RT inhibitors and do not exhibit antiviral effect in chronically-infected H9 cells *in vitro*, where RT activity is not required for virus replication.

Compounds that interfere with the maturation, infectivity, and release of the progeny virus (e.g. protease and glycosidase inhibitors) or compounds that inhibit replication of viral genome (antisense oligonucleotides or Tat inhibitors) are more likely to show some antiviral effect in these cells. Hence, all compounds which have unknown targets (except RT inhibitors and compounds inhibiting gp120/CD4 interaction), are tested for their ability to inhibit infectious virus yield from H9 cells. The same basic principles described in *Protocol 1* apply here. 100 μl dilutions of compounds are mixed with 100 μl of cells at 4×10^5/ml. The antiviral activity is monitored by titrating infectious virus yield on C8166 cells. It is important to ensure that the supernatant from H9 cells is sufficiently diluted such that antiviral activity of compound in C8166 cells is negligible (below its EC_{50} value).

4.5 Evaluation of compounds in HeLa-CD4 pBK-LTR-*lac* cells

The emergence of AZT-resistant variants of HIV-1 and their cross-resistance to other nucleoside analogues makes it necessary to test potential antiviral agents particularly those derived from AZT and related structures against

AZT-resistant strains. The choice of cell line sensitive to infection with resistant strains is critical. The best results are achieved in a cell line which was used initially to isolate these strains. The resistant strains used here are provided by the ADP Reagent Project which were isolated and propagated in HeLa cells expressing CD4 (10). Attempts to grow these strains in various cell lines including highly sensitive C8166 were unsuccessful; only MT2 and HeLa CD4-positive have proven useful for their propagation.

A modification of the plaque reduction assay (10) is described here for testing compounds against AZT-resistant strains (*Protocol 3*). The indicator cell line; HeLa-CD4 pBK-LTR-*lac* clone (11), is readily infected with AZT-resistant strains. After infection of these cells, clearly visible blue foci can be detected by staining with X-gal. This assay is similar in principle to the cell fusion assay described in Chapter 8.

Protocol 3. Evaluation of compounds against AZT-resistant strains

Equipment and reagents

- HIV-1 105A, sensitive to AZT
- MT2 cells, human T cell transformed by co-cultivating with leukaemia lymphocytes, harbouring HTLV-1, sheds the virus in culture supernatant (12)
- Indicator cell line; HeLa-CD4 pBK-LTR-*lac*, carrying the *E. coli* β-galactosidase gene under the control of the HIV LTR promoter (11)
- RPMI 10% containing 1 mg/ml Geneticin (selection for CD4-positive antigen) and 100 μg/ml hygromycen (selection for the BKLTRLac episome)
- X-gal, 5-bromo-4-chloro-3-indolyl-β-D-galactopyranoside (Sigma Chemical)
- HIV-1 105F, resistant to AZT (mutations in RT residues 67, 70, 215, and 219)

Method

1. Seed the indicator cells in 96-well plates and allow them to grow to confluence in RPMI 10% containing antibiotics.
2. Remove the medium, add 100 μl of a predetermined dilution of 105A and 105F-infected MT2 cells and supernatant; better foci are produced with the mixture.
3. Incubate the plates at 37°C.
4. Examine the plates for evidence of syncytia and cytotoxicity, four days post-infection.
5. Remove supernatant for confirmatory tests (gp120 measurement, *Protocol 2*).
6. Stain with X-gal. Wash the cells with PBS. Replace with 100 μl PBS containing 200 μg/ml X-gal, 3 mM potassium ferricyanide, 3 mM potassium ferrocyanide, 1.3 mM $MgCl_2$ pH 7.4.
7. Incubate at 37°C for 4 h or more to allow blue foci to develop.

8. Fix with 10% formaldehyde in PBS for microscopic examination and counting.
9. Use XTT–Formazan method to measure cytotoxicity (*Protocol 1*).

(a) Both gp120 and blue foci counts are expressed as percentages of infected, untreated controls to facilitate comparison between the two strains of virus.

(b) Although the blue foci produced are clearly visible, quantifying the results by microscopic analysis is not always reliable, therefore an additional quantitative confirmatory assay measuring gp120 is necessary for the calculation of EC_{50} values.

5. Evaluation of combination of compounds for anti-HIV activity

Prolonged AIDS therapy using AZT and other non-nucleoside RT inhibitors has resulted in the emergence of resistant strains. Combination therapy using two drugs may delay or even prevent virus resistance. Compounds of different modes of action (e.g. RT and protease inhibitors) may also have synergistic effects on the efficacy of drugs. Therefore assessment of compounds in combination is an important part of our testing programme.

Primary screening using cell viability assay has proven particularly useful for the evaluation of compounds in combination for anti-HIV activity and cytotoxicity. Dilutions of compounds are tested alone, then in combination in the same 96-well plate with infected and uninfected controls. The procedure used for testing is described in *Protocol 1*. Combinations of compounds are tested as follows:

- six concentrations of compound 1
- six concentrations of compound 2
- six concentrations of compound 1 + six concentrations of compound 2
- six concentrations of compound 1 + one concentration (EC_{50}) of compound 2

It is necessary that the highest concentration of each compound used is lower than its EC_{90} value (EC_{90} is the concentration which reduces the antigen gp120 production by 90% in infected cell cultures). Uninfected cells are treated similarly to provide compounds' cytotoxicity.

The cell viability results are computed as described in *Figure 1* to facilitate comparisons between the various combinations. Any antiviral effects should be confirmed by gp120 or infectivity assays.

6. Specialized assays

The methods outlined above have described briefly the details of tests in various cells suitable for the assessment of different types of compounds. However, there are a variety of simple assays which can be used in conjunction with these assays to further confirm the mechanism of antiviral action. These are briefly described below:

(a) Compounds for which the target is unknown are studied for their ability to inhibit infection when added at different times after infection.

(b) Compounds inhibiting the gp120/CD4 interaction are studied for their ability to inhibit fusion (reversible or irreversible) (see Chapter 8).

(c) Virus neutralization assays can be performed (see Chapter 7, Volume 1). A high titre virus stock is incubated with or without compound at 37°C for 1–2 h, and is then is serially diluted to remove the compound, and the end-point for virus infectivity is determined.

(d) The CD4–gp120 interaction can be conveniently studied using gp120 bound to GNA (*Protocol 2*) and detecting bound CD4 in an ELISA format. This assay provides a convenient method to study the effects of compounds on this interaction (see Chapter 8 for an alternative CD4–envelope binding assay).

(e) Using monoclonal antibodies in a GNA-bound gp120 ELISA it is possible to identify putative binding sites for compounds which are tested for their binding to gp120 in competition with monoclonal antibodies of known specificities.

(f) *In vitro* inhibition of enzymes, e.g. reverse transcriptase and protease (see Chapters 2 and 5).

(g) The emergence of resistant variants in the presence of compounds reveals specific binding sites. Resistant mutants are selected by growing virus in the presence of increasing amounts of compounds over a period of several weeks and sequenced to determine the binding site.

References

1. Weislow, O. S., Kiser, R., Fine, D. L., Bader, J., Shoemaker, R. H., and Boyd, M. R. (1989). *J. Natl Cancer Inst.*, **81**, 577.
2. Pauwels, R., Balzarini, J., Baba, M., Snoeck, R., Schols, D., Herdewijn, P., *et al.* (1988). *J. Virol. Methods*, **20**, 309.
3. Salahuddin, S. Z., Markham, P. D., Wong-Staal, F., Franchini, G., Kalyanaraman, V. S., and Gallo, R. C. (1983). *Virology*, **129**, 51.
4. Mahmood, N. and Hay, A. J. (1992). *J. Immunol. Methods*, **151**, 9.
5. Hoffman, A. D., Banapour, B., and Levy, J. A. (1985). *Virology*, **147**, 326.
6. Gartner, S., Markovits, P., Markovitz, D. M., Kaplan, M. H., Gallo, R. C., and Popovic, M. (1986). *Science*, **223**, 215.

7. Richman, D. D., Kornbluth, R. S., and Carson, D. A. (1987). *J. Exp. Med.*, **166**, 1144.
8. Schwenk, H. H. and Schneider, U. (1975). *Blut*, **B31**, 299.
9. Sundstrom, C. and Nilsson, K. (1976). *Int. J. Cancer*, **17**, 565.
10. Larder, B. A., Darby, G., and Richman, D. D. (1989). *Science*, **243**, 1731.
11. Akrigg, A., Wilkinson, G. W. G., Angliss, S., and Greenaway, P. J. (1991). *AIDS*, **5**, 153.
12. Miyoshi, I., Kubonishi, I., Yoshimoto, S., Aai, T., Ohtsuki, Y., Shiraishi, Y., *et al.* (1981). *Nature*, **294**, 770.

17

Inhibition of HIV by ribozymes and antisense oligonucleotides

D. CASTANOTTO, E. BERTRAND, and J. ROSSI

1. Introduction

Since the early 1980s, when the immunodeficiency virus (HIV) was shown to be the etiological agent of AIDS, a considerable amount of information on its life cycle has been collected. Although this growing body of knowledge has allowed investigators to evolve rational drug designs, the complexity of HIV has so far made it impossible to develop a cure or a vaccine. The variety of tissues and cells which HIV is able to infect, along with the high mutation rate of the viral genome and the secondary diseases associated with AIDS, further restrict the possibilities for achieving any effective therapy. Moreover, agents that block viral function often have adverse effects on the host, especially after prolonged use. A relatively new approach in the treatment of viral infections is the use of antisense and/or ribozymes in the attempt to bind and block expression of viral transcripts. To date, they have been reported to be effective in cell cultures against viral targets, oncogenes, and transfected genes. RNA enzymes or ribozymes are receiving considerable attention because, unlike antisense, ribozymes are able to turn over multiple substrates and inactivate them irreversibly. The ribozyme technology is based on the design of RNA molecules in which the antisense and enzymatic functions are combined into a single transcript, which can specifically pair and cleave any target RNA. Because of their high specificity and the possibility of being endogenously synthesized within a cell, it is likely that ribozymes will be major players in the development of new drugs during the next several years.

2. Ribozymes

At present there are four RNA catalytic motifs that are potentially useful for *trans*-mediated cleavage of target RNAs (*Figure 1*).

(a) 26S RNA intervening sequence of *Tetrahymena thermophila* first characterized by Cech and colleagues (1) and is perhaps the most thoroughly

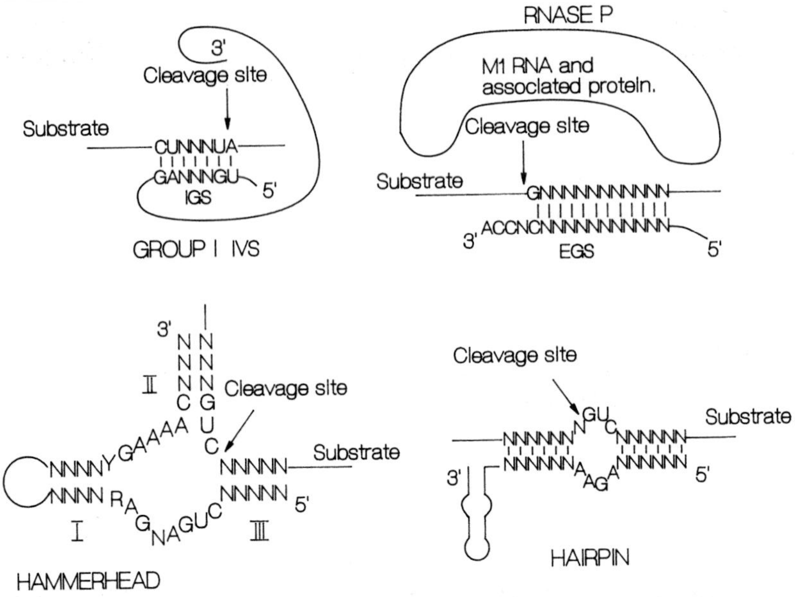

Figure 1. RNA Catalytic motifs. The group 1 intervening sequence (*Tetrahymena*), the *Escherichia coli* RNase P (comprised of RNase P proteins and M1 RNA to make up the holoenzyme), the hammerhead ribozyme, and the hairpin ribozyme are shown. IGS (internal guide sequence) is the guide sequence for the group 1 introns, and EGS the external guide sequence for the RNase P. The IGS is part of the ribozyme core and is the primary binding site for the substrate. Specific complementary base pairing interactions between this site and the substrate confer specificity to the ribozyme reaction. The EGS is an 'antisense' RNA with an appended CCA at the 3′ end, which is provided in *trans.* Complementary base pairing of the EGS with the target RNA results in a duplex substrate for RNase P, by mimicking the CCA stem of the natural tRNA substrate molecules. The Ns denote no specificity for nucleotides at these positions, and roman numerals indicate regions required for ribozyme function. All four ribozymes are shown as they would interact with substrates in *trans*. The substrates and the cleavage sites are indicated.

studied ribozyme. Although this RNA enzyme possesses a rather complex structure, it has been engineered to cleave a variety of substrates, including even single-stranded DNA (2, 3).

(b) RNase P is a ubiquitous enzyme that processes the 5′ leader sequences from pre-transfer RNAs and has an RNA component that has been shown to be the catalytic entity for the bacterial form of the enzyme (4). Forster and Altman (5) have characterized the substrate requirements for the *E. coli* enzyme, and have demonstrated that it can cleave a non-tRNA duplex provided that the target is in a helix and that the antisense to the target RNA has a CCA sequence at its 3′ terminus. Thus, in principle, any RNA can be targeted as a substrate for RNase P provided that the CCA-containing antisense RNA can be annealed to the target RNA.

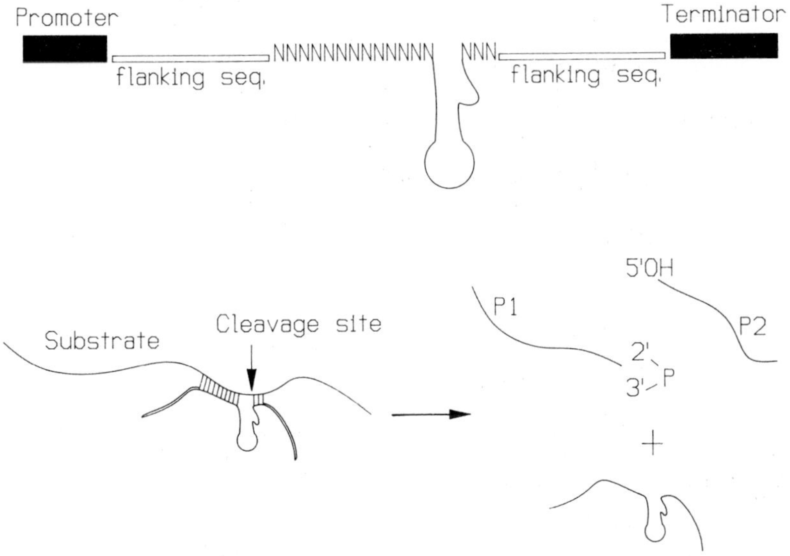

Figure 2. Schematic representation of a ribozyme expression vector and cleavage reaction. An asymmetric ribozyme gene is cloned between a promoter and an appropriate transcriptional terminator. The Ns indicate complementary nucleotides on both sides of the cleavage site. The adjacent open boxes represent the non-pairing flanking sequences. Below, a diagrammatic representation of the *trans*-acting ribozyme cleavage reaction. The substrate, cleavage site, and ribozyme generated cleavage products (P1 and P2), are shown. After product dissociation the ribozyme is recycled and can catalyse a new reaction. The cleavage reaction occurs only in presence of magnesium ions.

(c) Two RNA catalytic motifs that originate from a plant viroid and virusoids (single-stranded RNAs that infect flowering plants and replicate via a rolling circle mechanism) have received a great deal of attention for their use in antisense ribozymes. These are the hammerhead (6) and hairpin (7) self-cleavage domains (*Figure 1*). Uhlenbeck (8) and Haseloff and Gerlach (9) demonstrated that the hammerhead catalytic motif could be incorporated into *trans*-acting antisense RNAs, thereby imparting enzymatic activity on these molecules (for a schematic example of a ribozyme cleavage reaction see *Figure 2*). The hairpin catalytic domain was subsequently shown to be useful for incorporation into *trans*-acting ribozymes as well.

For applications where an antisense RNA is contemplated, it should also be possible to utilize a ribozyme, provided that there is a cleavable site in the target molecule. Since most of the research on therapeutic ribozymes has been performed with the hammerhead motif, we will focus on this type of catalytic molecule. However, the general criteria used in the design and applications of hammerhead ribozymes can be extended to other catalytic RNAs, as well as to antisense oligonucleotides.

3. Design of ribozymes and antisense oligonucleotides

The choice of whether to use a ribozyme or non-enzymatic antisense is the first decision to be made. The advantages of the ribozyme approach are still somewhat theoretical since it is very difficult to formally prove that a ribozyme is cleaving the target RNA or turning over multiple substrates in an intracellular environment. However, numerous experiments in which a standard antisense has been used as a control suggest that there is an added effectiveness if there is a catalytic centre included within the antisense domain.

3.1 Choice of the target

As a first step in the design of a ribozyme or antisense oligonucleotide, the target RNA needs to be selected:

(a) Theoretical considerations suggest that stable rather than unstable RNAs may be the best targets (10).
(b) RNAs encoding regulatory proteins are often preferred over the RNAs encoding structural proteins, because small variations in the expression of the former can result in a significant down-regulation of the genes with which these proteins interact.
(c) In the case of the human lentiviruses such as HIV, different RNA species are generated by alternative splicing. Thus, some sequences are present within two or more (sometimes all) RNA species. Their distribution in various transcripts makes these sequences potentially good targets because the same ribozyme or antisense molecules could inhibit the virus at different stages of its life cycle.

In choosing the region(s) to target within the selected RNA molecule, several aspects of mRNA biosynthesis in eukaryotic cells need to be considered:

1. RNAs are always associated with proteins inside the cell: in the nucleus, pre-mRNAs are complexed with heterogeneous nuclear ribonucleoproteins (hnRNPs) (11, 12) and snRNPs (13, 14), whereas in the cytoplasm, mRNAs are associated with the translational machinery. These factors are not distributed equally along the target RNA, so that some regions of the target may be highly accessible to the antisense or the ribozyme, and other regions may be totally inaccessible. The structure of the RNA can affect the availability of a targeted site. If the target sequence is sequestered by a secondary or tertiary structure, the accessibility of this sequence to degradation by a

ribozyme or binding by an antisense molecule will probably be impaired. The prediction of RNA secondary structure using computer-derived energy minimization programs is the most common method for target site selection (15), although the reliability of such computer generated structures is not always good. If variation in the sequences are known (such as for variants of HIV-1) a phylogenetic analysis can be a useful adjunct to the computer analysis for the prediction of the RNA structures. It is also useful to probe the structure of RNA *in vitro* using enzyme or chemical reagents (16). Although it is possible to directly determine the accessibility of a particular sequence *in vivo* (17) or in cell extracts, it requires intensive labour. Thus, accessible regions of the target should be found empirically by designing separate ribozymes or antisense oligonucleotides to target several potential sites and testing their relative effectiveness. One study which tested this problem directly found that only some regions of the HIV genome were accessible to inhibition by oligonucleotides (18).

(2) In the case of HIV, an additional concern in the choice of the target site is that the virus could become resistant (via mutation) to the antisense or the ribozyme cleavage, therefore, it is preferable to select a cleavage site in an essential region of the viral genome which can be identified by sequence comparisons of variants of closely related viruses. The regions with the lowest likelihood to vary are those that interact with essential viral or cellular proteins required for expression, such as the HIV TAR and REV sequences.

3.2 Antisense and ribozyme design

In the simplest design of the hammerhead motif, the cleavage site sequences are defined by XUY (where X is U, G, A, or C, and Y is U, C, or A), with cleavage occurring 3′ to Y. Different XUY sequences are cleaved with different efficiencies (19). Since the most commonly used site in nature is GUC (6), it is often the site of choice when present in the target RNA. When long and/or highly structured RNAs are targeted, the limiting step in the cleavage reaction is likely to be the binding/hybridization step (20, 21), thus the choice of the XUY trinucleotide may not be so critical. A potentially useful approach for target site selections is to incubate the RNA target (end-labelled) with a high molar excess of ribozymes with degenerate flanking sequences (22). This ribozyme population can select sites most accessible to binding and cleavage on the RNA substrate. The analysis of the timed appearance of cleavage products on a sequencing gel should result in the generation of a ladder of cleaved RNAs. Sites that are more accessible to ribozyme-mediated cleavage will preferentially show up as the strongest bands of the ladder.

The number of base pairs required for optimal ribozyme cleavage at different sites is variable and must be determined empirically for each specific site, because the stability of the ribozyme/substrate duplex will be affected by several factors including G–C content, temperature, and RNA structure (23).

(a) 12 to 14 base pairs is a good starting point, since it has been shown that fewer than 12 base pairs can result in poor binding and greater than 14 base pairs can reduce turnover by slowing dissociation of ribozyme and cleavage products (21).
(b) The inhibition that longer flanking sequences exert on product release may be overcome with the design of asymmetric ribozymes where the ribozyme/target pairing is extended on one side of the hybrid and shortened on the other, such that one of the two cleavage products is bound to the ribozyme by only a few bases (four or five: *Figure 2*).
(c) Ribozymes with long flanking arms might be modified and inactivated by the activity of the dsRNA unwinding/modifying enzyme, which requires only 15–20 base pairs of double-stranded RNA (24). Increasing the length of the hybrid will decrease ribozyme specificity by stabilizing binding and, thus, cleavage of mismatched targets (25).
(d) Since the complexity of human RNA is about 100-fold less than that for human DNA (26), target specificity can be achieved with as few as 12 base pairs.

4. Chemical synthesis of DNA and RNA

Antisense oligonucleotides and ribozymes can be efficiently synthesized on any DNA/RNA synthesizer. The major advantage that chemical synthesis offers over an endogenous expression system (from a transcriptional unit) is that base or backbone chemical modifications can be readily included in the molecules. Modifications such as phosphorothioates and methylphosphonates are more resistant to nuclease degradation than unmodified oligonucleotides (27), and can also be effective inhibitors of HIV-1 replication in infected cells (28, 29).

Several chemical modifications introduced into ribozymes have also been shown to increase stability without impairing catalytic capability.

(a) The 2′-ribose hydroxyl group renders RNA more sensitive to nucleases than DNA, and modification of this group can increase ribozyme stability. Only the 2′-OH required for catalysis must be preserved (30). For example, a ribozyme with DNA flanking sequences is catalytically active and two to three times more stable in the cell (31).
(b) Ribozymes containing 2′-fluorocytidine and 2′-fluorouridine, or 2′-aminouridines, are considerably more stable in serum and maintain catalytic activity (32). Modification of the 2′-OH in all but the six nucleotides of the hammerhead's conserved catalytic core to 2′-*O*-allylribonucleotides gave similar results (33).

(c) Phosphorothioate backbone substitutions can also render ribozymes more resistant to nucleases (21). In addition to increased stability, these types of modifications can be particularly informative for mechanistic, functional, and structural studies (30).

4.1 *In vitro* transcription of ribozymes

RNA molecules may be enzymatically synthesized using RNA polymerases (T7, T3, and SP6 are commonly employed).

(a) The DNA bearing the RNA molecule (antisense, ribozyme, or the target for the *in vitro* assays, see Chapter 9), should be cloned in a transcriptional unit. It should be inserted as close to the promoter as possible, and the plasmid should be linearized immediately downstream of the ribozyme gene in order to minimize vector-derived flanking sequences.

(b) *In vitro* transcription can yield RNA up to 50 times the amount of DNA template utilized in the reaction.

(c) To produce labelled transcripts use only 0.01–0.5 mM cold UTP, and 10 mCi [α-^{32}P]UTP (3000 Ci/nmol). Over 90% of the radioactivity can be incorporated into the transcripts.

(d) The addition of spermidine increases the efficiency of the reaction. This step should be performed at room temperature to avoid DNA precipitation.

4.2 *In vitro* analysis of ribozyme activity

If a ribozyme has been chosen as a viral inhibitor, its cleavage activity of the selected target should be directly assayed in an *in vitro* cleavage reaction.

(a) Standard cleavage reactions are performed by first heating to 90°C for 2 min, two separate tubes containing either the ribozyme or substrate (^{32}P-labelled target) in a solution of 20 mM Tris–HCl pH 7.5 and 0–140 mM KCl (or NaCl). A non-cleaving mutant version of the ribozyme should be used as a control.

(b) Renature the samples for 5 min at the temperature chosen for the cleavage reaction (37–55°C) in the presence of 10 mM $MgCl_2$.

(c) Mix different amounts of ribozyme and target (depending on the purpose of the analysis, at either equimolar, excess of ribozyme, or excess of target) at the desired temperature. Different time points should be taken, from 0–3 h; the reactions can be stopped by adding an equal volume of formamide gel loading dye containing 20 mM EDTA.

(d) Denature the samples by heating to 90°C for 2 min, and analyse the cleavage products on a 7 M urea polyacrylamide gel. Values for k_{cat}/K_m can be determined by incubating a constant concentration of substrate (around 1 nM) with increasing excess of ribozyme for a constant time. The k_{cat}/K_m value is derived using the equation:

$$-\ln (Frac\ S)/t = k_{cat}/K_m \times [\text{Ribozyme}]$$

where *Frac S* is the fraction of remaining substrate and *t* is time (21).

5. *In vivo* analysis of ribozyme activity

5.1 Exogenous delivery

To examine the effects of a ribozyme or antisense oligonucleotide *in vivo*, the first obstacle to overcome is the delivery of these molecules into the cells. A major drawback of exogenous delivery is that the inhibitory effects of the nucleic acids so delivered are transient and require repeated administrations. Despite this, exogenously delivered molecules can incorporate chemical modifications which increase stability; although these modified bases and sugar–phosphate backbones can contribute to toxicity of the molecules. Further investigation is needed to establish whether the advantages of chemical modifications will overcome the general disadvantages of an exogenous delivery.

RNA synthesized *in vitro* by T7 RNA polymerase can also be exogenously delivered. An advantage of this approach is that the antisense or ribozyme can be inserted in larger transcripts containing structural features (stem–loops) which confer stability. A disadvantage is that these extra sequences can negatively affect the catalytic activity of the ribozyme (34).

Many techniques have been developed to introduce functional, naked, DNA and some of these may also be applicable to the delivery of synthetic RNA.

(a) DNA is complexed with various compounds (e.g. polylysine) or to lipophilic groups (35–37) which increase cellular uptake.
(b) DNA can be complexed with receptor ligands for specific targeting and localization to defined types of cells (38).
(c) Conjugates of DNA and lipophilic derivatives can have higher anti-viral activity as shown in the case of HIV-1 (37, 39).

These procedures introduce nucleic acids into the cytoplasm, but access to the nucleus is remarkably poor, as a consequence of the degradation of the nucleic acids in the cytoplasm or endocytotic vesicles, and because of the physical barrier of the nuclear membrane. Some delivery techniques address this issue by including non-histone nuclear proteins (40), phage particles (41), or adenovirus protein in the nucleic acid-containing complex (42). These proteins can mediate to some extent the migration of nucleic acids to the nucleus either by protecting the DNA from degradation, or by facilitating the transfer through the nuclear membrane.

Antisense oligonucleotides can also be introduced into cells by direct addition to culture medium. RNA molecules, however, are extremely sensitive to degradation in the media, and require protection from serum ribonucleases. To date, the most commonly utilized technique for delivering pre-synthesized RNA molecules into cells in culture is via liposome encapsulation.

5.2 Liposomes

Liposomes are comprised of one or more concentric phospholipid bilayers (which can incorporate lipid soluble substances) surrounding an aqueous compartment that can incorporate water soluble substances. Size and lipid composition can vary, and different liposomes exhibit different characteristics as *in vivo* delivery systems (*Protocol 1*).

(a) Negatively charged lipids can increase the efficiency of cellular uptake; saturated lipids and the presence of cholesterol can increase liposome stability.
(b) Liposomes can be covalently attached to antibody molecules, resulting in specific binding to cellular antigens (43) and allowing specific targeting to different types of cells.
(c) pH-sensitive liposomes upon exposure to the low pH environment of the endosomes, fuse with the endosome membranes.
(d) Immunoliposomes (pH-sensitive liposomes conjugated to monoclonal antibodies) were successfully targeted to cell surface receptors *in vitro* and *in vivo* (44).
(e) The primary mechanism for cellular uptake of liposomes seems to be endocytosis (45). Once in the cytoplasm, the liposomes are degraded, and the nucleic acids contained inside are released.

Protocol 1. Introduction of synthetic DNA and RNA to cells via cationic liposomes[a]

Equipment and reagents

- Vortex mixer
- CO_2 incubator
- Laminar flow hood
- Lipofectin (Gibco–BRL or other cationic lipid analogue reagents): the reagent consists of 0.5 mg/ml DOTMA and 0.5 mg/ml dioleoylphosphatidylethanolamine in sterile water (reagents functionally similar to the Lipofectin agent, such as Lipofect ACE (Gibco–BRL), Transfectase (Gibco–BRL), Lipofect-AMINE (Gibco–BRL), Transfectam RM (Promega), and DOTAP (Boehringer Mannheim Corp.) can all be used in this procedure)
- Opti-MEM I (Gibco) medium
- Serum (dependent upon cell lines)
- DMEM high glucose (Irvine)
- PBS (1 ×) (Irvine)
- Fungi Bact (Irvine)
- Pen–Strep (Irvine)
- Fungizone (Irvine)
- β-mercaptoethanol
- Sodium pyruvate (Irvine)
- Trypsin (1 ×) (Irvine)
- Sodium bicarbonate (7.5%) solution
- L-glutamine (200 mM)
- Plasmid DNA containing ribozyme gene transcriptional unit
- RNA produced from *in vitro* transcription or chemically synthesized
- Synthetic antisense DNA oligonucleotides
- Polystyrene tubes, sterile, 17 × 100 mm (Myriad)
- 60 mm culture dish (Costar)

Protocol 1. *Continued*

Method

1. Plate exponentially growing cells in tissue culture dishes at 5×10^5 cells/well and grow overnight in a CO_2 incubator at 37°C to 80% confluency.
2. Dilute the pre-synthesized DNA or RNA molecules, and the Lipofectin reagent (BRL) with Opti-MEM I (Gibco) medium. The amounts of nucleic acids and liposome suspension need to be optimized for each cell type.
3. Mix the diluted reagent from the previous step, vortex gently, and incubate 5–10 min at room temperature (prepare this complex in a polystyrene tube because it can stick to polypropylene). If RNA is used, you may perform this step on ice to avoid chemical and enzymatic degradation.
4. Wash the cells three times with serum-free medium.
5. Add the liposome complex and incubate the cells at 37°C in a CO_2 incubator (5–10% CO_2) for 3–6 h. In general, transfection efficiency increases with time, although after 8 h toxic conditions may develop.
6. Add 3 ml medium with 20% of serum (the serum is dependent on the cell type, fetal calf serum may be used).
7. Incubate the cells 24–48 h at 37°C in a CO_2 incubator.
8. Harvest cells and assay for gene activity.

[a] J. Felgner, F. Bennett, and P. L. Felgner. (1993). *Methods: A companion to methods in enzymology*, Vol. **5**, 67–75.

From a therapeutic perspective, intravenously administered liposomes can only leave the circulation if there is a discontinuous or damaged endothelium. They thus have a tendency to accumulate in inflamed areas, tumours, bone marrow, and especially in the liver and spleen. This tendency as well as leakage and lack of control in the rate of release of the liposomal contents once inside the cell are among the major concerns for liposome delivery.

5.3 Endogenous delivery

Endogenous delivery involves the expression of an antisense RNA or a ribozyme from a DNA template permanently maintained within the cell. For an example of an expression vector see *Figure 2*.

(a) Expression of these molecules can be directed by Pol II or Pol III promoters. Pol II promoters include those promoters of viral origin, the long terminal repeat (LTR) promoter sequences of retroviruses, and strong cellular promoters (e.g. the actin promoter).

(b) Use of a strong promoter is advisable but high level expression may be difficult to achieve.

(c) Tandem repeats of the antisense or ribozyme genes, under the control of the same promoter, may help to alleviate this obstacle by increasing the effective concentration from each transcript. Inducible, repressible, or tissue-specific promoters can be used to confer temporal cell type and specific expression and may temper other problems, such as cellular toxicity generated by high levels of expression within the cells.

Endogenous expression from a Pol II promoter, with the exception of few specialized cases such as the human U1 snRNA Pol II promoter, necessitates a polyadenylation signal, which allows addition of a poly(A) tail. This, along with the 5′-m^7GpppG cap, common to Pol II transcripts, may prolong the intracellular half-life of the RNA molecules. Expressing these molecules under the control of Pol III promoters affords additional advantages.

Pol III driven gene expression seems to occur at high levels in all tissues. The size of Pol III transcribed genes is smaller, presenting a more defined transcript. Other expression strategies are possible. For instance, an snRNA transcription unit (46), which incorporates portions of the snRNA structural sequence and protein binding sites, could be used. This can facilitate targeting to the nucleus. Another option is to insert a ribozyme into the acceptor arm or the anticodon loop of a tRNA gene, which has resulted in higher levels of expression and stability of the ribozyme (47). However, this tRNA expression system has been shown to alter post-transcriptional processing and cellular transport (47).

5.4 Viral vectors

Although the ribozyme or antisense expression vector can be delivered with the previously described techniques, integration of the foreign DNA into the host genome does not occur with a high frequency. More promising and efficient technologies employ viral vectors.

Different viral vectors have the capacity to infect a variety of cell types with high efficiency. Several classes of viral vectors are being exploited for delivery of genes *in vivo*, including DNA (adenoviruses, herpes virus, adeno-associated virus), and RNA retroviruses. General concerns persist with the use of viral pathogens such as residual infectivity, toxicity, and rescue of infectivity by recombination. Additionally, each viral vector has its own set of advantages and disadvantages which ultimately dictate its use in a specific application.

To date the most extensively utilized viral vectors have been retroviruses. This class of viruses can infect a wide variety of cell types resulting in long-term persistence as a consequence of integration into the host chromosome. However, the integration process requires cell replication, thereby restricts retroviral use to actively dividing cells. Other potential concerns are low vector titres, lack of specific integration sites, the possibility of activating proto-oncogenes, and the potential for infectious helper virus rescue due to recombination. None the less, retroviruses possess considerable potential for efficient and effective *in vivo* delivery, and are currently the method of choice.

Another promising vector system is adeno-associated virus (AAV). The adeno-associated viruses are non-pathogenic, integrating viruses that require helper viruses for replication of their genome. The AAV vector can exist autonomously at high copy number within a cell, and can integrate into a specific site within chromosome 19 (48).

The intracellular transfer of nucleic acids is continuously improving, but many problems remain to be solved. The efficiency of cellular transformation, targeting to specific tissue and organs, subcellular localization of the RNAs, and maintenance of ribozyme activity within the intracellular environment are some of the major issues to be addressed. Other obstacles are the timing of expression and regulating the level of endogenously synthesized molecules inside the target cells.

5.5 *In vivo* assays

For *in vivo* analyses, standard techniques are performed at the levels of RNA (Northern gels, primer extension, PCR), and proteins. Assays based on virus titre, reduction of infectivity, and proviral DNA are good indicators of viral inhibition. The RNA analyses should reveal a reduced amount of viral RNA and, if a ribozyme is used, the presence of cleavage products.

(a) RNA analysis is not by itself conclusive because it is possible that the cleavage can occur during the RNA extraction.

(b) Use a mutant, non-cleaving ribozyme as control to establish that any effect seen *in vivo* is a result of a specific ribozyme activity.

(c) If antisense oligonucleotides are employed, the corresponding sense oligomers, or scrambled oligomer sequences with the same base composition as the antisense, should be selected to control for non-specific effects and nucleotide breakdown toxicities.

(d) The reduction in the level of viral RNA should be confirmed by a reduction in the level of the encoded viral proteins.

The following steps are used for *in vivo* assays of ribozyme or antisense genes directed against HIV:

(a) Clone the ribozyme gene or the antisense oligonucleotide in a mammalian expression vector, downstream of a strong promoter (e.g. human β-actin promoter). The vector must contain a marker for the isolation of transfected clones.

(b) Transfect the resulting vector into CD4-positive cells and isolate stable clones. It is also possible to work with pools of stable clones, so that clonal variability can be eliminated.

(c) Examine these clones for the presence of the ribozyme or antisense RNA by Northern or PCR analysis.

(d) Challenge the cell lines expressing the antisense (or catalytic) RNA, as well as untransformed parental cells, with HIV (see Chapters 2 and 3, Volume 1), Chapter 16.

6. Conclusions

The use of antisense and ribozymes as surrogate genetic tools and as therapeutic reagents for the treatment of disease is developing rapidly. Ribozymes have the inherent advantage over conventional antisense molecules of not only binding to the target RNA, but also cleaving it, thereby ensuring its permanent inactivation. Because ribozymes base pair with their substrates, antisense effects may contribute to a decrease in steady state levels of the targeted RNA encoded product. Although this is advantageous from a practical point of view, the optimal design of ribozymes for *in vivo* use requires the development of assays in which the antisense effects of ribozymes can be quantitated separately from the cleavage effects. To date, despite the potential for catalytic activity, great excesses of ribozyme over substrate are often required to achieve an inhibition of expression. We have a great deal to learn about maximizing intracellular targeting and stability of ribozymes, as well as learning more about the mechanisms governing intracellular localization before the maximal effectiveness of ribozymes can be achieved. The effectiveness of antisense RNAs are unpredictable as well. Several reports indicate that, for unknown reasons, stable and abundant expression of antisense RNA may have no detectable effect on expression of the targeted RNA (49–51).

Since the first demonstrations that ribozymes can cleave several different HIV-1 RNA targets and can effectively protect cells from HIV infection (52, 53), numerous experiments have been reported, the results of which appear to be very promising. Despite some of the pitfalls associated with these technologies, sufficient evidence has now accumulated to suggest that ribozymes and antisense oligonucleotides will be valuable tools for future therapeutic application.

Acknowledgements

This work was supported by American Foundation for AIDS Research (AmFAR) Grant 01917–14-RG, the American Foundation for Pediatric AIDS Research (PAF/AmFAR) Grant 500331–14-PG, and the National Institutes of Health Grants AI25959 and AI29329.

References

1. Kruger, K., Grabowski, P. J., Zaug, A. J., Sands, J., Gottschling, D. E., and Cech, T. R. (1982). *Cell,* **31**, 147.
2. Herschlag, D. and Cech, T. R. (1990). *Nature,* **344**, 405.
3. Robertson, D. L. and Joyce, G. F. (1990). *Nature,* **344**, 467.
4. Guerrier-Takada, C., Gardiner, K., Marsh, T., Pace, N., and Altman, S. (1983). *Cell,* **35**, 849.
5. Forster, A. C. and Altman, S. (1990). *Science,* **249**, 783.
6. Forster, A. C. and Symons, R. H. (1987). *Cell,* **49**, 211.
7. Hampel, A. and Tritz, R. (1989). *Biochemistry,* **28**, 4929.
8. Uhlenbeck, O. C. (1987). *Nature,* **328**, 596.
9. Haseloff, J. and Gerlach, W. L. (1988). *Nature,* **334**, 585.
10. Bertrand, E., Grange, R., and Pictet, R. (1992). In *Gene regulation biology of antisense RNA and DNA* (ed. R. P. Erickson and J. G. Ivant), pp. 71–81. Raven Press, New York.
11. Conway, G., Wooley, J., Bibring, T., and Lestrourgeon, W. M. (1988). *Mol. Cell. Biol.,* **8**, 2884.
12. Dreyfuss, G., Swanson, M. S., and Pinol-Roma, S. (1988). *Trends Biochem. Sci.,* **13**, 86.
13. Rinke, J., Bernd, A., Blocker, H., Frank, R., and Lurhmann, R. (1984). *Nucleic Acids Res.,* **12**, 4111.
14. Black, D. L., Chabot, B., and Steitz, J. A. (1985). *Cell,* **42**, 737.
15. Jaeger, J. A., Turner, D. A., and Zucker, M. (1989). *Proc. Natl Acad. Sci. USA,* **86**, 7706.
16. Ehresmann, C., Baudin, F., Mougel, M., Romby, P., Ebel, J. P., and Ehresmann, B. (1987). *Nucleic Acids Res.,* **15**, 9101.
17. Bertrand, E., Fromont-Racine, M., Pictet, R., and Grange, T. (1993). *Proc. Natl Acad. Sci. USA,* **90**, 3496.
18. Goodchild, J., Agrawal, S., Civeira, M. P., Sarin, P. S., Sun, D., and Zamecnik, P. C. (1988). *Proc. Natl Acad. Sci. USA,* **85**, 5507.
19. Ruffner, D. E., Stormo, G. D., and Uhlenbeck, O. C. (1990). *Biochemistry,* **29**, 10695.
20. Xing, A. and Whitton, J. L. (1992). *Biochemistry,* **66**, 1361.
21. Heidenreich, O. and Eckstein, F. (1992). *J. Biol. Chem.*, **267**, 1904.
22. Rittner, C., Burmester, C., and Sczakiel, G. (1993). *Nucleic Acids Res.,* **21**, 1381.
23. Freier, S. M., Kierzek, R., Jaeger, J. A., Sugimoto, N., Caruthers, M. H., Neilson, T., *et al.* (1986). *Proc. Natl Acad. Sci. USA,* **83**, 9373.
24. Nishikura, K., Yoo, C., Kim, U., Murray, J. M., Estes, P. A., Cash, F. E., *et al.*, (1991). *EMBO J.,* **10**, 3523.
25. Hershlag, D. (1991). *Proc. Natl Acad. Sci. USA,* **88**, 6921.
26. Lewin, B. (1983). In *Genes*, p. 295. John Wiley & Sons, New York.
27. Stein, C. A. and Cohen, J. S. (1990). In *Oligonucleotide antisense inhibitors of gene expression* (ed. J. S. Cohen), pp. 97–117. CRC Press, Boston.
28. Matsukura, M., Zon, G., Shinozuka, K., Robert-Guroff, M., Shimada, R., Stein, C. A., *et al.* (1989). *Proc. Natl Acad. Sci. USA,* **86**, 4244.
29. Shibahara, S., Mukai, S., Morisawa, H., Nakashima, H., Kobayashi, S., and Yamamoto, N. (1989). *Nucleic Acid Res.,* **17**, 239.

30. Heidenreich, O., Pieken, W., and Eckstein, F. (1993). *FASEB*, **7**, 90.
31. Taylor, N. R., Kaplan, B. E., Seiderski, O. P., Li, H., and Rossi, J. J. (1992). *Nucleic Acids Res.*, **20**, 4559.
32. Pieken, W. A., Olsen, D. B., Benseler, F., Aurup, H., and Eckstein, F. (1991). *Science*, **253**, 314.
33. Paolella, G., Sproat, B. S., and Lamond, A. I. (1992). *EMBO J.*, **11**, 1913.
34. Taylor, N. and Rossi, J. J. (1991). *Antisense Res. Dev.*, **1**, 173.
35. Leonetti, J. P., Degols, G., and Lebleu, B. (1990). *Bioconjugate Chem.*, **1**, 149.
36. Boutorin, A. S., Guskova, L. V., Inanova, E. M., Kobetz, N. D., Zarytova, V. F., Ryte, A. S., *et al.* (1989). *FEBS Lett.*, **254**, 129.
37. Letsinger, R. L., Zhang, G., Sun, D. K., Ikeuchi, T., and Sarin, P. S. (1989). *Proc. Natl Acad. Sci. USA*, **86**, 6553.
38. Wu, G. Y., Wilson, J. M., Shalaby, F., Grossman, M., Shafritz, D. A., and Wu, C. H. (1991). *J. Biol. Chem.*, **266**, 14338.
39. Abromova, T. V., Blinov, V. M., Vlassov, V. V., Gorn, V. V., Zarytova, V. F., Ivanova, E. M., *et al.* (1991). *Nucleotides Nucleosides*, **10**, 419.
40. Kaneda, Y., Kunimitsu, I., and Uchida, T. (1989). *Science*, **243**, 375.
41. Sugawa, H., Uchida, T., Yoneda, Y., Ishiura, M., and Okada, Y. (1985). *Exp. Cell Res.*, **159**, 410.
42. Curiel, D. T., Wagner, E., Cotten, M., Birnxtiel, M. L., Agarwal, S., Li, C. M., *et al.* (1992). *Hum. Gen. Ther.*, **3**, 147.
43. Wright, S. and Huang, L. (1989). *Adv. Drug Delivery Rev.*, **3**, 343.
44. Ho, R. J. Y., Rouse, B. T., and Huang, L. (1987). *J. Biol. Chem.*, **262**, 13973.
45. Alving, C. R. (1988). *Adv. Drug Delivery Rev.*, **2**, 107.
46. Guthrie, C. and Patterson, B. (1988). *Annu. Rev. Genet.*, **22**, 387.
47. Cotten, M. and Birnstiel, M. L. (1989). *EMBO J.*, **12**, 3861.
48. Kotin, R. M., Siniscalo, M., Samulski, J., Zhu, X., Hunter, L., Laughlin, C., *et al.* (1990). *Proc. Natl Acad. Sci. USA*, **87**, 2211.
49. Salmons, B., Groner, B., Friis, R., Muellener, D., and Jaggi, R. (1986). *Gene*, **45**, 215.
50. Gunning, P., Leavitt, J., Muscat, G., Ng, S., and Kedes, L. (1987). *Proc. Natl Acad. Sci. USA*, **84**, 4831.
51. Kerr, M. S., Stark, G. R., and Kerr, I. M. (1988). *Eur. J. Biochem.*, **175**, 65.
52. Sarver, N., Cantin, E. M., Chang, P. S., Zaia, J. A., Ladne, P. A., Stephens, D. A., *et al.* (1990). *Science*, **247**, 1222.
53. Dropulic, B., Lin, N. H., Martin, M. A., and Kuan-Teh, J. (1992). *J. Virol.*, **66**, 1432.

18

Gene therapy

C. SMITH, W. A. MARASCO, and E. BÖHNLEIN

1. Introduction

Somatic gene therapy was independently proposed in 1988 by D. Baltimore and J. C. Sanford as an alternative treatment of AIDS (1, 2). This new approach is based on the stable introduction of genetic material into cells of HIV-1-infected patients. The therapeutic genes can be delivered directly into the target cells of HIV in the peripheral blood (CD4-positive lymphocytes, macrophages) using techniques analogous to those used for the first human gene therapy trial for the treatment of ADA deficiency. Differentiated peripheral blood cells have a limited life expectancy and repeated treatment would be required. Alternatively, the antiviral genes can be introduced into the haematopoietic stem cell from which all HIV-1 target cells are derived. In this case, a single treatment could suffice to obtain the therapeutic effect.

HIV-1 offers at least ten functional genes as targets for somatic gene therapies:

(a) The introduced antiviral gene products (RNA, protein) should interfere with replication to eliminate HIV-1 or reduce the viral burden. Expression of *trans*-dominant (td) Gag proteins (3) which interfere with infectious particle formation and HIV LTR-directed suicidal genes (4, 5) could accomplish this goal.

(b) Some of the suggested strategies could protect CD4-positive cells against HIV-induced cytopathic effects. RNA decoys (6, 7), antisense strategies (8), ribozymes (9), td mutants of Rev (10) and Tat (11), and intracellular expression of anti-Env single chain monoclonal antibodies (12) fall into this category.

Reliable test systems are critical for the development of effective antiviral somatic gene therapy approaches. The same viral parameters which proved valuable for the development of antiviral drugs (e.g. RT activity, syncitia formation, p24 antigen expression, infectivity) and have been discussed in previous chapters in this and Volume 1 can also be used to evaluate somatic gene therapy approaches:

(a) 'Proof of principle' for a particular antiviral strategy can be achieved very quickly in transient transfection studies (10). However, cells amenable to these procedures are normally not infected by HIV. Quantitative evaluation is difficult in these test systems and bears little relevance for long-term interference with HIV-1 replication.

(b) More accurate model systems are based on CD4-positive cell lines which can be infected with HIV-1 (adapted strains and primary isolates).

Stable introduction of antiviral genes into CD4-positive cells is very difficult. Recombinant amphotropic retroviral vectors based on murine leukaemia virus (MLV) proved very helpful to overcome this technical hurdle (13):

(a) A retroviral expression plasmid containing the antiviral gene and a selectable marker, such as neomycin phosphotransferase, is transfected into an amphotropic 'packaging cell' (14) which provides all MLV helper functions (Gag, Pol, and Env).

(b) The transfected cells secrete infectious recombinant retroviral particles carrying the antiviral gene plus a selectable marker.

(c) Established CD4-positive cells are then infected with these recombinant particles and transduced cells can be selected in the presence of the drug (G418).

(d) The established, stably transduced CD4-positive cell population can be characterized molecularly and subjected to HIV-1 infection experiments to evaluate the antiviral potential.

CD4-positive peripheral lymphocytes, the natural target cells for HIV infection, are also amenable to retroviral transduction and represent the best available HIV *in vitro* model system. However, in contrast to established CD4-positive cell lines, they can only be kept in culture for a short time which complicates the quantitative assessment of antiviral effects.

2. Examples of antiviral strategies

2.1 Intracellular expression of an anti-Env single chain monoclonal antibody

The envelope glycoproteins of HIV-1 have been implicated in the cytopathic effect of the virus, which is specific for cells bearing high levels of the CD4 receptor (15). In tissue culture, the cytopathic effects of HIV-1 consist of:

- multinucleated giant cell (syncytium) formation
- lysis of single cells (16)

Syncytium formation is mediated solely by the HIV-1 envelope expressed on the infected cell surface (17). The envelope binds to the CD4 receptor present on adjacent cells and then, via a fusion analogous to that involved in virus entry, the apposed cell membranes are fused to form heterokaryons (see Chapter 8). In the presence of HIV-1 gp160, a ternary complex is formed between gp160–CD4 and $p56^{lck}$, a tyrosine kinase that binds to the cytoplasmic tail of CD4. This complex is retained in the ER and this mislocalization of tyrosine kinase to the cytoplasmic side of the ER may play a role in single cell lysis (18). Blockage of the envelope protein transport, therefore, may be an important target for AIDS gene therapy.

Recent advances in antibody engineering have allowed antibody genes to be manipulated and antibody molecules to be re-shaped (19). These advances raise the possibility that antibodies can be made to function intracellularly to bind and to inactivate molecules within cells. The single chain monoclonal antibody sFv105 was constructed from RT–PCR amplified DNA fragments and has been described in detail (12, 20). sFv105 is derived from the human monoclonal antibody F105, which recognizes the CD4 binding region of the HIV-1 envelope protein (21) and was designed to react with the nascent folded envelope protein within the ER (22) and to prevent transit of the envelope antibody complex to the cell surface.

Protocol 1. Construction of stably transformed cell lines by lipofection

Equipment and reagents

- 35 mm tissue culture plates (Nunc)
- COS-1 cells
- Dulbecco's modified Eagle's medium (DMEM; Gibco–BRL)
- Fetal bovine serum (FBS)
- G418 (BRL)
- Lipofectin™ (BRL)
- pCMV-sFv105
- Sterile water

Method

1. Plate COS-1 cells on 35 mm dishes and grow to 50–70% confluence.
2. Wash three times in serum-free media.
3. Dilute 10 μg of supercoiled plasmid DNA in 50 μl sterile water.
4. Dilute 10 μl of Lipofectin™ reagent in 50 μl sterile water.
5. Combine DNA and Lipofectin and allow to stand at room temperature for 20 min.
6. Add mixture to cells. Incubate at 37°C agitating briefly every 10 min.
7. Replace media with 1.5 ml of DMEM supplemented with 10% fetal bovine serum 2 h after transfection.
8. Replace media with DMEM with 10% fetal bovine serum containing 500 μg/ml G418 after 48 h.

Protocol 1. *Continued*

9. Continue to replace media regularly until G418-resistant colonies appear.
10. Pick colonies and grow in flasks for further assay. This will take approximately four to six weeks, depending on transfection efficiency.

The ability of the sFv105 and sFv105-KDEL proteins to inhibit the function of the envelope protein was determined by measurement of the ability of transformed COS sFv105 and COS sFv105-KDEL cells transfected with the HIV envelope gene to induce syncytium formation of CD4-positive cells (*Protocols 1* and *2*).

(a) The parental COS vector cells as well as the COS sFv105 and COS sFv105-KDEL cells were transfected with a plasmid that expresses a functional envelope glycoprotein.
(b) Two days post-transfection the cells were mixed at a ratio of about one to ten with a human CD4-positive T cell line, SupT1, that is susceptible to envelope-mediated fusion.
(c) Envelope-mediated syncytium formation was reduced by 80–90% in cells which express either the sFv105 or sFv105-KDEL proteins (*Table 1*).
(d) Inhibition of envelope processing was shown to occur and similar amounts of gp160 were made in all three lines as determined by metabolic labelling and precipitation of the transfected cultures (12).

Protocol 2. Syncytium formation assay for CD4-positive SupT1 cells

Equipment and reagents

- DMEM/10% FBS
- COS cells
- Phosphate-buffered saline (PBS)/50 mM EDTA
- pSVIIIENV

Method

1. Transfect the transformed COS cells (*Protocol 1*) with 3 μg of plasmid producing HIV-1 envelope (pSVIIIENV) (23).
2. Incubate cells for 48 h. Rinse cells in PBS, and then incubate with PBS containing 50 mM EDTA at 37 °C for 40 min.
3. Remove cells from the plate, wash with PBS, and then resuspend in 2 ml of DMEM supplemented with 10% fetal calf serum.
4. Add 2×10^6 SupT1 lymphocytes and incubate at 37 °C.
5. Score syncytia after 12 h. Syncytia are counted in five high power fields (HPF; magnification × 200) in each well.

Table 1. Syncytia formation and infectious virus production in transformed COS cells

Transforming gene	% syncytia[a]	Capsid proteins (day to release)	Virus titre ($TCID_{50}$)
None	100	—	NB[b]
Vector	~98	5	~10^5
sFv105	~10	10	~10^5
sFv105-KDEL	~20	10	ND

[a] Syncytia counted in five high power fields (magnification ×200).

To examine the ability of the sFv105 proteins to inhibit production of infectious virus, COS vector, COS sFv105, and COS sFv105-KDEL cells were transfected with a plasmid that contains a copy of the entire viral genome (23). Virus growth and virus infectivity (see Chapters 2, 3, and 7, Volume 1) was assayed by p24 assay (see Chapter 7, Volume 1). The results are shown in *Table 1*.

(a) Four days post-transfection, the virus in the culture supernatant fluids was used to initiate infection of the sensitive indicator cell line SupT1. The supernatants of all three transfected cell lines were shown to contain similar amounts of the viral capsid protein, p24.

(b) Release of capsid proteins into the cell supernatant has previously been shown to occur in the absence of synthesis of the envelope glycoprotein as well as in the presence of envelope glycoproteins that contain processing defects and are therefore retained in the ER (25).

(c) Virus replication in SupT1 cells initiated by supernatants from the transfected COS sFv105 or COS sFv105-KDEL cells is delayed about five days relative to that initiated by virus produced by a control COS-1 cell line that contains the vector without the sFv105 sequences.

(d) When serial dilutions of the supernatants were used to infect SupT1 cells, there was a greater than 10^3-fold reduction in syncytium formation.

(e) The delay in replication of virus produced by COS sFv105 cells, and the decrease in infectious titre, is due to the low infectivity of the virus produced in the antibody-expressing cells relative to that of virus produced by the control cell line.

(f) These studies illustrate the feasibility of designing antibodies that bind and inactivate molecules intracellularly and offer another strategy for gene therapy of AIDS.

2.2 Inhibition of HIV-1 replication in CEM cells transduced with a dominant-negative Rev mutant protein

The HIV-1 Rev protein interacts with a highly structured RNA sequence (Rev-response element; RRE) located in the *env* gene intron (see Chapter 9, and 11; ref. 26). Formation of this complex facilitates the transport of the incompletely spliced HIV transcripts to the cytoplasm where they are translated into structural proteins (27). Hence, HIV-1 Rev function is crucially important for infectious virion formation (28).

Mutant genes have been generated which interfere with Rev wild-type function in a dominant way in transient assays (10). We constructed a recombinant retroviral vector (*Figure 1*) to generate CD4-positive cell lines which stably express Rev protein. These cell lines were used to test whether the dominant-negative RevM10 mutant could interfere with HIV replication. CD4-positive CEM cells were transduced with the amphotropic recombinant retrovirus (*Protocol 3*) and the G418-resistant cell population was analysed for CD4 expression by FACS analysis (see Chapter 13, Volume 1) and transgene expression by Northern blotting.

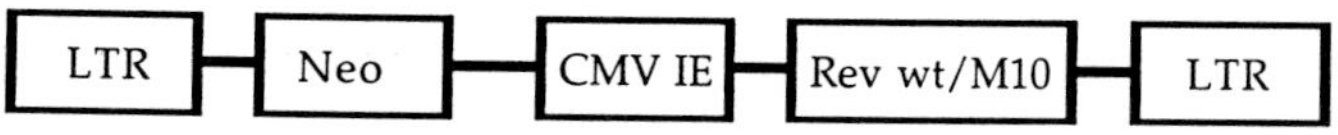

Figure 1. Organization of the integrated, proviral recombinant retroviral vector. Expression of the neomycin phosphotransferase gene (Neo) is directed by the Moloney murine leukaemia virus (MoMLV) promoter located in the long terminal repeat (LTR). Expression of the Rev genes (wild-type and dominant-negative mutant) is directed from the immediate early human cytomegalovirus promoter (CMV IE).

Protocol 3. Retroviral transduction of CEM cells

Equipment and reagents

- Retroviral expression plasmid
- PA 317 cells
- CEM cells
- Lymphoprep (Nycomed)
- DMEM/H21/10% fetal calf serum (FCS) (Gibco)
- RPMI 1640/10% FCS (Gibco)
- G418 (BRL)
- 2 M $CaCl_2$
- 10 mM Tris–HCl pH 7.6
- 2 × HBS: 50 mM Hepes, 1.5 mM Na_2HPO_4, 280 mM NaCl, pH 7.13
- TBS/3 mM EGTA

Method

1. Seed 1 × 10^6 PA317 cells/100 mm tissue culture dish in DMEM/H21/10% FCS.
2. Replace medium the next day.

3. Mix 500 μl solution containing 20 μg retroviral plasmid, and 62 μl of 2 M $CaCl_2$, 10 mM Tris–HCl pH 7.6
4. Add mix dropwise to 0.5 ml 2 × HBS while gently shaking on vortex.
5. Incubate for 30 min at room temperature.
6. Add precipitate to cells and distribute evenly.
7. Incubate 12–16 h at 37°C.
8. Wash cells three times with TBS/3 mM EGTA.
9. Add 10 ml RPMI 1640 medium containing 10% FCS.
10. Add 2 × 10^6 CEM cells, and co-culture overnight (16–24 h).
11. Transfer the CEM cells into a T75 flask in 10 ml RPMI 1640/10% FCS, and incubate overnight.
12. Transfer CEM cells into a new flask to remove any accidentally transferred PA317 cells. Repeat transfer three times every 2 h.
13. Add RPMI 1640 medium supplemented with 10% FCS and 800 μg/ml G418, and incubate at 37°C.
14. Centrifuge CEM cells on Lymphoprep™ (Nycomed) cushions at 800 *g*.
15. Repeat step 14 every three days or until no live cells are detectable in control cultures.

Protocol 4. HIV infectious centre assay

Equipment and reagents

- Coulter counter
- Multiwell tissue culture slides
- Heat inactivated human serum from HIV seropositive donors
- HIV $HTLV_{IIIB}$ stock
- CEM cells
- RPMI 1640/10% FCS
- PBS
- Acetone/methanol (1:1)
- FITC labelled donkey anti-human immunoglobulin (Kallestad Diagnostics)
- Evans blue (0.03% in PBS)
- Sterile water

Method

1. Infect 2 × 10^6 CEM cells with 200 TCID of HIV $HTLV_{IIIB}$.
2. Add 9 ml of RPMI 1640 medium with 10% FCS (f.c. 2 × 10^5 cells/ml).
3. Determine cell number using a Coulter counter every seven days and adjust to 2 × 10^5 cells/ml.
4. Collect samples for indirect immunofluorescence assay every seven days. Pellet 2–4 × 10^4 cells for 10 min at 500 *g*. Wash twice with PBS. Resuspend in 10 μl PBS and plate on multiwell slides.
5. Air dry slides for approx. 30 min.

Protocol 4. *Continued*

6. Fix dried cells with acetone/methanol (1 : 1). At this stage, the samples can be stored at −20°C for several weeks.
7. Incubate cells with a 1 : 20 dilution of HIV-1 infected patient sera in a humid chamber at 37°C for 30 min.
8. Wash cells three times or more with PBS.
9. Incubate for 30 min with a FITC labelled donkey anti-human immunoglobulin (Kallestad Diagnostics).
10. Wash three more times with PBS.
11. Counterstain with Evans blue (0.03% in PBS) for 2 min.
12. Rinse cells three times with distilled water and prepare for microscopic evaluation.

The CEM-Rev WT and CEM-RevM10 cells were infected with 200 tissue culture infectious doses (TCID) of HIV HTLV$_{IIIB}$ (see Chapter 2, Volume 1). The results are shown in *Table 2*:

(a) Clonal analysis shows that about 60% of the RevM10 transduced clonal lines did not support HIV-1 replication under these experimental conditions (29).

(b) 65% of the cells (CEM-Rev WT, CEM-RevM10) express HIV-1 antigens after two weeks. After three weeks, almost all CEM-Rev WT cells are HIV-1 antigen-positive and syncitia are readily detectable. A similar result is observed after 43 days.

Table 2. HIV-1 challenge experiment of transduced CEM pools

	Cell line			
	CEM-Rev WT		CEM-RevM10	
Time (days)	% cells expressing HIV antigens	Cell density (10E5 cells/ml)	% cells expressing HIV antigens	Cell density (10E5 cells/ml)
0	<2[a]	2	<2[a]	2
7	<2[a]	43	<10	32
14	65	22	65	14
21	>90	1.1[c]	65	0.7[c]
28	ND[b]	1[c]	ND[b]	9.2[c]
35	ND[b]	1[c]	ND[b]	9.2[c]
43	>90	1[c]	15	20

[a] <2% indicates background levels.
[b] ND: not determined.
[c] Cells were further diluted at this time point.

(c) Between days 14 and 43, the CEM-Rev WT cells are effectively killed by the replicating HIV $HTLV_{IIIB}$. In the CEM-RevM10 cultures, cell counts also decrease but these cells appear to be more resistant against the HIV-induced cytopathic effects than the CEM-Rev WT cells as indicated by the ten times higher cell densities at days 28 and 35.

(d) The number of HIV-1 antigen-positive cells decreases between day 21 and day 43 from 65% to 15% and no syncitia can be detected in the CEM-RevM10 cultures.

These results indicate that even in a pre-selected retrovirally transduced cell line, expression levels of the RevM10 genes are not sufficient to prevent HIV-1 replication in every cell. Nevertheless, the protected CEM-RevM10 cells proliferate in the presence of actively replicating HIV in the unprotected fraction of the transduced cells which is comparable to the situation in the patient.

2.3 RNA decoy inhibition of HIV in primary CD4-positive peripheral blood lymphocytes

The RNA decoy strategy is based on using gene transfer to introduce vectors which express high levels of critical portions of the HIV TAR and RRE RNA sequences. These oligonucleotides bind *tat* and *rev* (see Chapter 9) and competitively inhibit binding to the authentic TAR and RRE sequences of HIV (7, 30, 31).

(a) The RNA decoys were expressed from the $tRNA^{met}$ Pol III promoter which yields 10–100-fold higher RNA expression levels relative to Pol II promoters.

(b) Optimal expression was dependent on cloning the Pol III expression cassette into the 3′LTR of the retroviral vector (i.e. a double copy vector) presumably to avoid inhibition by Pol II transcription originating in the vector LTR (32).

Greater than 99% inhibition of the HIV p24 antigen, HIV reverse transcriptase (RT) activity, and infectious HIV virion production for more than 21 days was observed in cultures of selected CEM subclones expressing TAR decoys (7). In addition, HIV RNA transcription in TAR decoys transduced CEM cells was substantially reduced compared to transcription in control cells (7). The activity required an intact binding site on the TAR decoy, suggesting that *tat* was sequestered (31). Because the TAR sequence is highly conserved among HIV-1 and related viruses, intracellular expression of TAR decoys effectively protected CEM cells challenged with retroviral strains as disparate as SIV (7). Similarly, but less effectively, greater than 95% inhibition of HIV was observed in selected CEM subclones expressing high intracellular levels of RRE decoys (6). More recently, variations of RRE decoys have been constructed which appear to inhibit HIV significantly more effectively (33).

Table 3. Transduction of peripheral blood lymphocytes

Transducing vector	Survival of CD4-positive cells (% of uninfected control)		P24 (ng/ml)	
	Day 18	Day 21	Day 18	Day 21
DCT: M3[a]	60	26	89	942
DCT: Tar	90	85	>10	98

[a] M3: negative control.

The effectiveness of RNA decoys has also been evaluated in primary CD4-positive peripheral blood T cells (*Protocol 5* and *Table 3*). Unselected bulk populations of CD4-positive PBLs were challenged with HIV-1 to mimic as closely as possible the clinical setting. G418 selection was found to produce substantial artefacts in the growth of the surviving T cells which could have a significant impact on the kinetics of HIV spread. Following transduction, 10–15% of primary CD4-positive PBLs contained the vector as determined by DNA blot analysis and DNA quantitative competitive PCR (QC–PCR). RNA blot analysis revealed that transduced CD4-positive PBL expressed TAR decoy RNA at levels equivalent to selected CEM cells on a per cell basis for more than 28 days in culture. Bulk unselected populations of CD4-positive PBL transduced with TAR decoys had a survival advantage relative to control cells after infection by the cytolytic HIV-1 SF13 strain. TAR decoy transduced PBLs also produced lower amounts of p24 than control vector transduced PBL (*Table 3*), fewer HIV-infected cells could be identified by immunofluorescence staining at comparable time points, and the HIV infectious titre was reduced (data not shown). However, the observed inhibitory effect was transient and a population of cells resistant to HIV could not be recovered even after prolonged culture periods.

Protocol 5. Transduction of primary CD4-positive PBL

Equipment and reagents

- FACS-Scan (Becton Dickinson)
- T25 anti-CD4 MicroCellector Flasks (Applied Immune Sciences)
- Lymphoprep™ (Nycomed)
- Gamimmune (Cutter Labs)
- PBS/1 mM EDTA
- PBS
- Polybrene
- RPMI 1640/10% FCS
- Heat inactivated human AB serum (Pel-Freeze, Rogers)
- Recombinant human IL-2 (Sandoz)
- gpAM12-derived retroviral producer cell lines
- PHA-M (Sigma)

Method

1. Collect > 70 ml of peripheral blood into lithium heparin-containing collection tubes by phlebotomy from consenting individuals.
2. Dilute blood (1:1) with PBS and centrifuge on a Lymphoprep™ (Nycomed) cushion for 15 min at 800 *g*, 4°C.
3. Collect the mononuclear band, wash twice with PBS, and resuspend at 5×10^6 cells/ml in PBS [1 mM EDTA/0.5% Gamimmune (Cutter Lab, NL)] for 20 min at room temperature.
4. Seed 2×10^7 treated peripheral blood mononuclear cells (PBMC) on to a T25 anti-CD8 MicroCellector Flask (Applied Immune Sciences (AIS)) which was pre-primed with three washes with PBS/1 mM EDTA.
5. Remove non-adherent enriched CD4 cells.
6. Add 5 ml of RPMI 1640/10% heat inactivated human AB serum (Pel-Freeze)/50 U rh-IL-2 (Sandoz) (lymphocyte media) plus 5 μg/ml PHA-M (Sigma) per flask.
7. Incubate PBMCs at 37°C for three days.
8. Pool cells from multiple CD8 Cellector Flasks into a T75 flask and adjust to a concentration of 2×10^6 cells/ml with fresh lymphocyte media (see above).
9. Determine the proportion of CD4-positive cells by FACS analysis (typically > 90%).
10. Seed CD4-positive PBLs on to gpAM12/RNA decoy amphotropic producer line irradiated with 3000 Gy. (Producer lines have previously been characterized to have transducing titres of > 5×10^5/ml and to lack replication competent recombinant retrovirus.)
11. Co-cultivate CD4-positive PBL cells at 2×10^6 cells/ml with 8 μg/ml Polybrene for 24 h in T75 flask (10 ml medium).
12. Transfer cells at a concentration of 2×10^6 cells/ml three times into tissue culture flasks with fresh lymphocyte medium to remove residual producer cells.
13. Determine viable cell number every three days and adjust cell concentration to 2×10^6 cells/ml by addition of fresh lymphocyte medium.

In addition to examining the efficacy of RNA decoys in primary CD4-positive PBLs, a number of safety issues have been assessed:

(a) We could not detect replication competent murine retrovirus, alteration in growth kinetics or morphology, generation of RNA decoy escape mutants, or generations of IL-2-independent T cells in long-term PBL cultures.

(b) No alterations in the production of CD4-positive T cell-specific cytokines including IFN-γ, IL-6, IL-4, IL-3, IL-1, and GM-CSF were observed following transduction.

(c) No acute or chronic pharmacologic toxicity was observed in SCID mice inoculated with 10–50 × 10^6 RNA decoy transduced CD4-positive PBLs relative to control mice transplanted with non-transduced cells.

(d) No alteration in CTL activity was observed in G418 selected cells isolated from a CD4-positive CTL clone which had been transduced with TAR decoys.

In summary, transduction of CD4-positive PBL with RNA decoys appears to be safe and confers a transient survival advantage to transduced cells. These observations confirm the experiments performed in immortalized cell lines which proved efficacy against HIV. Improvements in gene transfer efficiency into CD4-positive PBL as well as improvements in the effectiveness of RNA decoy strategy may enhance the protection effect.

2.4 Quantitative PCR assays

Semi-quantitative PCR (see Chapter 9, Volume 1) has not been reliable for quantitating gene transfer efficiency into cells. Quantitative competitive PCR (QC–PCR) remedies this by including a known amount of a standard which co-linearly amplifies with the experimental sequence (34–36). The amount of experimental sequence is determined by comparing the ratio of standard to experimental amplified products. This procedure (*Protocol 6*) is very accurate, requires less than 1 × 10^6 cells per sample, and can detect 1/1000 cells containing vector.

Protocol 6. Quantitative competitive PCR assay

A. *Equipment and reagents*

- 10 × PCR buffer: 100 mM Tris–HCl pH 8.3, 500 mM KCl
- 25 mM $MgCl_2$
- 10 mM dATP, dCTP, dGTP, dTTP (−20°C)
- 25 μM primers (resuspended in H_2O and stored at 4°C)
  upstream (within tRNA sequence): 5′ CTT GGC AGA ACA GCA GAG TGG 3′
  downstream (within *gag*): 5′ ACA GAG ACA ACA CAG AAC GAT 3′
- HPLC grade water
- AmpliTaq polymerase (Cetus)
- Ampliwax gems (Cetus)
- [^{32}P]dCTP (3000 Ci/mM)
- X-ray films (Kodak)
- Internal standard genomic DNA/RNA isolated from a clonal cell line containing one copy per cell (the standard shares PCR primer binding sites with the experimental sample but contains either an insertion or deletion so that a different size product is produced by PCR)
- Isoquick Extractor Kit (MicroProbe)
- Cytoprobe Hoechst-DNA Assay Kit (Millipore)
- Agarose

Method

1. Isolate standard and experimental DNA (Isoquick Extractor Kit which reproducibly yields 1–2 μg of DNA/1 × 10^6 cells).

2. Determine the experimental and standard DNA concentration with the Cytoprobe Hoechst-DNA Assay Kit.
3. Confirm the integrity of DNA and concentration on a 1% agarose gel.
4. Prepare a Hot-Start PCR reaction per instructions in the PCR Ampliwax kit. Mix lower layer of PCR reaction:

• 10 × PCR buffer	2.5 μl	(1.25 ×)
• $MgCl_2$	8.0 μl	(2 mM)
• dNTPs	2.0 μl each	(200 μM each)
• primers	2.5 μl each	(0.5 μM)
• [^{32}P]dCTP	0.5 μl	

5. Add an Ampliwax gem, heat for 10 min at 80°C, and allow the wax to cool 1 min at room temperature.
6. Add the upper layer which consists of:

• water	9–54 μl	
• 10 × PCR buffer	10 μl	(1.25 ×)
• AmpliTaq polymerase	0.5 μl	(2.5 U)

7. Add equal amounts of the experimental DNA and the standard DNA. Also prepare a series of tubes with constant amounts of experimental DNA and a serial dilution of the standard cell line.
8. Perform PCR at 95°C 5 min; 64°C 1 min; 73°C 1 min for one cycle, followed by 95°C 1 min; 64°C 1 min; 73°C 1 min for 27 cycles (previously determined to be in the linear portion of the amplification reaction).
9. Remove one-tenth vol. of reaction and run on a 1.5% gel.
10. Remove the gel, dry, and autoradiograph. The lane where the experimental sample and the standard sample are equivalent indicates the proportion of transduced cells in the experimental sample.

3. Conclusions

All three examples described above provide experimental evidence that gene intervention strategies can effectively interfere with HIV-1 replication *in vitro*. In two cases, the genetically modified cells were significantly more resistant to HIV-induced cytopathic effects than in the control cells. However, gene transfer will probably be less efficient for PBMCs and, even less so for human haematopoietic stem cells. It is unlikely that these target cell populations can be selected extensively *ex vivo*. Because of this, establishment of a successful gene therapy will greatly depend on improved gene delivery systems for haematopoietic stem cell populations. An added complication is that it is currently unclear whether the transduction procedure interferes with the viability and activity of the target cells. *In vitro* transduction studies using PBLs indicate that this is not the case but animal studies in suitable model systems (e.g. SCID–hu mouse) are needed to determine safety and *in vivo* efficacy of these procedures. However refined the model system, eventually AIDS gene therapy approaches will be put to the real test in HIV-infected patients. Because of the severity of the disease, Phase 1 clinical AIDS gene therapy trials have been initiated based on *in vitro* data obtained with peripheral blood lymphocytes. An important outcome of these trials will be a determination of whether genetic intervention allows selection of HIV variants with resistance to the particular treatment, as has been seen with pharmacological therapies. Time will tell whether any of the currently pursued promising approaches can actually keep in check the variety of HIV species which are routinely detected in patients and whether this will eventually result in measurable clinical improvements of the infected individuals.

Acknowledgements

The authors would like to thank Dr J. Hauber, Dr M. Fung, Dr S.-Y. Chen, and Ms J. Bagley for their critical comments and helpful suggestions.

References

1. Baltimore, D. (1988). *Nature*, **335**, 395.
2. Sanford, J. C. (1988). *J. Theor. Biol.*, **130**, 469.
3. Trono, D., Feinberg, M. B., and Baltimore, D. (1989). *Cell*, **59**, 113.
4. Harrison, G. S., Maxwell, F., Long, C. J., Rosen, C. A., Glode, L. M., and Maxwell, I. H. (1991). *Hum. Gene Ther.*, **2**, 53.
5. Venkatesh, L. K., Arens, M., Subramanian, T., and Chinnadurai, G. (1990). *Proc. Natl Acad. Sci. USA*, **87**, 8746.
6. Lee, T. C., Sullenger, B. A., Gallardo, H. F., Ungers, G. E., and Gilboa, E. (1992). *New Biol.*, **4**, 66.
7. Sullenger, B. A., Gallardo, H. F., Ungers, G. E., and Gilboa, E. (1990). *Cell*, **63**, 601.
8. Rittner, K. and Sczakiel, G. (1991). *Nucleic Acids Res.*, **19**, 1421.
9. Sarver, N., Cantin, E. M., Chang, P. S., Zaia, J. A., Ladne, P. A., Stephens, D. A., *et al.* (1990). *Science*, **247**, 1222.
10. Malim, M. H., Böhnlein, S., Hauber, J., and Cullen, B. R. (1989). *Cell*, **58**, 205.
11. Pearson, L., Garcia, J., Wu, F., Modesti, N., Nelson, J., and Gaynor, R. (1990). *Proc. Natl Acad. Sci. USA*, **87**, 5079.
12. Marasco, W. A., Haseltine, W. A., and Chen, S.-Y. (1993). *Proc. Natl Acad. Sci. USA*, **90**, 7889.
13. Gilboa, E., Eglitis, M. A., Kantoff, P. W., and Anderson, W. F. (1986). *Biotechniques*, **4**, 504.
14. Markowitz, D., Goff, S., and Bank, A. (1988). *Virology*, **167**, 400.
15. DeRossi, A., Franchini, G., Aldióvini, A., Del Mistro, A., Chieco-Bianchi, L., Gallo, R. C., *et al.* (1986). *Proc. Natl Acad. Sci. USA*, **83**, 4297.

16. Popovic, M., Sarngadharan, M. C., Read, E., and Gallo, R. C. (1984). *Science*, **224**, 497.
17. Sodroski, J., Goh, W. C., Campbell, K., and Haseltine, W. A. (1986). *Nature*, **322**, 470.
18. Crise, B. and Rose, J. K. J. (1992). *J. Virol.*, **66**, 2296.
19. Winter, G. and Milstein, C. (1991). *Nature*, **349**, 293.
20. Marasco, W. A., Bagley, J., Zani, C., Posner, M., Cavacini, L., Haseltine, W. A., *et al.* (1992). *J. Clin. Invest.*, **90**, 1467.
21. Posner, M. R., Hideshima, T., Cannon, T., Mukherjee, M., Mayer, K. H., and Byrn, R. A. (1991). *J. Immunol.*, **146**, 4325.
22. Munro, S. and Pelham, H. R. B. (1987). *Cell*, **48**, 899.
23. Helseth, E. M., Kowalski, M., Gabuzda, D., Olshevsky, U., Haseltine, W., and Sodroski, J. (1990). *J. Virol.*, **64**, 2416.
24. Johnson, V. A. and Byngton, R. E. (1990). In *Techniques in HIV research* (ed. A. Aldovini and B. D. Walker), pp. 71–6, Stockton Press, New York.
25. McCune, J. M., Rabin, L. B., Feinberg, M. B., Lieberman, M., Kosek, J. C., Reyes, G. R., *et al.* (1988). *Cell*, **53**, 55.
26. Daly, T., Cook, K. S., Gray, G. S., Maione, T. E., and Rusche, J. R. (1989). *Nature*, **342**, 816.
27. Malim, M. H., Hauber, J., Le, S.-Y., Maizel, J. V., and Cullen, B. R. (1989). *Nature*, **338**, 254.
28. Cullen, B. R. (1991). *J. Virol.*, **65**, 1053.
29. Bevec, D., Dobrovnik, M., Hauber, J., and Böhnlein, E. (1992). *Proc. Natl Acad. Sci. USA*, **89**, 9870.
30. Sullenger, B., Lee, T., Smith, C., Ungers, G., and Gilboa, E. (1990). *Mol. Cell. Biol.*, **10**, 6152.
31. Sullenger, B., Gallardo, H. F., Ungers, G. E., and Gilboa, E. (1991). *J. Virol.*, **65**, 6811.
32. Hantzopoulos, P. A., Sullenger, B. A., Ungers, G., and Gilboa, E. (1989). *Proc. Natl Acad. Sci. USA*, **86**, 3519.
33. Lee, S., Gallardo, H. F., Gilboa, E., and Smith, C. (1994). *J. Virol.*, **68**, 8254.
34. Wang, A., Doyle, M., and Mark, D. (1989). *Proc. Natl Acad. Sci. USA*, **86**, 9717.
35. Ferre, F. (1992). *PCR Methods Applications*, **2**, 1.
36. Gilliland, G., Perrin, S., and Bunn, H. F. (1990). In *PCR protocols* (ed. M. A. Innis, D. H. Gelfland, J. J. Sninsky, and T. J. White), pp. 60–6. Academic Press, New York.

A1

List of suppliers

Ambion, 2130 Woodward Street #200, Austin, Texas 78744–1832, USA.

Amersham Corporation, 2636 South Clearbrook Drive, Arlington Heights, Illinois 60005, USA.

Amersham International, Amersham Place, Little Chalfont, Buckinghamshire HP7 9NA, UK; 2636 South Clearbrook Drive, Arlington Heights, Illinois 60005, USA.

Amicon Inc., 72 Cherry Hill Drive, Beverly, Massachusetts 01915, USA.

Apothekerns Laboratorium A.S., 3 Harbizallen, N-0275 Oslo, Norway.

Bachem BioSciences, 3700 Horizon Drive, King of Prussia, Pennsylvania 19406, USA.

Beckman Instruments Inc., 2500 Harbor Boulevard, PO Box 3100, Fullerton, California 92634–3100, USA.

Becton Dickinson Labware, 2 Bridgewatter Lane, Lincoln Park, New Jersey 07035, USA.

Biochrom Seromed KG, Leonorenstrasse 2–6, D-1000 Berlin 46, Germany.

Bio-Rad Laboratories, 2000 Alfred Nobel Drive, Hercules, California 94547, USA.

Boehringer Mannheim GmbH, Biochemica, Sandhöferstrasse 116, Postfach 310120, D-68000 Mannheim 31, Germany. 9115 Hague Road, PO Box 50414, Indianapolis, Indiana 46250–0414, USA.

Brinkman Instruments, Inc. (Eppendorf), Cantiague Road, Westbury, NY 11590, USA.

Calbiochem-Novabiochem Corp., 10394 Pacific Center Court, San Diego, CA 92121, USA.

Clonetech, 4030 Fabian Way, Palo Alto, California 94303–9605, USA.

Costar Corporation, One Alewife Center, Cambride, MA 02140, USA.

Difco Laboratories, PO Box 331058, Detroit, Michigan 48232–7058, USA.

Dupont (Sorvall Division) Co. Inc., 31 Peck's Lane, PO Box 5509, Newtown, CT 06470-5509, USA.

Dupont de Nemours, Barley Mill Plaza P-24, Wilmington, DE 19898, USA.

Eastman Kodak Corp., 343 State, Building 701, Rochester, New York 14652, USA.

FisherBiotech, 585 Alpha Drive, Pittsburgh, Pennsylvania 15238, USA.

Fisher Scientific, 711 Forbes Avenue, Pittsburgh, Pennsylvania 15219, USA.

5 Prime->3 Prime Inc., 5603 Arapahoe Boulevard, Boulder, Colorado 80303, USA.

Flow Laboratories: now **ICN Biochemicals Inc.**—see below).

Gibco–BRL (Life Technologies): now part of **Life Technologies Ltd.**—see below.

Gibco–BRL Inc.: now part of **Life Technologies Inc.**—see below.

Hewlett Packard S.A., 150 Route du Nant-d'Avril, CH-1217 Meyrin 2, Geneva, Switzerland.

Hi-Tech Scientific Ltd., Brunel Road, Salisbury SP2 7PU, UK.

Hoefer Scientific Instruments, 654 Minnesota Street, PO Box 77387, San Francisco, California 94197–0387, USA.

ICN Biomedicals Inc., PO Box 19536, Irvine, California 92713–9921, USA.

Imperial Laboratories, West Portway, Andover, Hampshire SP10 3LF, UK.

Invitrogen Corp., 3985 B Sorrento Valley Blvd, San Diego, California 92121, USA.

Jandel, Shimmelbuschstraβe 25, 4006 Erkrath, Germany.

Jandel Scientific, 2591 Kerner Boulevard, San Rafael, CA 94901, USA.

Janseen Chimica, Janseen Pharmaceuticalaan, 3–2440 Geel, Belgium.

JRH-Sera-Lab UK Ltd., Hophurst Lane, Crawley Down, Crawley, Sussex RH10 4BR, UK.

KinTek Instruments, 106 Althouse Laboratory, University Park, Pennsylvania 16802, USA.

Life Technologies Inc., Gibco BRL Division, 8400 Helgerman Court, PO Box 6009, Gaithersburg, MD 20884-9980, USA.

Life Technologies Ltd., PO Box 35, Renfrew Road, Paisley, Scotland, PA3 4EF.

Medical Research Council AIDS Reagent Repository, National Institute for Biological Standards and Control, Blanche Lane, South Mimms, Potters Bar, Hertfordshire EN6 3QG, UK.

E. Merck, Frankfurterstrasse 250, D-6100, Darmstadt 1, Germany.

Merck (BDH), Broom Road, Poole, Dorset DH12 4NN, UK.

Microfluidics Corp., 90 Oak Street, Newton, Massachusetts 02164, USA.

Millipore Corp., 80 Ashby Road, Bedford, MA 01730, USA.

Millipore UK Ltd., The Boulevard, Blackmoor Lane, Watford, Herts WD1 8YW, UK.

Molecular Devices Corp., Menlo Oaks Coporate Center, 4700 Bohannon Drive, Menlo Park, California 94025, USA.

Molecular Kinetics, PO Box 2475 C. S., Pullman, Washington 99165–2475, USA.

Moravek Biochemicals Inc., 577 Mercury Lane, Brea, California 92621–4890, USA.

National Institutes of Health AIDS Research and Reference Reagent, Program, 685 Lofstrand Lane, Rockville, MD 20850, USA.

New England Biolabs, 32 Tozer Road, Beverly, Massachusetts 01915–5599, USA.

New England Nuclear Research Products, Dupont Company, 549 Albany Street, Boston, Massachusetts 02118, USA.

Novagen Inc., 597 Science Drive, Madison, Wisconsin 53711, USA.

Nunc A/S, Kamstrupvej 90, Kamstrup, DK-4000 Roskilde, Denmark.

Nunc Inc., 2000 North Aurora Road, Naperville, Illinois 60566, USA. (Now part of **Life Technologies Inc.**—see above.)

Perkin Elmer, 761 Main Avenue, Norwalk, Connecticut 06859–0156, USA.

Pharmacia Biotech Co., 23 Grosvenor Road, St Albans, Herts AL1 3AW, UK.

Pharmacia Biotech Co. Inc., 800 Centennial Avenue, Piscataway, New Jersey 08855–1327, USA.

PharMingen, 11555 Sorrento Valley Road, San Diego, California 92121, USA.

Pierce Chemical Co., 3747 North Meridian Road, PO Box 117, Rockford, Illinois 61105, USA.

Promega Biotech Corp., 2800 South Fish Hatchery Road, Madison, Wisconsin 53711, USA.

Qiagen Inc., 9259 Eton Ave., Chatsworth, California 91311, USA.

Quality Biological Inc., 7581 Lindeburgh Drive, Gaithersburg, Maryland 20879, USA.

Repligen Corp., One Kendall Square, Building 700, Cambridge, Massachusetts 02139, USA.

Research Products International Corp., 410 North Business Center Drive, Mount Prospect, Illinois 60056–2190, USA.

Röhn Pharma GmbH, Postfach 4347, D-61000, Darmstadt 1, Germany.

Schleicher & Schuell Inc., 10 Optical Avenue, PO Box 2012, Keene, New Hampshire 03431, USA.

Sepracor Inc., 33 Locke Drive, Marlborough, Massachusetts 01752, USA.

Sigma-Aldrich Co. Inc., 1001 West Saint Paul Avenue, Milwaukee, Wisconsin 53233, USA.

Sigma Chemical Company, Fancy Road, Poole, Dorset BH17 7NH, UK; 3050 Spruce Street, PO Box 14508, St Louis, Missouri 63103, USA.

Spectrum Medical, 1100 Rankin Road, Houston, Texas 77073, USA.

Stratagene, 11011 North Torrey Pines Road, La Jolla, California 92037, USA.

Toso Haas GmbH, Zettachring 6, D-70567, Stuttgart, Germany.

University of Kent, Department of Chemistry, Canterbury, Kent, UK.

United States Biochemical Corp.: now part of **Amersham Corporation**—see above.

Whatman LabSales, 5285 N. E. Elam Young Parkway, Suite A-400, Hillsboro, Oregon 97124–9981, USA.

Worthington Biochemicals, Halls Mill Road, Freehold, New Jersey 07728, USA.

Index